Dirac Operators: Yesterday and Today

Dirac Operators: Yesterday and Today

Proceedings of the Summer School & Workshop
Center for Advanced Mathematical Sciences
American University of Beirut, Lebanon
August 27 to September 7, 2001

edited by

Jean-Pierre Bourguignon
Thomas Branson
Ali Chamseddine
Oussama Hijazi
Robert J. Stanton

International Press
www.intlpress.com

Dirac Operators: Yesterday and Today

Editors:
Jean-Pierre Bourguignon
Thomas Branson
Ali Chamseddine
Oussama Hijazi
Robert J. Stanton

ISBN 978-1-57146-184-1

Paperback reissue 2010. Previously published in 2005 under ISBN 1-57146-175-2 (clothbound).

Typeset using the LaTeX system.

Preface

The Summer School "Dirac Operators: Yesterday and Today" was an ambitious endeavour to show to a public of advanced students a tableau of modern mathematics. The idea was to make available a topic, of considerable importance for mathematics itself because of its deep impact on many mathematical questions, but also for its interactions with one of mathematics historic partners, physics.

The Dirac operator was undoubtedly created by Paul Adrien Maurice Dirac to deal with an important physical question. That it now has a deep presence in mathematics is a long story, for which Michael F. Atiyah and Isadore M. Singer are the main players. Indeed, if the basic objects involved in the definition of the Dirac operator, namely the spinor fields, had been introduced for their own sake by Élie Cartan, and then firmly developed by Richard Brauer and Hermann Weyl, the remarkable insight that there is a very distinguished operator acting on them was Dirac's. Fifty years later Atiyah and Singer turned it into a mathematical cornerstone by one of the far-reaching developments of their monumental work on the index theorem. In some sense one can say that they rediscovered the Dirac operator in a Riemannian setting. As a consequence of their work, this operator now has a central role in modern mathematics, at the crossroads of many disciplines, and in particular as a unifying tool bringing together analysis and geometry under the umbrella of what is now called "global analysis."

To understand problems connected with the Dirac operator, one needs to present its many facets. This set of notes covering courses and advanced seminars that accompanied the courses presents basic ingredients that constitute the foundations of the theory. It is exemplary of the recent evolution of mathematics, and of the strengthening of its historic unity which occurs in a very dynamic way. One needs to bring together many components of mathematics, often considered as separate: algebra, analysis and geometry in its broadest context. Indeed, at the basis of the study of Dirac operators one finds representation theory of the orthogonal groups, bundle theory and the theory of elliptic operators on manifolds. Last but not least, one of course cannot forget the link to theoretical physics, which has originally inspired most of the developments, and which recently broadened considerably the scope of the subject.

As it often is the case with multi-faceted objects, one never knows what should come first. This set of notes proposes an itinerary, going from the simplest notions to more advanced ones. There is always the risk that the reader loses patience from frustration of waiting too long before she or he can get a global view of the landscape. The discovery of this breath-taking view, when all pieces have been properly put in place, will be one of the high points of the journey, a very valuable pay-off. The hope of course is soon after to gain new viewpoints on classical questions. In the case at hand, some anticipate that parts of Riemannian Geometry, undoubtedly the classical backbone of the theory of Dirac operators, will soon be much enriched by a systematic spinorial approach. In a sense particle physicists have led the way in establishing long ago, almost as soon as Dirac came up with his model for relativistic electron motion, that particle physics must associate spinor fields and tensor fields. Throughout the twentieth century, it got richer and richer as a result of one of the most extraordinary endeavours of human history. In this context spinor fields are the wave functions of half-spin particles, while tensor fields are the wave functions of spinless particles or particles having an integral spin.

It took many years for mathematicians to really accept the vision that spinorial objects and tensorial objects have to be considered on an equal footing. In the late 70s and more systematically in the 80s, physicists went one step further. They started to focus their attention on theories that would associate, in an even more intimate way spinorial and tensorial fields, the so-called supersymmetric theories. Centering around the idea that special transformations actually exchanging spinorial and tensorial particles, the notion of supersymmetry would dramatically impact physical models in several parts of physics. It still has to prove realistic by some experimental evidence. When formulated in mathematical terms, this notion brings to the forefront special geometric structures in a very natural way. These, in turn, seem to have a physical bearing in string theory, something difficult to guess a priori. These notes do give a few hints of these important developments as one will read in some seminars.

Another way in which concepts at the heart of this lecture series find an exemplary domain of illustration is non-commutative geometry. Indeed, in order to introduce the notion of a metric in this context, Alain Connes proposed very successfully that this be done via the definition of an object with is the fundamental solution of a would-be "Dirac operator." This is a remarkable realisation of the principle this preface means to stress, namely that in modern Geometry the functional point of view, and very concretely the consideration of special operators related to classical geometric situations, is fundamentally fruitful. A very deep lesson this set of notes will hopefully help to digest.

It is our great pleasure to thank the following institutions for their generous financial support: the International Mathematical Union (IMU); the National Science Foundation (NSF) and the Clay Mathematics Institute (CMI) in the United States; and the International Center for Theoretical Physics (ICTP), Trieste, Italy. We are also grateful for the financial and logistical support of the following French Institutions: the Centre International de Mathématiques Pures et Appliquées (CIMPA); the Institut des Hautes Études Scientifiques (IHÉS); the Institut Élie Cartan (IÉCN), Nancy; and the Ministère des Affaires Étrangères. Special thanks go to the American University of Beirut (AUB) and the National Council for Scientific Research (CNRS), Lebanon.

We acknowledge the financial support of the Hariri Foundation and the Audi Bank, Beirut. Particular thanks go to Dr. Marwan Ghandour, Chairman and Managing Director, Lebanon Invest SAL, for his enthusiasm and active support for the promotion of Science in Lebanon. We would also like to thank the College of Liberal Arts and Sciences and the Department of Mathematics of the University of Iowa for their assistance with the distribution of the conference proceedings volume.

The organizers would like to recognize John Waterbury, President of the AUB, and Mrs. Bahia Hariri, Parliamentary deputy and president of the Parliamentary Commission for Culture and Education, Lebanon, for their supportive Opening Ceremony speeches and for their hospitality.

Finally, we express our deep appreciation to the Center for Advanced Mathematical Sciences (CAMS) for the exceptional atmosphere and facilities we enjoyed during the meeting; to Rola Ghalayini, Assistant to the Director, for her efficient assistance to the Organizing Committee and to Bertrand Morel for his assistance with LATEX2ε.

The Editors

Contents

Part 1

Summer School

CHAPTER 1

Introduction to the Dirac Operator

H. BLAINE LAWSON, JR.

The point of this article is to give a brief history of the emergence of the Dirac Operator in physics and mathematics in the middle years of the twentieth century. Most of the more technical points touched on here will be fully elaborated in subsequent lectures.

The story begins in the 1920's when physicists were trying to develop a relativistic version of quantum mechanics. Recall that the basis of quantum mechanics is the **Schrödinger Equation**

$$i\hbar\frac{\partial\Psi}{\partial t} = H\Psi$$

for the wave function Ψ where H is a differential operator determined canonically by the classical energy in the system. Here are some examples.

Example 1. A free particle. Here $E = \frac{p^2}{2m}$ is the free particle energy and the Schrödinger Equation becomes

$$i\hbar\frac{\partial\Psi}{\partial t} = -\frac{\hbar^2}{2m}\Delta\Psi.$$

Example 2. A particle in a scalar potential field.

$$i\hbar\frac{\partial\Psi}{\partial t} = \left(-\frac{\hbar^2}{2m}\Delta + V(x)\right)\Psi$$

Example 3. A charged particle in an electromagnetic field.

$$i\hbar\frac{\partial\Psi}{\partial t} = \left\{\frac{1}{2m}\left(\frac{\hbar}{c}\nabla - \frac{e}{c}\vec{A}\right)^2 + e\phi\right\}\Psi$$

This formulation of quantum mechanics was based on classical Newtonian mechanics. However, Einstein's new **Principle of Special Relativity** stated that physical laws should be invariant under *Lorentz transformations* of space-time coordinates. By definition these are the linear transformations of $\mathbb{R}^4$ which preserve the indefinite quadratic form:

$$(x, x) = x_0^2 - x_1^2 - x_2^2 - x_3^2$$

where $x_0 = ct$ is the time-like coordinate and the constant c is the speed of light *in vacuo*.

Now the principle of Lorentz Invariance led Einstein to the conclusion that the energy of a free particle is given by

$$E = \sqrt{p^2c^2 + m^2c^4}$$

and not by the formula appearing in Example 1. Unfortunately, because of the square root it is impossible to convert this expression into a classical differential operator as we did above. The first solution to this difficulty was to apply the Schrödinger Equation twice. This gives a term corresponding to E^2 on the right-hand side and yields the **Klein-Gordon Equation**

$$-\hbar^2\frac{\partial^2\Psi}{\partial t^2} = -\hbar^2c^2\Delta\Psi + m^2c^4\Psi$$

which can be rewritten as

$$\left\{\Box + \left(\frac{mc}{\hbar}\right)^2\right\}\Psi = 0$$

where

$$\Box = \frac{1}{c^2}\frac{\partial^2}{\partial t^2} - \Delta = \frac{\partial^2}{\partial x_0^2} - \frac{\partial^2}{\partial x_1^2} - \frac{\partial^2}{\partial x_2^2} - \frac{\partial^2}{\partial x_3^2}$$

This brings us roughly to the year 1928 and the physicist P. A. M. Dirac, who was fundamentally dissatisfied with the Klein-Gordon approach. According to Dirac, the principle of causality required the Schrödinger Equation to be *first-order* in time. Lorentz invariance then implied that the Schrödinger operator should be *first-order in all variables.* This was impossible to achieve for scalar wave functions, so Dirac decided to consider vector-valued wave functions

$$\Psi = (\psi_1, \ldots, \psi_N)$$

and to search for a first-order linear Lorentz-invariant operator D with

$$D^2 = \Box$$

Writing

$$D = \sum_{\mu=0}^{3} \gamma_\mu \frac{\partial}{\partial x_\mu}$$

where the γ_μ are $N \times N$-matrices, we see that

$$D^2 = \sum_{\mu,\nu=0}^{3} \frac{1}{2}(\gamma_\mu\gamma_\nu + \gamma_\nu\gamma_\mu)\frac{\partial}{\partial x_\mu}\frac{\partial}{\partial x_\nu} = \Box$$

implies the equations

$$\gamma_\mu\gamma_\nu + \gamma_\nu\gamma_\mu = \pm 2\delta_{\mu\nu} \tag{1.1}$$

These are easily solved for small values of N. For example when $N = 4$:

$$\sigma_1 = \begin{pmatrix} 0 & i \\ i & 0 \end{pmatrix} \quad \sigma_2 = \begin{pmatrix} 0 & -1 \\ 1 & 0 \end{pmatrix} \quad \sigma_3 = \begin{pmatrix} i & 0 \\ 0 & -i \end{pmatrix}$$

$$\gamma_0 = \begin{pmatrix} 0 & Id \\ Id & 0 \end{pmatrix} \quad \gamma_\mu = \begin{pmatrix} \sigma_\mu & 0 \\ 0 & -\sigma_\mu \end{pmatrix}, \quad \mu = 1, 2, 3$$

Using such matrices Dirac wrote down a relativistic Schrödinger equation for an electron in a magnetic field which eventually gave spectacular accord with experiment. However there was a significant problem: the existence of negative energy solutions, i.e., eigenfunctions of the operator with negative eigenvalues.

Dirac, believing in the beauty of the equations, asserted that such particles should exist. He proposed the idea that the negative states are filled in general, but when a state is empty, we see it. Such empty states were called "antiparticles". The antielectron was called a *positron*. If an electron encounters a positron, it will be pulled into the empty state causing an annihilation of the pair

$$\text{electron} + \text{positron} \quad \longrightarrow \quad \emptyset$$

In 1932 the positron was found in the laboratory.

Underlying Dirac's discovery was an important mathematical fact: Lorentz tranformations of space-time coordinates yield linear transformations of $\Psi = (\psi_1, \ldots, \psi_N)$ which were determined only up to sign ± 1. The additional property determining the sign was thought of as an internal *spin* of the particle. The transformations of the spinor wave functions Ψ did not constitute a representation of the Lorentz group L but rather a representation of its simply-connected 2-fold covering group $\widetilde{L} \longrightarrow L$.

The importance of Dirac's work led to renewed mathematical interest in representations of the universal covering groups of the orthogonal groups. The first construction of these representations was due to Élie Cartan in 1913 [1]. However, in 1935 Richard Brauer and Herman Weyl revisited and expanded Cartan's theory.[2] Two years later Cartan wrote a fundamental monograph[3] aimed at introducing spinors into Riemannian geometry and thereby into general relativity. However, the full realization of the importance of spinors to geometry would not come for some decades.

Let us briefly review Cartan's theory of spinors in the case of positive definite quadratic forms. Let $\|x\|^2 = \sum_{i=1}^n x_i^2$ be the standard norm on $\mathbb{R}^n$ and denote by $\mathfrak{M}_{n\times n}$ the set of all $n \times n$ matrices with real coefficients. The orthogonal group and its Lie algebra are then defined by

$$\begin{aligned} \mathrm{SO}_n &= \{g : \mathbb{R}^n \to \mathbb{R}^n : \|gx\|^2 = \|x\|^2 \ \ \forall x\} \\ &= \{g \in \mathfrak{M}_{n\times n} : \ g^t g = Id\} \\ \mathfrak{so}_n &= \{g \in \mathfrak{M}_{n\times n} : \ g^t + g = 0\}. \end{aligned}$$

It is a basic fact that for $n > 1$

$$\pi_1 \mathrm{SO}_n \ = \ \mathbb{Z}_2.$$

and there is a universal covering group which sits in a short exact sequence

$$0 \to \mathbb{Z}_2 \to \mathrm{Spin}_n \to \mathrm{SO}_n \to 1.$$

The representations of Spin_n (which correspond bijectively to representations of $\mathfrak{so}_n$) are generated via direct sums, tensor products and exterior powers, by:

The *basic* representation on $\mathbb{R}^n$, and

The *spinor* representations given by Clifford multiplication.

These latter representations are **not** representations of SO_n. To understand them systematically we must first understand Clifford algebras and their classification.

[1] É. Cartan, Les groupes projectifs qui ne laissent invariante aucune multiplicité plane, Bull. Soc. Math. France, **41** (1913), 53-97

[2] R. Brauer and H. Weyl, Spinors in n-dimensions, Amer. J. Math. **57** (1935),425-449.

[3] É. Cartan, La Théorie des Spineurs, Hermann, Paris, 1937.

Clifford Algebras

Let V be a real vector space with a quadratic form q. The associated **Clifford algebra** is the quotient

$$C\ell(V,q) \equiv \bigoplus_{k\geq 0} V^{\otimes k}/I$$

of the tensor algebra of V by the ideal

$$I = \langle v\otimes v + q(v)\cdot 1 \;:\; v\in V\rangle$$

Example: $C\ell_n \equiv C\ell\left(\mathbb{R}^n, \|\cdot\|^2\right)$. Let $e_1,\ldots,e_n$ be an o.n. basis of $\mathbb{R}^n$. Then the algebra $C\ell_n$ can be written in terms of generators and relations as (cf. (1.1))

$$C\ell_n = \langle 1, e_1,\ldots,e_n \;:\; e_ie_j + e_je_i = -2\delta_{ij}\rangle.$$

Note the isomorphisms:

$$C\ell_0 = \mathbb{R} \quad C\ell_1 = \mathbb{C} \quad C\ell_2 = \mathbb{H},$$

Every Clifford algebra has the structure of a $\mathbb{Z}_2$-graded algebra

$$C\ell(V,q) = C\ell(V,q)^0 \oplus C\ell(V,q)^1 \qquad \text{where}$$

$C\ell(V,q)^0$ is generated by $V^{\otimes 2*}$ and $C\ell(V,q)^1$ is generated by $V^{\otimes(2*+1)}$

Spinor Groups

For any Clifford algebra $C\ell(V,q)$ we can consider the subgroups of the units in $C\ell(V,q)$ given by

$$\mathrm{Pin}(V,q) = \{v_1\cdots v_m \in C\ell(V,q) : v_i\in V \text{ and } q(v_i)=1\}$$
$$\mathrm{Spin}(V,q) = \mathrm{Pin}(V,q)\cap C\ell(V,q)^0$$

called the **Pinor** and **Spinor** groups of (V,q). Note that for $v,w\in V$ with $q(v)=1$

$$\begin{aligned} v\cdot w\cdot v &= v\cdot(-v\cdot w - 2\langle v,w\rangle)\\ &= w - 2\langle v,w\rangle v\\ &= \rho_v(w)\end{aligned}$$

where $\langle\cdot,\cdot\rangle$ is the polarization of q and ρ_v denotes orthogonal reflection through the hyperplane perpendicular to v. Therefore there are group homomorphisms

$$\mathrm{Pin}(V,q)\to \mathrm{O}(V,q), \qquad \mathrm{Spin}(V,q)\to\mathrm{SO}(V,q)$$

defined by

$$v_1\ldots v_m \mapsto v_1\ldots v_m w v_m\ldots v_1 = \rho_{v_1}\circ\cdots\circ\rho_{v_m}(w).$$

When $V=\mathbb{R}^n$ with $q=\|\cdot\|^2$, this gives a direct construction of the universal covering group

$$\mathrm{Spin}_n \to \mathrm{SO}_n.$$

mentioned above. The kernel of this homomorphism is generated by -1. Note that

$$\mathrm{Spin}_n \subset C\ell_n^0.$$

Consequently, *any representation of the Clifford algebra $C\ell_n$, in fact any representation of the algebra $C\ell_n^0 \cong C\ell_{n-1}$, gives a representation of the group* Spin_n. Furthermore, since in an algebra representation, the element -1 must act by $-$Identity, one sees that *no such representation of $Spin_n$ can be a pull-back of a representation of SO_n.*

Now the algebras $C\ell_n$ are classified. Each is simple or the direct sum of two simple algebras and therefore has exactly one irreducible representation S_n (or in the second case two S_n, S'_n) . The facts are given by the following table and periodicity result.

n	$C\ell_n$	S_n
0	$\mathbb{R}$	$\mathbb{R}$
1	$\mathbb{C}$	$\mathbb{C}$
2	$\mathbb{H}$	$\mathbb{H}$
3	$\mathbb{H} \oplus \mathbb{H}$	$\mathbb{H}, \mathbb{H}'$
4	$\mathbb{H}(2)$	$\mathbb{H}^2$
5	$\mathbb{C}(4)$	$\mathbb{C}^4$
6	$\mathbb{R}(8)$	$\mathbb{R}^8$
7	$\mathbb{R}(8) \oplus \mathbb{R}(8)$	$\mathbb{R}^8, \mathbb{R}^{8'}$
8	$\mathbb{R}(16)$	$\mathbb{R}^{16}$

Basic Periodicity Theorem: *For all n there are graded algebra isomorphisms*

$$C\ell_{n+8} \cong C\ell_n \otimes_{\mathbb{R}} C\ell_8$$

and therefore module isomorphisms

$$S_{n+8} \cong S_n \otimes_{\mathbb{R}} S_8 \cong S_n \otimes_{\mathbb{R}} \mathbb{R}^{16}$$

The Index Theorem

This brings us to 1960 when Michael Atiyah and Isador Singer were searching for a general topological formula expressing the index of an elliptic operator on a compact manifold. Such a formula would vastly generalize the classical Riemann-Roch Theorem from algebraic geometry and have far-reaching consequences in mathematics.

It was known from work of Milnor and Borel-Hirzebruch that manifolds X which were *Spin*, i.e., whose structure group could be simplified from SO_n to Spin_n, had special properties. Among them was the fact that a certain Pontrjagin number $\widehat{A}(X)$, called the $\widehat{A}$-genus of X, which was generally not an integer (for example, $\widehat{A}(\mathbb{P}^2(\mathbb{C})) = \frac{1}{8}$), was always an integer in this case.

Atiyah and Singer speculated: *$\widehat{A}$ might be the index of some operator defined only on Spin manifolds.* Singer, recalling Chevalley's book on spinors, used a Dirac-type construction to define an elliptic first-order operator

$$\not{D} : C^\infty(\not{S}^+) \to C^\infty(\not{S}^-)$$

with

$$\not{D}^2 \quad = \quad \nabla^*\nabla + \text{zero-order terms}$$

where $\nabla^*\nabla$ is the connection Laplacian. From here the index formula was quicky found.

Essentially every elliptic operator on a manifold is given by the Dirac construction. On a Spin manifold every elliptic operator is the Atiyah-Singer operator twisted by a coefficient bundle. The following is an important example.

Hirzebruch-Riemann-Roch: Let $E \to X$ be a holomorphic vector bundle over a compact complex manifold. Then one has the Dolbeault complex

$$\mathcal{E}^{0,0}(X,E) \xrightarrow{\overline{\partial}} \mathcal{E}^{0,1}(X,E) \xrightarrow{\overline{\partial}} \cdots \xrightarrow{\overline{\partial}} \mathcal{E}^{0,n}(X,E)$$

which, by taking the adjoint of $\overline{\partial}$, can be assembled into an elliptic operator

$$D_E \equiv \overline{\partial} + \overline{\partial}^* : \mathcal{E}^{0,\mathrm{ev}}(X,E) \to \mathcal{E}^{0,\mathrm{odd}}(X,E)$$

whose index is given by the formula

$$\begin{aligned} \mathrm{ind}(D_E) &= \sum_{q=0}^{n} (-1)^q \dim H^{0,q}(X;\, E) \\ &= \{\mathrm{ch}(E) \cdot Td(X)\}\, [X] \end{aligned}$$

where $\mathrm{ch}(E)$ and $Td(X)$ are certain explicit polynomials in the Chern classes of E and X respectively, and $\{\Phi\}[X]$ essentially means "integration of Φ over X".

Atiyah-Singer Dirac operator: The index of the Dirac operator (twisted by a vector bundle E)

$$\not{D}_E : C^\infty(\not{S}^+ \otimes E) \to C^\infty(\not{S}^- \otimes E),$$

which Atiyah and Singer constructed on any compact spin manifold, has a similar form

$$\mathrm{ind}(\not{D}_E) = \left\{\mathrm{ch}(E) \cdot \widehat{A}(X)\right\} [X] \tag{1.2}$$

The general Atiyah-Singer Index Theory is quite broad. It provides formulas for the index of:

(1) Elliptic operators (on compact manifolds), where the index is an integer.
(2) G-invariant elliptic operators, where the index lies in the representation ring R_G.
(3) Families of elliptic operators, where the index lies the K-theory of the base.
(4) $C\ell_k$-linear elliptic operators, where the index lies in KO-Theory.

This constitutes one of the fundamental results of the 20th century with important applications in analysis, geometry, topology, number theory, and mathematical physics.

The Atiyah-Singer Dirac Construction

Let (X, g) be a Riemannian manifold with tangent bundle $TX \to X$. By the **Clifford bundle** of (X, g) we mean the bundle of algebras over X:

$$C\ell(X) \to X$$

whose fibre at a point $x \in X$ is the Clifford algebra

$$C\ell_x(X) = C\ell\,(T_xX, g_x)$$

Let $S \to X$ be a bundle of modules over this bundle of algebras, i.e., suppose there exists a bundle mapping

$$C\ell(X) \otimes S \stackrel{\cdot}{\to} S$$

which makes each S_x a $C\ell_x(X)$-module. Note that since $TX \subset C\ell(X)$, this map defines "Clifford muliplication" by tangent vectors of X on S. We assume that S

carries a connection ∇ which is compatible with this Clifford multiplication and the Levi-Civita connection on TX, i.e., so that $\nabla(v \cdot s) = (\nabla v) \cdot s + v \cdot (\nabla s)$. Then the Atiyah-Singer Dirac operator

$$D : C^\infty(S) \to C^\infty(S)$$

is defined by

$$\boxed{D\sigma = \text{trace}\{\bullet \cdot \nabla_\bullet \sigma\} = \sum_{i=1}^{n} e_i \cdot \nabla_{e_i}\sigma}$$

where $e_1, \ldots, e_n$ is an orthonormal basis of T_xX at each x.

Basic Fact: If X is a Spin manifold, then there exists a bundle

$$\$ \to X$$

of $C\ell(X)$-modules such that $\$_x$ is an **irreducible** module at each point x. It is given by

$$\$ \equiv (P_{\text{Spin}} \times S_n)/\text{Spin}_n$$

where S_n is the representation of Spin_n produced by Clifford multiplication and P_{Spin} is a principal Spin_n-bundle covering the principal SO_n-bundle of othonormal frames in TX.

Now the operator D is self-adjoint, and so its index = dim(ker D) - dim(coker$D$) is zero. However, when $n = 2m$ the bundle can be split by using the *volume form*

$$\omega \equiv i^m e_1 \ldots e_{2m}$$

which satisfies $\omega^2 = 1$. Let

$$\$^+ = \frac{1}{2}(1+\omega)\cdot\$ \qquad\qquad \$^- = \frac{1}{2}(1-\omega)\cdot\$$$

be the ± 1-eigenbundles under multiplication by ω. Then $\$ = \$^+ \oplus \$^-$ and restriction of D gives

$$\not{D} : C^\infty(\$^+) \to C^\infty(\$^-).$$

This operator has an interesting index. It can be extended by taking the tensor product with any bundle E with unitary connection to obtain

$$\not{D}_E : C^\infty(\$^+ \otimes E) \to C^\infty(\$^- \otimes E).$$

whose index is computed by (4.2). This construction yields virtually all elliptic operators on X up to homotopy. Therefore, its discovery established the desired index formula at least for spin manifolds. Passing from there to general manifolds was not difficult.

This construction plays a fundamental role in many areas of mathematics. For example, the Seiberg-Witten equations which lead to profound results in 4-dimensional topology involve $\not{D}_E$ for various connections on E. In Riemannian geometry the operators $\not{D}_E$ are essential in the study of spaces of positive scalar curvature. We shall discuss this topic next.

Riemannian Geometry

The Spinor bundle $\$$ and its connection are constructed **canonically** out the Riemannian metric. In fact, the metric is required for its existence. So one might guess that spinors play a fundamental role in Riemannian geometry. They do! (as you will see in subsequent lectures.)

Example: Computing the square of a Dirac operator gives

$$\begin{aligned} D\!\!\!\!/^2 &= \sum_{i,j} e_i e_j \nabla^2_{e_i,e_j} \\ &= -\sum_i \nabla^2_{e_i,e_i} + \sum_{i<j} e_i e_j \{\nabla^2_{e_i,e_j} - \nabla^2_{e_j,e_i}\} \\ &\equiv \nabla^*\nabla + \mathcal{R} \end{aligned}$$

where $\mathcal{R}$ is a unversal expression in curvature. For the Spinor bundle $S\!\!\!/$ we have

$$\mathcal{R} = \frac{1}{4}\kappa \, \mathrm{Id}$$

where κ is the *scalar curvature* of X. Together with the index theorem this proves the following.

Theorem. *If X is a spin manifold which admits a metric of positive scalar curvature, then*

$$\widehat{A}(X) = 0.$$

There are many spin manifolds for which $\widehat{A}(X) \neq 0$. For example, let $X^{4n}(d)$ be the complex hypersurface of complex projective space $\mathbb{P}^{2n+1}(\mathbb{C})$ defined in homogeneous coordinates by the equation $z_0^d + z_1^d + \cdots + z_{2n+1}^d = 0$. Then $X^{4n}(d)$ is spin iff d is even, and $\widehat{A}(X^{4n}(d))$ is a polynomial in d which does not vanish for $d > 2n$.

Topology

There are many relations to topology even at the most basic level. Here is an example. Let $\mathbb{R}^{N+1}$ be a module for the Clifford algebra $C\ell_n$. We can suppose that Clifford multiplication by unit vectors in $\mathbb{R}^n \subset C\ell_n$ preserves the standard inner product on $\mathbb{R}^{N+1}$. Each unit vector $e \in \mathbb{R}^n$ gives a nowhere-vanishing vector field $V(x) = e \cdot x$ on the unit sphere $S^N = \{x \in \mathbb{R}^{N+1} : \|x\| = 1\}$. In fact the vector fields $V_1, ..., V_n$ corresponding to an orthonormal basis $e_1, ..., e_n$ of $\mathbb{R}^n$ are pointwise orthonormal on S^N, and in particular everywhere linearly independent. A deep theorem of Frank Adams states that the maximal number of everywhere linearly independent vector fields on a sphere is the number constructed in this fashion.

There is also an important relationship to Bott periodicity. Let $\mathfrak{M}_n$ denote the Grothendieck group of real representations of the Clifford algebra $C\ell_n$. Let $i^*\mathfrak{M}_{n+1} \subset \mathfrak{M}_n$ be those which are restrictions of representations of $C\ell_{n+1}$ and set

$$\widetilde{\mathfrak{M}}_n = \mathfrak{M}_n / i^*\mathfrak{M}_{n+1}$$

Then one finds that:

n	0	1	2	3	4	5	6	7
$\widetilde{\mathfrak{M}}_n$	$\mathbb{Z}_2$	$\mathbb{Z}_2$	0	$\mathbb{Z}$	0	0	0	$\mathbb{Z}$

That is we have

$$\widetilde{\mathfrak{M}}_n \cong \pi_n \mathrm{O} \cong KO_{n+1}(\mathrm{pt}) \tag{1.3}$$

for $n \leq 7$. By the Basic Periodicity Theorem above, we see that

$$\widetilde{\mathfrak{M}}_n \cong \widetilde{\mathfrak{M}}_{n+8}$$

extending (4.4) to all n by Bott Periodicity. There is in fact a geometrically contructed map from $\widetilde{\mathfrak{M}}_n$ to $\pi_n \mathrm{O}$, and one might imagine that it provides a proof of the Bott Periodicity Theorem. However, to date this is not the case. The only known proof that this map is an isomorphism uses the Bott theorem. Nevertheless this isomorphism is of fundamental importance in understanding the index of certain Dirac operators.

H. Blaine Lawson, Jr., Mathematics Department, State University of New York, Stony Brook, NY 11794, USA.
E-mail address: blaine@math.sunysb.edu

CHAPTER 2

Introduction to Differential Manifolds

Christian Bär

In this lecture we give a brief introduction to the theory of manifolds and related basic concepts of differential geometry. Since a lot of material will have to be covered in a very short time we can hardly give any proofs. Most proofs are fairly straightforward anyway. The reader interested in details may consult [**Spi79**] or any other introduction to differential geometry.

Definition 2.1. A topological space X is called an *n-dimensional topological manifold* if and only if

i) X is Hausdorff and has a countable basis.
ii) X is locally homeomorphic to $\mathbb{R}^n$, i. e., for every point $p \in X$ there exists an open neighborhood U of p in X, an open subset $V \subset \mathbb{R}^n$ and a homeomorphism $x : U \to V$.

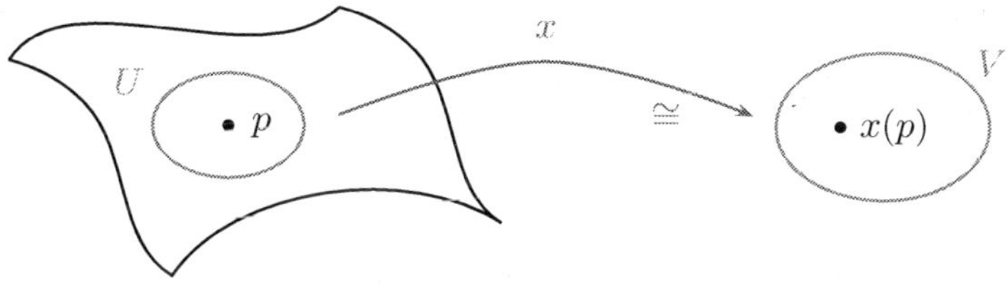

Fig. 1

The first condition is of technical nature and sometimes omitted. For subsets of $\mathbb{R}^N$ it is automatically satisfied. The crucial condition for a topological space to be a manifold is the second one. Topological manifolds are precisely those spaces that look locally like $\mathbb{R}^n$.

The homeomorphisms $x : U \to V$ are called *charts*. Given a chart x the standard coordinates on $\mathbb{R}^n$ can be used to specify points in U. Namely, a point p in U is uniquely determined by the n numbers $(x^1(p), \ldots, x^n(p))$. A collection of charts $x_\alpha : U_\alpha \to V_\alpha$ is called an *atlas* of M if the U_α cover M, i. e., $M = \bigcup_\alpha U_\alpha$.

Examples 2.2. 1. Euclidean space $\mathbb{R}^n$ itself is of course an n-dimensional topological manifold. Here we can use an atlas consisting of only one chart, namely $x = \mathrm{Id} : \mathbb{R}^n \to \mathbb{R}^n$.

2. The sphere S^n of unit vectors in $\mathbb{R}^{n+1}$ is an n-dimensional topological manifold. To find a chart containing a given point $p \in S^n$ one can project the hemisphere about p onto the n-dimensional disk perpendicular to p and then rotate this disk into $\mathbb{R}^n$.

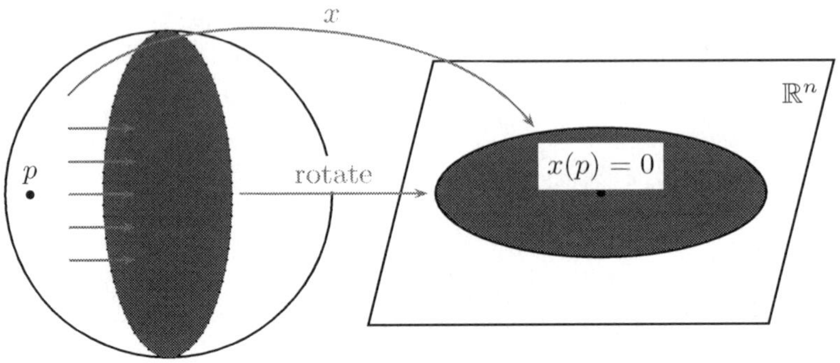

FIG. 2

3. Surfaces are 2-dimensional topological manifolds. Compact connected oriented[1] surfaces are classified up to homeomorphism by their *genus* (which can be any number in $\mathbb{N}_0$).

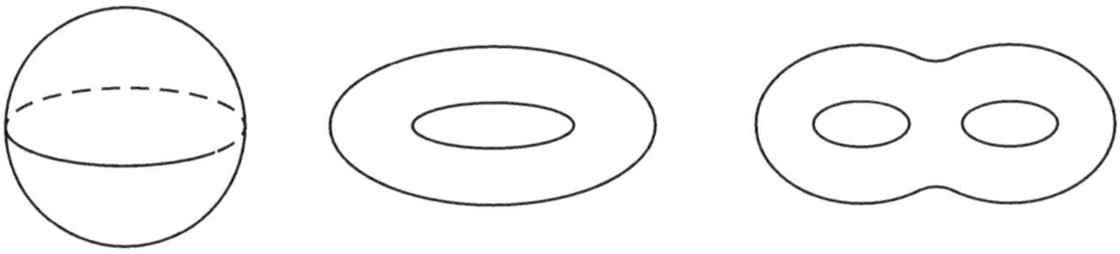

FIG. 3

To get a better feeling for the concept of a manifold let us also look at some topological spaces which are not manifolds.

NON-EXAMPLES.

1. The union M of the x-axis and the y-axis in the plane is not a manifold. The problems arise at the origin where the two lines intersect. In fact, if we remove the origin, then the remaining space $M - \{(0,0)\}$ is a 1-dimensional topological manifold. That M itself is not a manifold can be seen as follows. If M were a 1-dimensional topological manifold, then there would be a chart about $(0,0)$, i. e., a homeomorphism $x : U \to V$ where U is an open neighborhood of $(0,0)$ and V is some open subset of $\mathbb{R}$. By restricting x to a smaller open neighborhood if necessary we may assume that V is an open interval. If we now remove $x(0,0)$ from V, then

[1]see Definition 2.31.

we are left with two components whereas $U - \{(0,0)\}$ has at least four components, a contradiction to x being a homeomorphism.

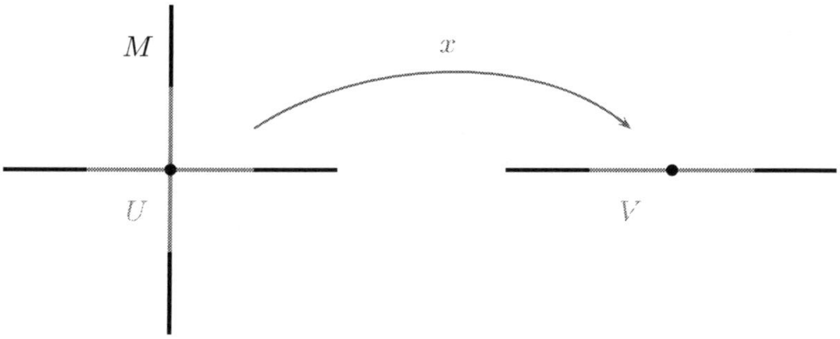

FIG. 4

2. Now let us look at the upper half-plane $\mathcal{H} = \{(x^1, x^2) \in \mathbb{R}^2 \mid x^2 \geq 0\}$. This time the boundary points $(x^1, 0)$ are the ones which are not contained in a chart. $\mathcal{H}$ is what is called a *manifold with boundary*, a concept which we will meet again when we get to the Stokes theorem.

Manifolds with boundary are defined like manifolds except that the target set V of a chart $x : U \to V$ is now an open subset of the upper half-space $\mathbb{R}^{n-1} \times [0, \infty)$ instead of $\mathbb{R}^n$. Note that a manifold is also a manifold with (empty) boundary whereas a manifold with boundary is in general not a manifold. All the concepts for manifolds to be introduced in this section can with slight modifications also be defined for manifolds with boundary.

There are some standard methods to construct new manifolds out of given ones.

EXAMPLES FOR CONSTRUCTION METHODS.

1. If M is an m-dimensional topological manifold and N an n-dimensional one, then $M \times N$ is an $(n+m)$-dimensional topological manifold. To obtain charts for $M \times N$ one can simply take the products of charts $x : U \to V$ for M and $\tilde{x} : \tilde{U} \to \tilde{V}$ for N,
$$x \times \tilde{x} : U \times \tilde{U} \to V \times \tilde{V} \subset \mathbb{R}^n \times \mathbb{R}^m = \mathbb{R}^{n+m}.$$
2. If M is an n-dimensional manifold and $U \subset M$ an open subset, then U is also an n-dimensional manifold. Charts for U are obtained from charts for M by restricting their domain to the intersection with U.

For a topological space X it makes sense to talk about *continuous* functions
$$f : X \longrightarrow \mathbb{R}\,.$$
Since we plan to do *differential* geometry we would like to know:

QUESTION 2.3. *What are* differentiable *functions?*

We do know what differentiability of a function means if it is defined on an open subset of $\mathbb{R}^n$. Since by definition manifolds look locally like $\mathbb{R}^n$ the natural attempt would be to call a function $f : M \to \mathbb{R}$ defined on a manifold M differentiable at $p \in M$ if and only if $f \circ x^{-1} : V \to \mathbb{R}$ is differentiable at $x(p)$ for some chart $x : U \to V$ whose domain U contains p. But then we have to check if this definition depends on the choice of chart x. For a second chart y we have near $y(p)$

$$f \circ y^{-1} = (f \circ x^{-1}) \circ (x \circ y^{-1}).$$

If we knew that $x \circ y^{-1}$ is a *diffeomorphism* and not just a homeomorphism we could conclude differentiability of $f \circ y^{-1}$ from differentiability of $f \circ x^{-1}$. For this reason we make the following definitions.

Let M be an n-dimensional topological manifold.

DEFINITION 2.4. Two charts $x : U_1 \to V_1$ and $y : U_2 \to V_2$ of a topological manifold are called C^∞*-compatible* if and only if $y \circ x^{-1} : x(U_1 \cap U_2) \to y(U_1 \cap U_2)$ is a C^∞-diffeomorphism.

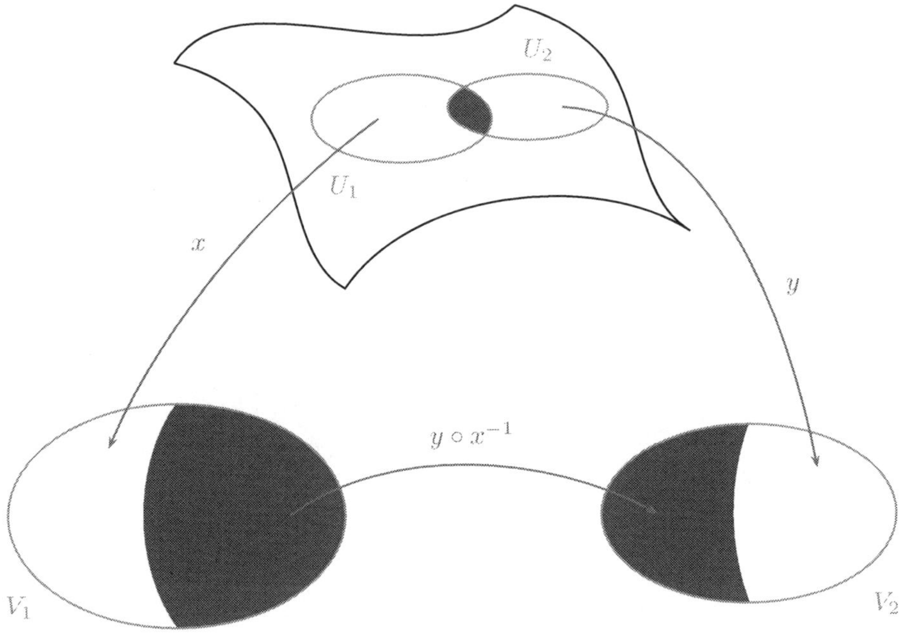

FIG. 5

DEFINITION 2.5. A collection $\mathcal{A} = \{x_\alpha : U_\alpha \to V_\alpha\}$ of charts of M is called a C^∞*-atlas* if and only if

$$\bigcup_\alpha U_\alpha = M$$

and all charts in $\mathcal{A}$ are mutually C^∞-compatible.

DEFINITION 2.6. A pair $(M, \mathcal{A})$ consisting of an n-dimensional topological manifold M and a maximal C^∞-atlas $\mathcal{A}$ of M is called an n-dimensional *differentiable manifold.*

Here maximality of the atlas means that if x is a chart of M which is C^∞-compatible with all charts in $\mathcal{A}$, then x itself also belongs to $\mathcal{A}$.

All examples of topological manifolds we have seen so far are in fact differentiable manifolds. In dimensions larger than 3 there exist topological manifolds which do not have a C^∞-atlas but we will not be concerned about them.

DEFINITION 2.7. Let $(M, \mathcal{A})$ be a differentiable manifold. A function $f : M \to \mathbb{R}$ is called *differentiable* (or C^k, $C^\infty, \ldots$) if and only if

$$f \circ x^{-1} : V \longrightarrow \mathbb{R}$$

is differentiable (or C^k, $C^\infty, \ldots$) for all charts $(x : U \to V) \in \mathcal{A}$.

More generally,

DEFINITION 2.8. Let $(M, \mathcal{A})$ and $(\widetilde{M}, \widetilde{\mathcal{A}})$ be two differentiable manifolds. A continuous map $\Phi : M \to \widetilde{M}$ is called *differentiable* (or C^k, $C^\infty, \ldots$), if and only if

$$\widetilde{x} \circ \Phi \circ x^{-1} : x(U \cap \Phi^{-1}(\widetilde{U})) \longrightarrow \widetilde{V}$$

is differentiable (or C^k, $C^\infty, \ldots$) for all charts $(x : U \to V) \in \mathcal{A}$ and all charts $(\widetilde{x} : \widetilde{U} \to \widetilde{V}) \in \widetilde{\mathcal{A}}$.

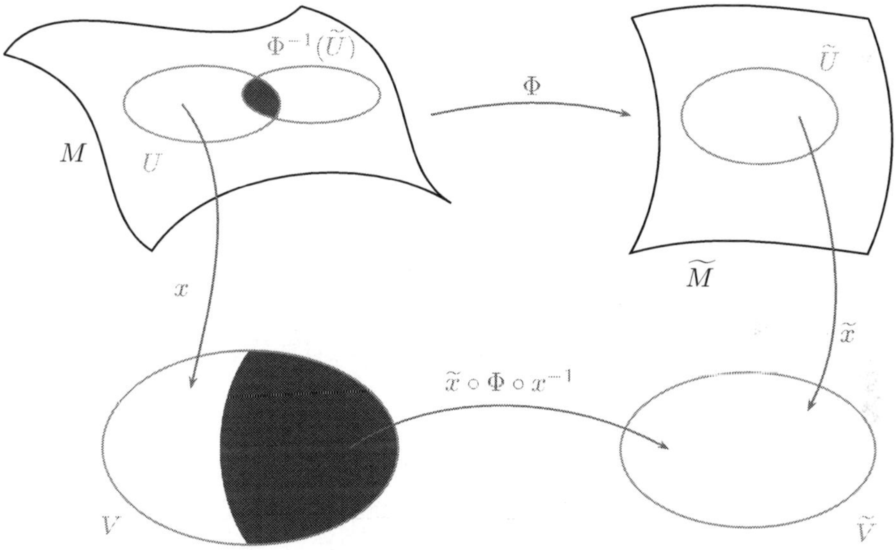

FIG. 6

Now that we know what differentiable maps are the next question arises:

QUESTION 2.9. *What is the derivative of a differentiable map?*

When dealing with differentiable maps between open subsets of Euclidean spaces many people think of the derivative as the Jacobi matrix containing the partial derivatives of the component functions of the map. This is certainly not a good point of view for generalizing the concept of derivative to maps defined between manifolds because it heavily uses the standard coordinates of $\mathbb{R}^n$. When working with manifolds we should think in a coordinate independent way. The right interpretation of the derivative is to consider it as the linear approximation of the map at a given point. But linear maps are defined between linear spaces (vector spaces) and manifolds are not vector spaces. So the question that has to be answered first is

QUESTION 2.10. *What is the linear approximation of a manifold?*

For the sake of simplicity we will from now on use the following

CONVENTION. From now on *manifold* stands for *differentiable manifold*. A *chart* will mean a chart in the corresponding C^∞-atlas $\mathcal{A}$.

DEFINITION 2.11. Let M be a manifold. Fix a point $p \in M$. Let $\gamma : I \to M$ and $\widetilde{\gamma} : \widetilde{I} \to M$ be two differentiable curves with

$$\gamma(0) = \widetilde{\gamma}(0) = p ,$$

where I, $\widetilde{I}$ are open intervals containing 0. We call the two curves *equivalent* and write $\gamma \sim \widetilde{\gamma}$ if and only if

$$\frac{d}{dt}(x \circ \gamma)\bigg|_{t=0} = \frac{d}{dt}(x \circ \widetilde{\gamma})\bigg|_{t=0}$$

for some chart $x : U \to V$ with $p \in U$.

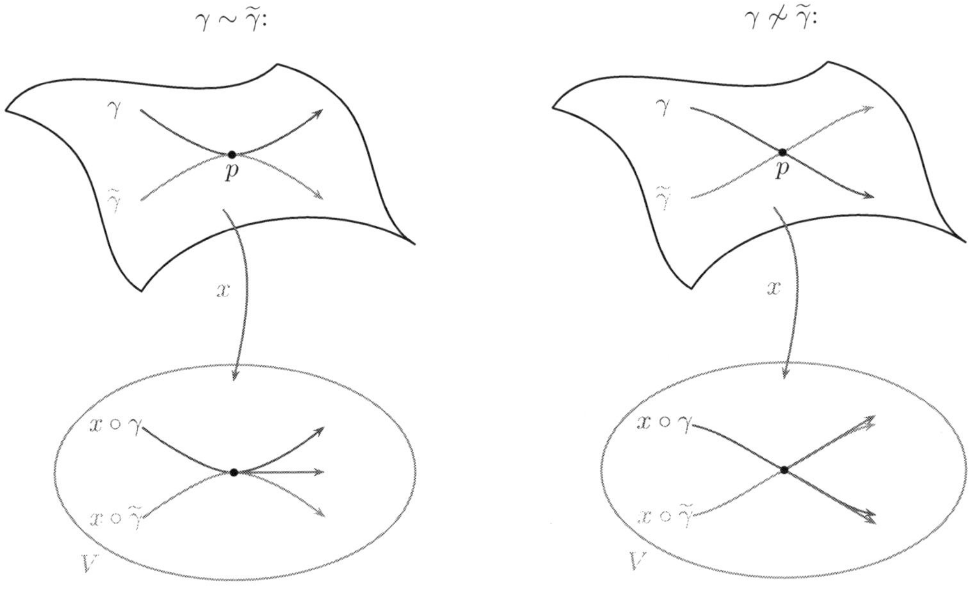

FIG. 7

Obviously this defines an equivalence relation on the set $\mathcal{C}_p$ of differentaible curves γ with $\gamma(0) = p$. We have to check that the definition of $\sim$ does not depend on the choice of chart x. If y is another chart containing p, then we get by the chain rule

$$\begin{aligned} \frac{d}{dt}(y \circ \gamma)\bigg|_{t=0} &= \frac{d}{dt}((y \circ x^{-1}) \circ (x \circ \gamma))\bigg|_{t=0} \\ &= D(y \circ x^{-1})(x(p))\,\frac{d}{dt}(x \circ \gamma)\bigg|_{t=0} . \end{aligned}$$

Hence $\frac{d}{dt}(x \circ \gamma)\big|_{t=0}$ and $\frac{d}{dt}(x \circ \widetilde{\gamma})\big|_{t=0}$ coincide if and only if $\frac{d}{dt}(y \circ \gamma)\big|_{t=0}$ and $\frac{d}{dt}(y \circ \widetilde{\gamma})\big|_{t=0}$ coincide. Here $D(y \circ x^{-1})(x(p))$ denotes the Jacobi matrix of the diffeomorphism $y \circ x^{-1}$ at the point $x(p)$.

Denote the equivalence class of γ by $\dot{\gamma}(0)$. We think of $\dot{\gamma}(0)$ as being the velocity vector of γ at p. Now we are ready to define the linear approximation of a manifold at a given point p.

DEFINITION 2.12. The set $T_pM := \{\dot{\gamma}(0) \mid \gamma \in \mathcal{C}_p\}$ is called the *tangent space* of M at p.

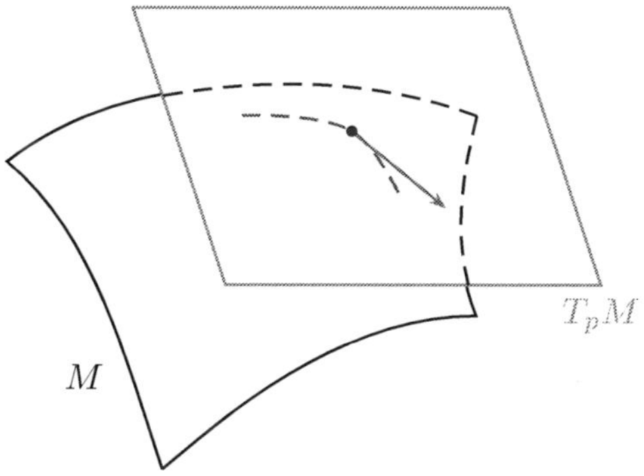

FIG. 8

In case the manifold happens to be a (finite dimensional real) vector space, then the tangent space of V at any point $p \in V$ can be *canonically* identified with V itself by the isomorphism

$$V \to T_pV, \quad v \mapsto \dot{\gamma}_{v,p}(0),$$

where $\gamma_{v,p}(t) = p + t \cdot v$.

DEFINITION 2.13. For $X = \dot{\gamma}(0) \in T_pM$ and a differentiable function $f : M \to \mathbb{R}$ define the *directional derivative* of f in direction X by

$$\partial_X f := \frac{d}{dt}(f \circ \gamma)\bigg|_{t=0}.$$

PROPERTIES.

1. $\partial_X f$ is well-defined, that is, it gives the same result for all $\gamma \in \mathcal{C}_p$ with $\dot{\gamma}(0) = X$.
2. ∂_X is local, i.e., if f and g coincide in a neighborhood of p, then $\partial_X f = \partial_X g$.
3. ∂_X is linear, i.e.,

$$\partial_X(\alpha \cdot f + \beta \cdot g) = \alpha \cdot \partial_X f + \beta \cdot \partial_X g$$

for all differentiable functions f and g and constants α and β.

4. ∂_X satisfies the Leibnitz rule,

$$\partial_X(f \cdot g) = \partial_X f \cdot g(p) + f(p) \cdot \partial_X g.$$

for all differentiable functions f and g.

An operator mapping differentiable functions defined near p to real numbers which satisfies properties 2., 3. and 4. is called a *derivation*. Hence ∂_X is a well-defined derivation. Write Der_p for the set of all derivations (at p).

FACTS.

1. The map $T_pM \to \mathrm{Der}_p$, $X \mapsto \partial_X$, is bijective.
2. For a chart $x : U \to V$ with $p \in U$ the derivations
$$\frac{\partial}{\partial x^i}(p) : f \longmapsto \frac{\partial f}{\partial x^i}(p) := \frac{\partial(f \circ x^{-1})}{\partial x^i}(x(p))$$
form a basis of the vector space Der_p, $i = 1, \ldots, n$, where n is the dimension of M.

COROLLARY 2.14. *Der_p is an n-dimensional vector space.*

Because of Fact 1 we can identify T_pM and Der_p which we will from now on do. This means that we will not distinguish between a tangent vector X and the corresponding derivation ∂_X. In particular, the tangent space T_pM inherits the structure of an n-dimensional vector space from Der_p. Very often in the literature T_pM is simply defined to be Der_p.

DEFINITION 2.15. Let $f : M \to N$ be a differentiable map. Fix $p \in M$. Then the *differential* of f at p is the linear map

$$df(p) : T_pM \to T_{f(p)}N, \quad \dot\gamma(0) \longmapsto \dot{\widehat{(f \circ \gamma)}}(0).$$

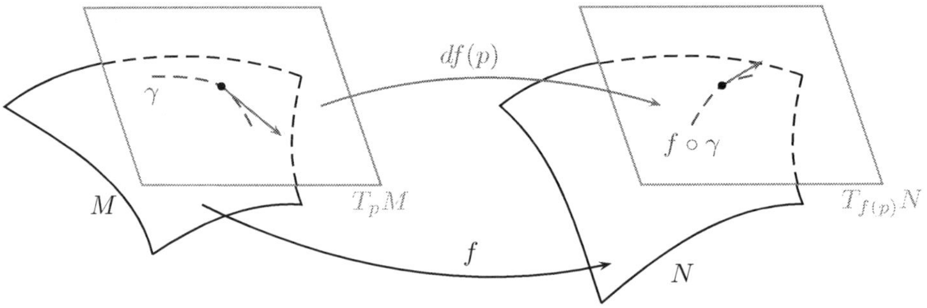

FIG. 9

Many other notations for the differential of a map are used in the literature, e. g. $df(p) = Df(p) = T_pf = f_{*,p}$.

EXPRESSION WITH RESPECT TO CHARTS. Given a chart x on M and a chart y on N one easily computes

$$df(p)\left(\frac{\partial}{\partial x^i}(p)\right) = \sum_j \frac{\partial(y^j \circ f \circ x^{-1})}{\partial x^i}(x(p)) \cdot \frac{\partial}{\partial y^j}(f(p)) .$$

Hence with respect to the basis $\frac{\partial}{\partial x^i}(p)$ of T_pM and the basis $\frac{\partial}{\partial y^j}(f(p))$ of $T_{f(p)}N$ the linear map $df(p)$ is given by the Jacobi matrix of $y \circ f \circ x^{-1}$ at $x(p)$.

DEFINITION 2.16. A *vector field* on M is a map X mapping each point $p \in M$ to a tangent vector $X(p) \in T_pM$.

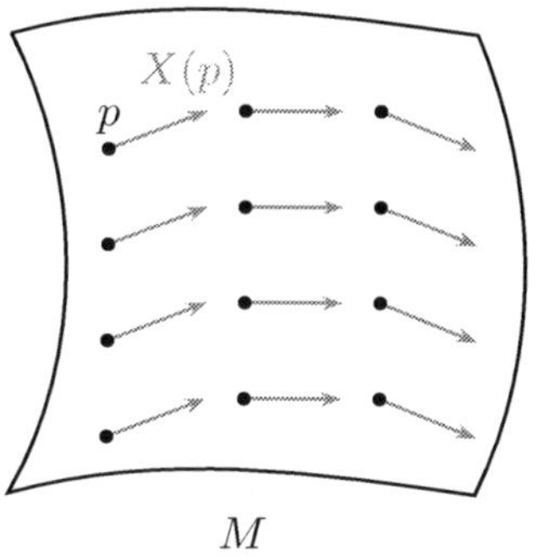

FIG. 10

EXPRESSION IN CHARTS. If $x : U \to V$ is a chart, then X can be written on U as

$$X(p) = \sum_{i=1}^{n} f_i(p) \cdot \frac{\partial}{\partial x^i}(p)$$

with uniquely determined functions

$$f_i : U \longrightarrow \mathbb{R}.$$

DEFINITION 2.17. A vector field is called *continuous* (or differentiable, $C^k, C^\infty, \dots$) if and only if for each chart the corresponding functions $f_1, \dots, f_n$ are continuous (or differentiable, $C^k, C^\infty, \dots$).

After so many definitions it is time for a result.

THEOREM 2.18 (Hedgehog Combing Theorem). *Every continuously combed hedgehog has at least one bald point.*

FIG. 11

Perhaps we should give a more mathematical formulation of this theorem. The surface of the hedgehog is modelled by a (fairly deformed) 2-dimensional sphere. When combed the pricks of the hedgehog form a continuous vector field. A bald point is a point at which the vector field has a zero. Hence the mathematical way of expressing Theorem 2.18 is

THEOREM 2.19. *Every continuous vector field on S^2 has at least one zero.*

In fact, this theorem is true for all even-dimensional spheres.

EXERCISE 2.20. Find continuous vector fields on S^n, n odd, which vanish nowhere.

DEFINITION 2.21. A *k-covector* at p is a multilinear alternating map

$$\underbrace{T_pM \times \cdots \times T_pM}_{k \text{ factors}} \longrightarrow \mathbb{R}.$$

We write $\Lambda^k T_p^* M := \{k\text{-covectors at } p\}$.

EXAMPLES 2.22. 1. By convention we put $\Lambda^0 T_p^* M := \mathbb{R}$.

2. $\Lambda^k T_p^* M = 0$ for $k > n$.
3. $\Lambda^1 T_p^* M = (T_pM)^* =$ dual space of T_pM.

3'. Let $f : M \to \mathbb{R}$ be differentiable. Then the differential of f at p, $T_pM \stackrel{df(p)}{\longrightarrow} T_{f(p)}\mathbb{R} \cong \mathbb{R}$, is a linear map from tangent space to $\mathbb{R}$, hence we may consider $df(p)$ as a 1-covector, $df(p) \in \Lambda^1 T_p^* M$.

For a chart x we know that $\frac{\partial}{\partial x^1}(p), \dots, \frac{\partial}{\partial x^n}(p)$ is basis of T_pM. Since the coordinate functions $x^1, \dots, x^n$ are smooth functions defined near p we can form the differentials $dx^1(p), \dots, dx^n(p) \in \Lambda^1 T_p^* M$. It turns out that they form the dual basis of $\Lambda^1 T_p^* M = (T_pM)^*$. Namely,

$$dx^i(p)\left(\frac{\partial}{\partial x^j}(p)\right) = \frac{\partial x^i}{\partial x^j}(p) = \delta^i_j .$$

Hence every 1-covector at p can be expressed in a unique way as a linear combination of the $dx^1(p), \dots, dx^n(p)$. For the differential of a function one checks

$$df(p) = \sum_{j=1}^{n} \frac{\partial f}{\partial x^j}(p) \cdot dx^j(p) .$$

WEDGE PRODUCT. For 1-covectors $\varphi_1, \dots, \varphi_k \in \Lambda^1 T_p^* M$ we define the k-covector $\varphi_1 \wedge \cdots \wedge \varphi_k \in \Lambda^k T_p^* M$ by

$$(\varphi_1 \wedge \cdots \wedge \varphi_k)(X_1, \dots, X_k) := \det\left(\varphi_i(X_j)\right)_{i,j=1,\dots,k} .$$

By the properties of the determinant we see that with this definition $\varphi_1 \wedge \cdots \wedge \varphi_k$ is indeed a k-covector, namely it is multilinear and alternating in the X_j. Moreover, it is also multilinear and alternating in the φ_j. It is a standard result of multilinear algebra [**Spi79**, p. 279] that suitable wedge products of basis vectors for the space of 1-covectors yields a basis for the space of k-covectors. More precisely,

$$dx^{i_1}(p) \wedge \cdots \wedge dx^{i_k}(p), \ 1 \le i_1 < \cdots < i_k \le n,$$

form a basis of $\Lambda^k T_p^* M$.

Hence every k-covector ω at p can be written in a unique manner in the form

$$\omega = \sum_{i_1 < \cdots < i_k} \underbrace{f_{i_1 \cdots i_k}}_{\in \mathbb{R}} dx^{i_1}(p) \wedge \cdots \wedge dx^{i_k}(p) .$$

DEFINITION 2.23. A *differential form of order* k (briefly k-*form*) is a map ω sending each point $p \in M$ to a k-covector $\omega(p)$ at p, $\omega(p) \in \Lambda^k T_p^* M$.

EXAMPLES 2.24. 1. 0-forms = functions.

2. For a differentiable function f, its differential df is a 1-form.

EXERCISE 2.25. Show that for all $X \in T_pM$

$$df(p)(X) = \partial_X f.$$

DEFINITION 2.26. A k-form is called *continuous* (or differentiable, $C^k, C^\infty, \dots$) if and only if for all charts the corresponding functions $f_{i_1\cdots i_k} : U \to \mathbb{R}$ are continuous (or differentiable, $C^k, C^\infty, \dots$).

Write $\Omega^k(M)$ for the space of all smooth k-forms defined on M. Differentiating functions is a map

$$d : \Omega^0(M) \longrightarrow \Omega^1(M)\,,\ f \mapsto df\ .$$

This has a generalization to forms of higher order,

$$d : \Omega^k(M) \longrightarrow \Omega^{k+1}(M)\,, \quad \omega \mapsto d\omega\ ,$$

defined with respect to a local chart x by

$$d\left(\sum_{i_1<\cdots<i_k} f_{i_1\cdots i_k}\, dx^{i_1} \wedge \cdots \wedge dx^{i_k}\right) := \sum_{i_1<\cdots<i_k} df_{i_1\cdots i_k} \wedge dx^{i_1} \wedge \cdots \wedge dx^{i_k}.$$

One checks that this definition of d is independent of the choice of the chart x. The operator $d : \Omega^k(M) \longrightarrow \Omega^{k+1}(M)$ is called *exterior differentiation.*

EXAMPLE 2.27. Let us look at the case $M = \mathbb{R}^2$. We work with the chart $x = \mathrm{Id} : \mathbb{R}^2 \to \mathbb{R}^2$. In other words, we work with the usual Cartesian coordinates. We compute the differential of the 1-form $\omega = x^1x^2dx^1$.

$$\begin{aligned} d\omega &= d\,(x^1x^2dx^1) = d\,(x^1x^2) \wedge dx^1 \\ &= (x^1dx^2 + x^2dx^1) \wedge dx^1 \\ &= x^1dx^2 \wedge dx^1 + x^2dx^1 \wedge dx^1 \\ &= -x^1dx^1 \wedge dx^2 + 0\ . \end{aligned}$$

Exterior differentiation has an important property, its square is zero, $d^2 = 0$. Note that d^2 increases the order by two,

$$\Omega^k(M) \xrightarrow{d} \Omega^{k+1}(M) \xrightarrow{d} \Omega^{k+2}(M)\ .$$

Property $d^2 = 0$ can also be expressed as $\mathrm{Im}(\Omega^{k-1}(M) \xrightarrow{d} \Omega^k(M)) \subset \mathrm{Ker}(\Omega^k(M) \xrightarrow{d} \Omega^{k+1}(M))$. This allows us to define *deRham cohomology* by taking the quotient space

$$H^k_{dR}(M) := \frac{\mathrm{Ker}(\Omega^k(M) \xrightarrow{d} \Omega^{k+1}(M))}{\mathrm{Im}(\Omega^{k-1}(M) \xrightarrow{d} \Omega^k(M))}\ .$$

For compact manifolds M these cohomology spaces turn out to be finite dimensional,

$$b_k(M) := \dim H^k_{dR}(M) < \infty\ .$$

The dimension $b_k(M)$ is called the kth *Betti number* of M.

Let $\Phi : M \to N$ be a smooth map. The map Φ can be used to pull back functions defined on N to functions defined on M as follows: For $f : N \to \mathbb{R}$ put

$$\Phi^* f := f \circ \Phi : M \to \mathbb{R}\ .$$

This has a generalization to differential forms: Define the *pull back*

$$\Phi^* : \Omega^k(N) \to \Omega^k(M)$$

by

$$(\Phi^* \omega)(p)\underbrace{(X_1, \ldots, X_k)}_{\in T_p M} := \omega(\Phi(p))\underbrace{(d\Phi(p)(X_1), \ldots, d\Phi(p)(X_k))}_{\in T_{\Phi(p)} N}.$$

Pull back has the following properties:

1. Φ^* is linear.
2. $(\Phi \circ \Psi)^* = \Psi^* \circ \Phi^*$ and $(\mathrm{Id}_M)^* = \mathrm{Id}_{\Omega^k(M)}$.
3. The diagram

$$\begin{array}{ccc} \Omega^k(N) & \xrightarrow{\Phi^*} & \Omega^k(M) \\ \downarrow d & & \downarrow d \\ \Omega^{k+1}(N) & \xrightarrow{\Phi^*} & \Omega^{k+1}(M) \end{array}$$

commutes.

Corollary 2.28. *Φ^* induces well-defined linear map*

$$\Phi^\sharp : H^k_{dR}(N) \longrightarrow H^k_{dR}(M)$$

mapping the cohomology class of ω to the cohomology class of $\Phi^*(\omega)$. This map has the properties

$$(\Phi \circ \Psi)^\sharp = \Psi^\sharp \circ \Phi^\sharp \quad \textit{and} \quad (\mathrm{Id}_M)^\sharp = \mathrm{Id}_{H^k_{dR}(M)}.$$

Corollary 2.29. *If M and N are diffeomorphic, then*

$$b_k(M) = b_k(N).$$

Proof. Let $\Phi : M \to N$ be a diffeomorphism. Then $\Phi^\sharp \circ (\Phi^{-1})^\sharp = (\Phi^{-1} \circ \Phi)^\sharp = (\mathrm{Id}_M)^\sharp = \mathrm{Id}_{H^k_{dR}(M)}$ and similarly $(\Phi^{-1})^\sharp \circ \Phi^\sharp = \mathrm{Id}_{H^k_{dR}(N)}$. Therefore $\Phi^\sharp : H^k_{dR}(N) \to H^k_{dR}(M)$ is an isomorphism with inverse $(\Phi^{-1})^\sharp$. QED

Example 2.30. If M is a compact surface of genus g, then it is known that

$$b_0(M) = b_2(M) = 1, \quad b_1(M) = 2g.$$

Hence surfaces of different genera cannot be diffeomorphic.

Definition 2.31. A C^∞-atlas of M is called *oriented* if and only if for any two charts x and y

$$\det D(y \circ x^{-1}) > 0.$$

A pair $(M, \mathcal{A})$ where M is a topological manifold and $\mathcal{A}$ is a maximal oriented atlas is called an *oriented differentiable manifold.*

Let M be an n-dimensional oriented manifold, let $\omega \in \Omega^n(M)$ (with compact support). Then

$$\int_M \omega \in \mathbb{R}$$

can be defined using oriented charts $x : U \to V$ by

$$\int_U \omega := \int_V (f \circ x^{-1})\, dx^1 \cdots dx^n,$$

where we have written

$$\omega = f \cdot dx^1 \wedge \cdots \wedge dx^n$$

and the right hand side $\int_V (f \circ x^{-1})\, dx^1 \cdots dx^n$ is the usual Lebesgue integral of a function defined on an open subset of $\mathbb{R}^n$.

In the zero dimensional case the underlying manifold is a discrete and countable set of points, $M = \{p_1, p_2, p_3, \ldots\}$. The orientation consists of attaching a sign $\varepsilon(p) = \pm 1$ to each point p. The integral of a function f is then given by

$$\int_M f = \sum_{p \in M} \varepsilon(p)\, f(p).$$

An orientation on a manifold with boundary induces an orientation on the boundary.

THEOREM 2.32 (Stokes). *Let M be an n-dimensional compact oriented manifold with boundary ∂M. Let $\omega \in \Omega^{n-1}(M)$. Then*

$$\int_M d\omega = \int_{\partial M} \omega$$

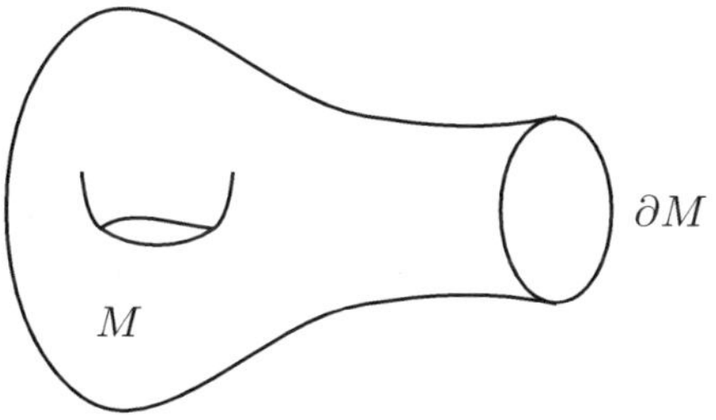

Fig. 12

EXAMPLE 2.33. Let us look at the Stokes Theorem in the simplest case where M is a 1-dimensional interval, $M = [a, b]$. Then the boundary consists of two points, $\partial M = \{a, b\}$. We pick a smooth 0-form, i. e. a function $\omega = f : [a, b] \to \mathbb{R}$. Its differential then is $d\omega = f' \cdot dx$. The Stokes Theorem now says

$$\int_a^b f'\, dx = f(b) - f(a)\,.$$

Hence the Theorem of Stokes is a generalization of the fundamental theorem of calculus to higher dimensions.

CHRISTIAN BÄR, UNIVERSITÄT POTSDAM, INSTITUT FÜR MATHEMATIK, PF 60 15 53, D-14415 POTSDAM, GERMANY
E-mail address: baer@math.uni-potsdam.de

CHAPTER 3

Clifford Algebras and Spinor Representations

Oussama Hijazi

Abstract. Clifford algebras are algebras naturally associated with vector spaces endowed with symmetric bilinear forms. In the case of the $n-$dimensional Euclidean vector space, these are algebras of square matrices with coefficients in $\mathbb{R}, \mathbb{C}$ or $\mathbb{H}$ or the sum of two such spaces. For simplicity, we mainly consider complex Clifford algebras and their representations. In the Clifford algebra of the n-dimensional euclidean space, we construct the spin group $\mathrm{Spin_n}$, the 2-fold covering of the n-dimensional special orthogonal group $\mathrm{SO_n}$. We then study the spinor representations and some of their basic properties.

Contents

For a general introduction, we refer to the excellent book of H.B. Lawson and M.L. Michelsohn on Spin Geometry [**LM89**].

3.1. Clifford Algebras

Definition 3.1. Let V be an n-dimensional vector space over a field $\mathbb{K} = \mathbb{R}$ or $\mathbb{C}$ and q a non-degenerate bilinear form on V. The *Clifford algebra* $\mathrm{C}\ell(V,q)$ associated with (V,q) is the associative algebra with unit, defined by

$$\mathrm{C}\ell(V,q) := V^{\otimes}/\mathcal{I}(V,q)$$

where $V^{\otimes} = \oplus_{i\geq 0}\ V^{\otimes i}$ is the tensor algebra of V, and $\mathcal{I}(V,q)$ is the two-sided ideal generated by all elements of the form $x \otimes x + q(x,x)1$, for $x \in V$. The product in the algebra $\mathrm{C}\ell(V,q)$ will be denoted by "$\cdot$".

The Clifford algebra shares with the tensor algebra the following important property:

PROPOSITION 3.2. *[Universal Property] Let $\mathcal{A}$ be an associative algebra with unit and $f\colon V \to \mathcal{A}$ a linear map such that for all $v \in V$,*

$$f(v) \circ f(v) = -q(v,v)1. \tag{3.1}$$

Then f uniquely extends to a $\mathbb{K}$–algebra homomorphism

$$\widetilde{f}\colon \mathrm{C}\ell(V,q) \to \mathcal{A}.$$

Furthermore, $\mathrm{C}\ell(V,q)$ is the unique associative $\mathbb{K}$–algebra with this property.

PROOF. Let $\iota : V \hookrightarrow V^{\otimes}$ be the natural inclusion. By the universal property of the tensor algebra, there exists an algebra morphism F such that $f = F \circ \iota$. It is then sufficient to show that $\mathcal{I}(V,q) \subset \mathrm{Ker}F$. In fact, for any $v \in V$, we have

$$\begin{aligned} F(v \otimes v + q(v,v)1) &= F(v) \cdot F(v) + q(v,v)1_{\mathcal{A}} \\ &= f(v) \cdot f(v) + q(v,v)1_{\mathcal{A}} \\ &= 0\,. \end{aligned}$$

□

REMARKS 3.3. (1) The Clifford algebra $\mathrm{C}\ell(V,q)$ is the algebra generated by V subject to the relations

$$x \cdot y + y \cdot x = -2q(x,y)1, \qquad \forall x, y \in V\,. \tag{3.2}$$

(2) If $\{e_1, \dots, e_n\}$ is a q–orthonormal basis of V, then the system

$$\{1,\, e_{i_1} \cdot \ldots \cdot e_{i_k} \mid 1 \leq i_1 < \ldots < i_k \leq n, \quad 0 \leq k \leq n\} \tag{3.3}$$

genrates the underlined vector space structure of $\mathrm{C}\ell(V,q)$, thus $\dim \mathrm{C}\ell(V,q) \leq 2^n$. We will see that in fact such a system is a basis of $\mathrm{C}\ell(V,q)$.

(3) The Clifford algebra $\mathrm{C}\ell(V,0)$ is the exterior algebra of V, $\Lambda^* V$.

We next prove that, as vector spaces, the Clifford algebra and the exterior algebra can be naturally identified.

PROPOSITION 3.4. *There is a canonical isomorphism of vector spaces between the exterior algebra and the Clifford algebra of (V,q) which is given by:*

$$\begin{array}{ccc} \Lambda^* V & \longrightarrow & \mathrm{C}\ell(V,q) \\ e_{i_1} \wedge \ldots \wedge e_{i_k} & \longmapsto & e_{i_1} \cdot \ldots \cdot e_{i_k}\,. \end{array} \tag{3.4}$$

This isomorphism does not depend on the choice of the orthonormal basis.

PROOF. For any vector $v \in V$, let $v\wedge$ and $v\lrcorner$ be respectively the exterior and interior multiplications by v and consider the map

$$\begin{array}{cccc} f : & V & \longrightarrow & \mathrm{End}(\Lambda^* V) \\ & v & \longmapsto & f(v)(\omega) = v \wedge \omega - v \lrcorner \omega\,. \end{array}$$

By definition of the exterior and interior products, it follows

$$\begin{aligned} f(v) \circ f(v)(\omega) &= f(v) \circ (v \wedge \omega - v \lrcorner \omega) \\ &= v \wedge (v \wedge \omega - v \lrcorner \omega) - v \lrcorner\, (v \wedge \omega - v \lrcorner \omega) \\ &= -v \wedge (v \lrcorner \omega) - v \lrcorner\, (v \wedge \omega) \\ &= -v \wedge (v \lrcorner \omega) - q(v,v)\omega + v \wedge (v \lrcorner \omega) \\ &= -q(v,v)\omega\,, \end{aligned}$$

hence f satisfies condition (3.1), with $\mathcal{A} = \mathrm{End}(\Lambda^* V)$. Hence, by the universal property, f uniquely extends to a homomorphism

$$\widetilde{f} : \mathrm{C}\ell(V, q) \longrightarrow \mathrm{End}(\Lambda^* V).$$

Now define the linear map

$$\begin{array}{rcll} F & : & \mathrm{C}\ell(V,q) & \longrightarrow & \Lambda^* V \\ & & \varphi & \longmapsto & \left(\widetilde{f}(\varphi)\right)(1_{\Lambda^* V}). \end{array}$$

If $\{e_1, \ldots, e_n\}$ is a q–orthonormal basis of V then for any $i_1 < \ldots < i_k$ we have

$$F(e_{i_1} \cdot \ldots \cdot e_{i_k}) = e_{i_1} \wedge \ldots \wedge e_{i_k},$$

i.e., F is surjective. On the other hand, Remark (2) above shows that $\dim \mathrm{C}\ell(V,q) \leq 2^n = \dim \Lambda^* V$, thus F is an isomorphism.

□

Definition 3.5. The Clifford algebra $\mathrm{C}\ell_n := \mathrm{C}\ell(\mathbb{R}^n, q^{\mathbb{R}})$ (resp. $\mathbb{C}\ell_n := \mathrm{C}\ell(\mathbb{C}^n, q^{\mathbb{C}})$) associated with the euclidean scalar product $q^{\mathbb{R}}$ (resp. $q^{\mathbb{C}}$) defined by $q^{\mathbb{R}}(x,y) = \sum_i x_i y_i$ (resp. $q^{\mathbb{C}}(z,w) = \sum_i z_i w_i$) is called the n–dimensional real (resp. complex) Clifford algebra.

Examples 3.6. If $\{e_1, \ldots, e_n\}$ denotes the canonical orthonormal basis of $\mathbb{R}^n$, then

$$e_i \cdot e_j + e_j \cdot e_i = -2\delta_{ij}\,. \tag{3.5}$$

Hence,

(1) a basis of $\mathrm{C}\ell_1$ is given by $\{1, e_1\}$. Since $e_1^2 = -1$, one has $\mathrm{C}\ell_1 \cong \mathbb{C}$.

(2) a basis of $\mathrm{C}\ell_2$ is given by $\{1,\, e_1,\, e_2,\, e_1{\cdot}e_2\}$. Since $e_1^2 = e_2^2 = (e_1{\cdot}e_2)^2 = -1$, one has $\mathrm{C}\ell_2 \cong \mathbb{H}$.

(3) a basis of $\mathrm{C}\ell_3$ is given by $\{1, e_1,\, e_2,\, e_3,\, e_1{\cdot}e_2,\, e_2{\cdot}e_3,\, e_3{\cdot}e_1,\, e_1{\cdot}e_2{\cdot}e_3\}$. With the help of the following natural linear identifications with the quaternions:

$$\{1,\, e_1 \cdot e_2,\, e_2 \cdot e_3,\, e_3 \cdot e_1\} \cong \{1, i, j, k\},$$

and

$$\{e_1, e_2, e_3, e_1 \cdot e_2 \cdot e_3\} \cong \{i, j, k, -1\},$$

it follows that $\mathrm{C}\ell_3 \cong \mathbb{H} \oplus \mathbb{H}$.

An immediate consequence of the universal property is the fact that the complex Clifford algebra is isomorphic to the complexification of the real Clifford algebra, i.e., $\mathbb{C}\ell_n \cong \mathrm{C}\ell_n \otimes_{\mathbb{R}} \mathbb{C}$. To see this, note that by the universal property, the inclusion $\mathbb{R}^n \to \mathbb{C}^n \subset \mathbb{C}\ell_n$ could be extented to a morphism $\imath : \mathrm{C}\ell_n \to \mathbb{C}\ell_n$ which induces the morphism

$$\begin{array}{rcll} \imath^{\mathbb{C}} & : & \mathrm{C}\ell_n \otimes_{\mathbb{R}} \mathbb{C} & \longrightarrow & \mathbb{C}\ell_n \\ & & a \otimes z & \longmapsto & \imath(a) z. \end{array}$$

On the other hand, the $\mathbb{C}$–linear map

$$\begin{array}{rcll} \jmath^{\mathbb{C}} & : & \mathbb{C}^n & \longrightarrow & \mathrm{C}\ell_n \otimes_{\mathbb{R}} \mathbb{C} \\ & & v & \longmapsto & \Re(v) \otimes 1 + \mathrm{Im}(v) \otimes i \end{array}$$

satisfies condition (3.1), i.e., $\jmath^{\mathbb{C}}(v)^2 = -q^{\mathbb{C}}(v,v)1$, hence it could be extended to a morphism $\mathbb{C}\ell_n \to \mathrm{C}\ell_n \otimes_{\mathbb{R}} \mathbb{C}$, which is the inverse of $\imath^{\mathbb{C}}$. We also have the following useful corollary:

COROLLARY 3.7. *Let $f\colon (V,q) \to (\overline{V},\overline{q})$ be an isometry between euclidean vector spaces. Then f uniquely extends to a $\mathbb{K}$–algebra homomorphism*

$$\widetilde{f}\colon \mathrm{C}\ell(V,q) \to \mathrm{C}\ell(\overline{V},\overline{q}).$$

REMARK 3.8. We denote by $\mathrm{O}(V,q)$ the space of isometric homomorphisms of an euclidean vector space (V,q). On the Clifford algebra $\mathrm{C}\ell(V,q)$, we have two basic automorphisms:

(1) The isometry $-\mathrm{Id} \in \mathrm{O}(V,q)$ gives rise to

$$\begin{array}{rccc} \alpha : & \mathrm{C}\ell(V,q) & \longrightarrow & \mathrm{C}\ell(V,q) \\ & e_{i_1}\cdot \ldots \cdot e_{i_k} & \longmapsto & (-1)^k e_{i_1}\cdot \ldots \cdot e_{i_k} \end{array}$$

As $\alpha^2 = \mathrm{Id}$, we get the decomposition

$$\mathrm{C}\ell(V,q) = \mathrm{C}\ell^0(V,q) \oplus \mathrm{C}\ell^1(V,q),$$

where

$$\mathrm{C}\ell^i(V,q) := \{\, \varphi \in \mathrm{C}\ell(V,q) \mid \alpha(\varphi) = (-1)^i \varphi \,\}. \tag{3.6}$$

Clearly, for $i,\, j \in \mathbb{Z}_2$, we have

$$\mathrm{C}\ell^i(V,q)\cdot \mathrm{C}\ell^j(V,q) \subset \mathrm{C}\ell^{i+j}(V,q)\,.$$

Thus the Clifford algebra $\mathrm{C}\ell(V,q)$ is a $\mathbb{Z}_2$–graded algebra. The subspace $\mathrm{C}\ell^0(V,q)$ (resp. $\mathrm{C}\ell^1(V,q)$) is called the *even* (resp. *odd*) part of $\mathrm{C}\ell(V,q)$.

(2) Consider the $\mathbb{K}$–algebra anti-automorphism defined by

$$\begin{array}{rccc} {}^t(\,) : & V^{\otimes} & \longrightarrow & V^{\otimes} \\ & x_{i_1}\otimes \ldots \otimes x_{i_k} & \longmapsto & x_{i_k}\otimes \ldots \otimes x_{i_1}\,. \end{array}$$

Since ${}^t(\mathcal{I}(V,q)) \subset \mathcal{I}(V,q)$, there is an induced homomorphism

$$\begin{array}{rccc} {}^t(\,) : & \mathrm{C}\ell(V,q) & \longrightarrow & \mathrm{C}\ell(V,q) \\ & x_{i_1}\cdot \ldots \cdot x_{i_k} & \longmapsto & x_{i_k}\cdot \ldots \cdot x_{i_1}. \end{array}$$

Another important consequence of the universal property of Clifford algebras is the following:

PROPOSITION 3.9. *The Clifford algebra $\mathrm{C}\ell_n$ is isomorphic to the even part of $\mathrm{C}\ell_{n+1}$, i.e.,*

$$\mathrm{C}\ell_n \cong \mathrm{C}\ell^0_{n+1}\,. \tag{3.7}$$

PROOF. Denote by $\{e_1,\ldots,e_n\}$ and $\{e_1,\ldots,e_{n+1}\}$ the canonical bases of $\mathbb{R}^n$ and $\mathbb{R}^{n+1}$. Note that we are identifying $\mathbb{R}^n$ with its canonical injection in $\mathbb{R}^{n+1}$. Define the linear map

$$\begin{array}{rccc} f : & \mathbb{R}^n & \longrightarrow & \mathrm{C}\ell^0_{n+1} \\ & e_i & \longmapsto & e_i\cdot e_{n+1}. \end{array}$$

By definition of f, we have $f(e_i)^2 = -1$, thus f extends to $\widetilde{f}\colon \mathrm{C}\ell_n \to \mathrm{C}\ell^0_{n+1}$. Clearly, $\widetilde{f}$ is an injective linear map between vector spaces of the same dimension, thus the map $\widetilde{f}$ is an algebra isomorphism.

□

In Proposition 3.4, we introduced the canonical vector isomorphism between $C\ell_n$ and $\Lambda^*\mathbb{R}^n$. The following proposition gives the relation between the product in the Clifford algebra in terms of the exterior and interior products in the exterior algebra.

PROPOSITION 3.10. *For all $v \in \mathbb{R}^n$ and all $\varphi \in \Lambda^p\mathbb{R}^n \subset \Lambda^*\mathbb{R}^n \cong C\ell_n$, we have*

$$v \cdot \varphi \cong v \wedge \varphi - v \lrcorner\, \varphi \qquad \text{and} \qquad \varphi \cdot v \cong (-1)^p (v \wedge \varphi + v \lrcorner\, \varphi). \tag{3.8}$$

PROOF. Let $v = e_j$ and $\varphi = e_{i_1} \cdot \ldots \cdot e_{i_p}$.

1. If there exists i_k such that $j = i_k$ then $v \wedge \varphi = 0$ and

$$\begin{aligned} v \lrcorner\, \varphi &= (-1)^{k-1} e_{i_1} \wedge \ldots \wedge e_{i_{k-1}} \wedge e_{i_{k+1}} \wedge \ldots \wedge e_{i_p} \\ &\cong (-1)^{k-1} e_{i_1} \cdot \ldots \cdot e_{i_{k-1}} \cdot e_{i_{k+1}} \cdot \ldots \cdot e_{i_p} \\ &= -v \cdot \varphi \\ &= (-1)^p \varphi \cdot v. \end{aligned} \tag{3.9}$$

2. If $j \notin \{i_1, \ldots, i_p\}$ then $v \lrcorner\, \varphi = 0$ and

$$\begin{aligned} v \wedge \varphi &= e_j \wedge e_{i_1} \wedge \ldots \wedge e_{i_p} \\ &\cong e_j \cdot e_{i_1} \cdot \ldots \cdot e_{i_p} \\ &= v \cdot \varphi \\ &= (-1)^p \varphi \cdot v. \end{aligned} \tag{3.10}$$

We then conclude by bilinearity.

□

We also need the following interesting feature of the Clifford product:

PROPOSITION 3.11. *For any $\varphi \in \Lambda^p\mathbb{R}^n \subset \Lambda^*\mathbb{R}^n \cong C\ell_n$, we have*

$$\sum_{j=1}^{n} e_j \cdot \varphi \cdot e_j = (-1)^{p+1}(n - 2p)\, \varphi\,. \tag{3.11}$$

PROOF. It is sufficient to consider $\varphi = e_{i_1} \cdot \ldots \cdot e_{i_p}$. Then by (3.5) and for a fixed $j \in \{1, \ldots, n\}$, it is straightforward to see

$$e_j \cdot \varphi \cdot e_j = \begin{cases} (-1)^p \varphi, & \text{if } j \in \{i_1, \ldots, i_p\} \\ (-1)^{p+1} \varphi, & \text{if } j \notin \{i_1, \ldots, i_p\}\,, \end{cases}$$

hence

$$\sum_{j=1}^{n} e_j \cdot \varphi \cdot e_j = \left[(-1)^p p + (-1)^{p+1}(n-p)\right] \varphi = (-1)^{p+1}(n - 2p)\, \varphi\,. \tag{3.12}$$

□

3.2. Classification of Clifford Algebras

We now give some basic facts which lead to the classification of the complex Clifford algebras. For the real Clifford algebras, we mention without proof the following proposition:

PROPOSITION 3.12. *For all $n \in \mathbb{N}$, one has: $C\ell_{n+8} \cong C\ell_n \otimes_{\mathbb{R}} C\ell_8$.*

The periodicity of the real Clifford algebras combined with Proposition 3.9 and the knowledge of $\mathbb{C}\ell_n$ for $1 \leq n \leq 8$, yield to the classification of the real Clifford algebras.

Denote by $\mathbb{C}(2^m)$ the $\mathbb{C}$-algebra of $2^m \times 2^m$ complex matrices. We have

PROPOSITION 3.13. *The complex Clifford algebras are either isomorphic to $\mathbb{C}(2^m)$, or to $\mathbb{C}(2^m) \oplus \mathbb{C}(2^m)$. In particular:*

$$\begin{aligned} \mathbb{C}\ell_{2m} &\cong \mathbb{C}(2^m) \cong \mathrm{End}_{\mathbb{C}}(\Sigma_{2m})\,, \\ \mathbb{C}\ell_{2m+1} &\cong \mathbb{C}(2^m) \oplus \mathbb{C}(2^m) \cong \mathrm{End}_{\mathbb{C}}(\Sigma_{2m}) \oplus \mathrm{End}_{\mathbb{C}}(\Sigma_{2m}), \end{aligned} \tag{3.13}$$

where Σ_{2m} is a complex vector space isomorphic to $\mathbb{C}^{2^m}$.

PROOF. We only give the proof for $\mathbb{C}\ell_{2m}$. An abstract proof of the isomorphism (3.13) will be given later. For this, let $\{e_1, \ldots, e_m, e_{m+1}, \ldots, e_{2m}\}$ be the canonical orthonormal basis of $\mathbb{R}^{2m}$ and $\{z_j, \overline{z}_j\}_{j=1,\ldots,m}$, a Witt basis of $\mathbb{R}^{2m} \otimes_{\mathbb{R}} \mathbb{C}$, i.e.,

$$z_j := \frac{1}{2}(e_j \otimes 1 - e_{j+m} \otimes i) \qquad \text{and} \qquad \overline{z}_j := \frac{1}{2}(e_j \otimes 1 + e_{j+m} \otimes i).$$

These vectors satisfy the relations

$$\begin{aligned} q^{\mathbb{C}}(z_j, z_k) = q^{\mathbb{C}}(\overline{z}_j, \overline{z}_k) &= 0\,, \\ q^{\mathbb{C}}(\overline{z}_j, z_k) = q^{\mathbb{C}}(z_j, \overline{z}_k) &= \tfrac{1}{2}\delta_{jk}\,, \end{aligned} \tag{3.14}$$

which yield

$$\begin{aligned} z_j \cdot_{\mathbb{C}} z_k + z_k \cdot_{\mathbb{C}} z_j &= 0 \\ \overline{z}_j \cdot_{\mathbb{C}} \overline{z}_k + \overline{z}_k \cdot_{\mathbb{C}} \overline{z}_j &= 0 \\ z_j \cdot_{\mathbb{C}} \overline{z}_k + \overline{z}_k \cdot_{\mathbb{C}} z_j &= -\delta_{jk}, \end{aligned} \tag{3.15}$$

since for $x, y \in \mathbb{C}^{2m} \subset \mathbb{C}\ell_{2m}$: $x \cdot_{\mathbb{C}} y + y \cdot_{\mathbb{C}} x = -2q^{\mathbb{C}}(x,y)$. For simplicity we write "$\cdot$" for the complex Clifford product "$\cdot_{\mathbb{C}}$".

Let $\overline{\omega} := \overline{z}_1 \cdot \ldots \cdot \overline{z}_m$ and $z_{L_r} := z_{l_1} \cdot \ldots \cdot z_{l_r}$, for $1 \leq l_1 < \ldots < l_r \leq m$. Note that $\overline{z}_k \cdot \overline{\omega} = 0$ for $1 \leq k \leq m$ by the relations (3.15). Then, define

$$\Sigma_{2m} = \mathrm{span}\{\, z_{L_r} \cdot \overline{\omega} \mid 1 \leq l_1 < \ldots < l_r \leq m,\ 0 \leq r \leq m\}.$$

Since Σ_{2m} and $\mathbb{C}^{2^m}$ have the same dimension, they are isomorphic as complex vector spaces. We define the linear map:

$$\begin{aligned} \rho:\quad \mathbb{C}\ell_{2m} &\longrightarrow \mathrm{End}(\Sigma_{2m}) \\ v = z_{J_p} \cdot \overline{z}_{K_q} &\longmapsto \rho(v) = \big(z_{L_r} \cdot \overline{\omega} \mapsto z_{J_p} \cdot \overline{z}_{K_q} \cdot z_{L_r} \cdot \overline{\omega}\big)\,. \end{aligned}$$

Obviously, ρ is a homomorphism of algebras. We show now that ρ is surjective. First note that, for $1 \leq j \leq m$:

$$\begin{aligned} \rho(z_j \cdot \overline{z}_1)(z_1 \cdot \overline{\omega}) &= z_j \cdot \overline{z}_1 \cdot z_1 \cdot \overline{\omega} \\ &= z_j \cdot (-z_1 \cdot \overline{z}_1 - 1) \cdot \overline{\omega} \\ &= -z_j \cdot z_1 \cdot \underbrace{\overline{z}_1 \cdot \overline{\omega}}_{0} - z_j \cdot \overline{\omega} \\ &= -z_j \cdot \overline{\omega} \end{aligned} \tag{3.16}$$

and for $2 \leq l \leq m$

$$\begin{aligned} \rho(z_j \cdot \overline{z}_1)(z_l \cdot \overline{\omega}) &= z_j \cdot \overline{z}_1 \cdot z_l \cdot \overline{\omega} \\ &= z_j \cdot (-z_l \cdot \overline{z}_1) \cdot \overline{\omega} \\ &= 0 \end{aligned} \tag{3.17}$$

Similarly, one computes that the image under $\rho(z_j \cdot \overline{z}_1)$ of a basis vector of Σ_{2m} not containing z_1 is zero, whereas the image of a basis vector containing z_1 is the same vector with z_1 being replaced by z_j or $-z_j$.

Let $\omega := z_1 \cdot \ldots \cdot z_m$. Then, for elements of the Clifford algebra of the form

$$v = z_{J_p} \cdot \overline{\omega} \cdot \omega \cdot \overline{z}_{K_q} := z_{j_1} \cdot \ldots \cdot z_{j_p} \cdot \overline{\omega} \cdot \omega \cdot \overline{z}_{k_1} \cdot \ldots \cdot \overline{z}_{k_q}$$

one has

$$\rho(v)\left(z_{k_1} \cdot \ldots \cdot z_{k_q} \cdot \overline{\omega}\right) = \pm z_{j_1} \cdot \ldots \cdot z_{j_p} \cdot \overline{\omega},$$

whereas $\rho(v)$ maps all other basis vectors of Σ_{2m} to zero. Thus ρ is surjective, and since $\dim \mathbb{C}\ell_{2m} = \dim \operatorname{End}(\mathbb{C}^{2^m})$, it is an isomorphism.

□

The above considerations yield to the classification of the complex Clifford algebras. For completeness, we also include the real Clifford algebras in the following table:

n	1	2	3	4	5	6	7	8
Cl_n	$\mathbb{C}$	$\mathbb{H}$	$\mathbb{H} \oplus \mathbb{H}$	$\mathbb{H}(2)$	$\mathbb{C}(4)$	$\mathbb{R}(8)$	$\mathbb{R}(8) \oplus \mathbb{R}(8)$	$\mathbb{R}(16)$
$\mathbb{C}l_n$	$\mathbb{C} \oplus \mathbb{C}$	$\mathbb{C}(2)$	$\mathbb{C}(2) \oplus \mathbb{C}(2)$	$\mathbb{C}(4)$	$\mathbb{C}(4) \oplus \mathbb{C}(4)$	$\mathbb{C}(8)$	$\mathbb{C}(8) \oplus \mathbb{C}(8)$	$\mathbb{C}(16)$

TABLE I

3.3. Spin Groups

Consider the following map

$$\begin{array}{rcl} N : \mathbb{C}\ell(V,q) & \longrightarrow & \mathbb{C}\ell(V,q) \\ a & \longmapsto & \alpha({}^t a) \cdot a\,. \end{array}$$

Note that $N(v) = |v|^2$, for all $v \in V$. Denote by $\mathbb{C}\ell^*(V,q)$ the group of invertible elements of $\mathbb{C}\ell(V,q)$. It could be easily checked that $N(\alpha({}^t(a^{-1}))) \cdot N(a) = 1$, this implies $N(\mathbb{C}\ell^*(V,q)) \subset \mathbb{C}\ell^*(V,q)$. Since, $v \cdot v = -|v|^2$ it is clear that any nonzero vector $v \in V \subset \mathbb{C}\ell(V,q)$ lies in $\mathbb{C}\ell^*(V,q)$, hence $v^{-1} = -v/|v|^2$. For each $a \in \mathbb{C}\ell^*(V,q)$, we consider the interior automorphism

$$\begin{array}{rcl} \mathrm{Ad}_a : \mathbb{C}\ell(V,q) & \longrightarrow & \mathbb{C}\ell(V,q) \\ b & \longmapsto & a \cdot b \cdot a^{-1} \end{array}$$

and the twisted interior automorphism defined by

$$\begin{array}{rcl} \widetilde{\mathrm{Ad}}_a : \mathbb{C}\ell(V,q) & \longrightarrow & \mathbb{C}\ell(V,q) \\ b & \longmapsto & \alpha(a) \cdot b \cdot a^{-1}. \end{array}$$

Observe that for any nonzero vector v, the vector space $V \subset \mathbb{C}\ell(V,q)$ is stable by $\widetilde{\mathrm{Ad}}_v$. In fact, the map $\widetilde{\mathrm{Ad}}_v|_V$ is the symmetry w.r.t the hyperplane $v^\perp$.

Define the group $P(V,q)$ by

$$P(V,q) = \{a \in \mathbb{C}\ell^*(V,q) \mid \widetilde{\mathrm{Ad}}_a(V) \subset V\},$$

and write $P_n := P(\mathbb{R}^n, q^{\mathbb{R}})$.

In the following, we take $(V,q) = (\mathbb{K}^n, q^{\mathbb{K}})$. For $\mathbb{K} = \mathbb{C}$, this choice is not restrictive, while in case $\mathbb{K} = \mathbb{R}$, this choice means that we will not consider the case of bilinear forms with nonzero signature.

LEMMA 3.14. (i) *The kernel of the map* $\widetilde{\mathrm{Ad}} : P(V,q) \longrightarrow \mathrm{Gl_n}(\mathbb{K})$ *is* $\mathbb{K}^*$.

(ii) *We have:* $N(P(V,q)) \subset \mathbb{K}^*$.

(iii) *The map* $N|_{P(V,q)} : P(V,q) \longrightarrow \mathbb{K}^*$ *is a morphism of groups.*

PROOF. (i) Consider $a \in C\ell(V,q)$ such that

$$\alpha(a) \cdot v = v \cdot a \ , \qquad \forall v \in \mathbb{K}^n \tag{3.18}$$

and, as above, denote by $\{e_1, \ldots, e_n\}$ the canonical basis of $\mathbb{K}^n$. By Proposition 3.4, for each fixed $i, 1 \leq i \leq n$, any element $a \in P(V,q)$ could be uniquely written as $a = b + e_i \cdot c$, where b and c are linear combinations of elements in the basis (3.3), not involving e_i. Condition (3.18), written for $v = e_i$, is equivalent to

$$\alpha(b) \cdot e_i = e_i \cdot b \qquad \text{and} \qquad \alpha(c) \cdot e_i = -e_i \cdot c. \tag{3.19}$$

The last relation implies $c = 0$. Since the argument holds for each e_i, it follows that a is a scalar and since $a \in C\ell^*(V,q)$, it should be a nonzero scalar.

(ii) For all $v \in \mathbb{K}^n$ and $a \in P(V,q)$ we have $\widetilde{\mathrm{Ad}}_a(v) \in \mathbb{K}^n$, hence $\widetilde{\mathrm{Ad}}_a(v) = {}^t(\widetilde{\mathrm{Ad}}_a(v))$, i.e., $\widetilde{\mathrm{Ad}}_{N(a)}(v) = v$. From (i) we conclude that $N(a) \in \mathbb{K}^*$.

(iii) By (ii) we get $N(a \cdot b) = \alpha({}^tb \cdot {}^ta)a \cdot b = \alpha({}^tb) \cdot N(a)b = N(a)N(b)$.

□

DEFINITION 3.15. (i) The spin group Spin_n is the subgroup of elements in P_n of the form $v_1 \ldots v_{2l}$, with $l \geq 1$ and $|v_i| = 1$, for $1 \leq i \leq 2l$.

(ii) The conformal spin group $\mathrm{CSpin_n}$ is the group $\mathrm{Spin}_n \times \mathbb{R}^*_+$.

PROPOSITION 3.16. *For* $n \geq 2$*, the morphism* $\xi := \widetilde{\mathrm{Ad}}|_{\mathrm{Spin}_n}$ *is a nontrivial double covering of the special orthogonal group* SO_n*. In particular for* $n \geq 3$*, the group* Spin_n *is the universal cover of* SO_n*.*

PROOF. We know that the image of a nonzero vector by the map $\widetilde{\mathrm{Ad}} : C\ell^*_n \to \mathrm{Gl_n}$ is the symmetry w.r.t the hyperplan orthogonal to this vector. The image of $\widetilde{\mathrm{Ad}}|_{\mathrm{Spin}_n}$ is then the group of even products of such symmetries, which by the Cartan-Dieudonné Theorem, is exactly the group SO_n. Using (iii) of Lemma 3.14, any element a of the spin group satisfies $N(a) = 1$. By (i) of Lemma 3.14, we conclude that the kernel of $\widetilde{\mathrm{Ad}} : \mathrm{Spin}_n \to \mathrm{SO}_n$ is $\{\pm 1\}$. To show that the covering is nontrivial it is sufficient to check that 1 and -1 belong to the same connected component of Spin_n. To see this, choose unit orthogonal vectors $v, w \in \mathbb{R}^n$ $(n \geq 2)$ and note that the curve

$$\begin{array}{rccl} c : & [0, \pi/2] & \longrightarrow & \mathrm{Spin}_n \\ & t & \longmapsto & c(t) = (v \sin t + w \cos t) \cdot (v \sin t - w \cos t) \end{array}$$

satisfies $c(0) = 1$, $c(\pi/2) = -1$. Finally, the last assertion is due to the fact that for $n \geq 3$, we have $\pi_1(\mathrm{SO}_n) = \mathbb{Z}_2$.

□

EXAMPLES 3.17. With the obvious notations, define the following classical groups:

(1) The unitary group $\mathrm{U}_n := \{A \in \mathbb{C}(n) \mid {}^tA\overline{A} = \mathrm{Id}_n\}$;

(2) The special unitary group $\mathrm{SU}_n := \{A \in \mathrm{U}_n \mid \det A = 1\}$;

(3) The symplectic group

$$\mathrm{Sp}_n := \{A \in \mathrm{Hom}_{\mathbb{H}}(\mathbb{H}^n, \mathbb{H}^n) \mid \forall (x,y) \in (\mathbb{H}^n, \mathbb{H}^n), (A(x), A(y)) = (x,y)\},$$

where $(x,y) := x\overline{y}$ and conjugation is defined, for any element $v = v_0 + v_1 i + v_2 j + v_3 k \in \mathbb{H} \cong \mathbb{R}^4$, by $\overline{v} := v_0 - v_1 i - v_2 j - v_3 k$;

(4) The unit sphere $S^n := \{x \in \mathbb{R}^{n+1} \mid (x,x) = 1\}$.

Then, it is not too difficult to check the following isomorphisms (see [**LM89**]):

(1) $\mathrm{Spin}_1 \subset C\ell_1^0 \cong C\ell_0 \cong \mathbb{R}$, hence $\mathrm{Spin}_1 \cong \{\pm 1\} = \mathbb{Z}_2 \mapsto \{1\}$;

(2) $\mathrm{Spin}_2 \subset C\ell_2^0 \cong C\ell_1 \cong \mathbb{C}$, hence $\mathrm{Spin}_2 \cong \mathrm{U}_1 \cong S^1 \mapsto S^1 \cong \mathrm{SO}_2$;

(3) $\mathrm{Spin}_3 \subset C\ell_3^0 \cong C\ell_2 \cong \mathbb{H} \cong \mathbb{C}^2 \cong \mathbb{R}^4$, hence $\mathrm{Spin}_3 \cong \mathrm{Sp}_1 \cong S^3 \cong \mathrm{SU}_2 \mapsto \mathrm{SO}_3$;

(4) $\mathrm{Spin}_4 \subset C\ell_4^0 \cong C\ell_3 \cong \mathbb{H} \oplus \mathbb{H}$, hence $\mathrm{Spin}_4 \cong \mathrm{Sp}_1 \times \mathrm{Sp}_1 \cong \mathrm{SU}_2 \times \mathrm{SU}_2 \mapsto \mathrm{SO}_4$.

We identify CO_n^+ with $\mathrm{SO}_n \times \mathbb{R}_+^*$ via the canonical isomorphism:

$$\begin{array}{rccc} \phi : & \mathrm{SO}_n \times \mathbb{R}_+^* & \longrightarrow & \mathrm{CO}_\mathrm{n}^+ \\ & (A,t) & \longmapsto & tA\,. \end{array}$$

Moreover, by Proposition 3.16 it is clear that, for $n \geq 3$, the morphism

$$\begin{array}{rccc} \zeta : & \mathrm{CSpin}_\mathrm{n} & \longrightarrow & \mathrm{CO}_\mathrm{n}^+ \\ & (a,t) & \longmapsto & t\xi(a)\,, \end{array}$$

is the universal cover of the conformal group. We have the following exact sequences

$$\{1\} \longrightarrow \mathbb{Z}_2 \longrightarrow \mathrm{Spin}_n \xrightarrow{\ \xi\ } \mathrm{SO}_n \longrightarrow \{1\}$$

and

$$\{1\} \longrightarrow \mathbb{Z}_2 \longrightarrow \mathrm{CSpin}_\mathrm{n} \xrightarrow{\ \zeta\ } \mathrm{CO}_\mathrm{n}^+ \longrightarrow \{1\}.$$

We now consider the group morphism

$$\begin{array}{rccc} \jmath : & \mathrm{Spin}_n \times \mathrm{U}_1 & \longrightarrow & \mathbb{C}\ell_n^* \\ & (a,z) & \longmapsto & \imath(a)z \end{array}$$

where $\imath : C\ell_n \to \mathbb{C}\ell_n \cong C\ell_n \otimes_{\mathbb{R}} \mathbb{C}$ is the standard inclusion of $C\ell_n$ in $\mathbb{C}\ell_n$. For simplicity, we shall identify $C\ell_n$ with its image by $\imath$. Since the kernel of $\jmath$ is $\{(1,1),(-1,-1)\} \cong \mathbb{Z}_2$, its image is isomorphic to $(\mathrm{Spin}_n \times \mathrm{U}_1)/\mathbb{Z}_2$, which will be *denoted* by $\mathrm{Spin}_n \times_{\mathbb{Z}_2} \mathrm{U}_1$.

DEFINITION 3.18. (i) The complex spin group is the group

$$\mathrm{Spin}_n^c := \mathrm{Spin}_n \times_{\mathbb{Z}_2} \mathrm{U}_1,$$

which via $\jmath$, is identified with a subgroup of $\mathbb{C}\ell_n^*$.

(ii) The conformal complex spin group is the group

$$\mathrm{CSpin}_\mathrm{n}^\mathrm{c} := (\mathrm{Spin}_\mathrm{n} \times_{\mathbb{Z}_2} \mathrm{U}_1) \times \mathbb{R}_+^*,$$

which via $\jmath \times \mathrm{Id}$, is identified with a subgroup of $\mathbb{C}\ell_n^* \times \mathbb{R}_+^*$.

EXAMPLES 3.19 (The group Spin_4^c). First recall the following identifications:

$$\begin{array}{ccccc} \mathbb{R}^4 & \longrightarrow & \mathbb{C}^2 & \longrightarrow & \mathbb{H}\subset\mathbb{C}(2) \\ \begin{pmatrix} x_1 \\ x_2 \\ x_3 \\ x_4 \end{pmatrix} & \longmapsto & \begin{pmatrix} u = x_1 + ix_4 \\ v = x_2 + ix_3 \end{pmatrix} & \longmapsto & \begin{pmatrix} u & -\overline{v} \\ v & \overline{u} \end{pmatrix} = \boldsymbol{x}, \end{array}$$

where $\boldsymbol{x} = x_1\mathbf{1} + x_2\boldsymbol{i} + x_3\boldsymbol{j} + x_4\boldsymbol{k}$ and

$$\mathbf{1} := \begin{pmatrix} 1 & 0 \\ 0 & 1 \end{pmatrix}, \quad \boldsymbol{i} := \begin{pmatrix} 0 & -1 \\ 1 & 0 \end{pmatrix}, \quad \boldsymbol{j} := \begin{pmatrix} 0 & i \\ i & 0 \end{pmatrix}, \quad \boldsymbol{k} := \begin{pmatrix} i & 0 \\ 0 & -i \end{pmatrix}.$$

Note that

$$\begin{cases} \det \boldsymbol{x} & = u\overline{u} + v\overline{v} = (\boldsymbol{x}, \boldsymbol{x}) \\ {}^t\boldsymbol{x}\,\overline{\boldsymbol{x}} & = (\boldsymbol{x}, \boldsymbol{x})\,\mathrm{Id}_2\,. \end{cases}$$

and define the map

$$\begin{array}{rcl} \xi\ :\ \mathrm{Spin}_4 \cong \mathrm{SU}_2\times\mathrm{SU}_2 & \longrightarrow & \mathrm{GL}(\mathbb{H}) \\ (a,b) & \longmapsto & \xi(a,b) : \boldsymbol{x} \longmapsto b\boldsymbol{x}a^{-1} = \boldsymbol{X}. \end{array}$$

Since the Euclidean scalar product on $\mathbb{H}\cong\mathbb{R}^4$ is left invariant by $\xi(a,b)$, i.e.,

$$(\boldsymbol{X}, \boldsymbol{X}) = \det \boldsymbol{X} = \det \boldsymbol{x} = (\boldsymbol{x}, \boldsymbol{x}),$$

it follows that the surjective map

$$\xi : \mathrm{Spin}_4 \longrightarrow \mathrm{SO}_4$$

has kernel $\mathbb{Z}_2 = \{(\mathrm{Id}_2, \mathrm{Id}_2), (-\mathrm{Id}_2, -\mathrm{Id}_2)\}$. By definition of the complex spinor group we have

$$\mathrm{Spin}_n^c := \mathrm{Spin}_n \times_{\mathbb{Z}_2} \mathrm{U}_1 \subset C\ell_n \otimes \mathbb{C}.$$

By the above identifications, the spinor group Spin_4 could be realized as

$$\mathrm{Spin}_4 = \left\{ \begin{pmatrix} a & 0 \\ 0 & b \end{pmatrix} \mid a, b \in \mathrm{SU}_2 \right\},$$

and for the complex spin group

$$\mathrm{Spin}_4^c = \left\{ \begin{pmatrix} za & 0 \\ 0 & zb \end{pmatrix} \mid a, b \in \mathrm{SU}_2 \quad \text{and} \quad z \in \mathrm{U}_1 \right\}.$$

Moreover, the representation ξ^c and the map det are given by

$$\begin{array}{rccl} \xi^c\ : & \mathrm{Spin}_4^c & \longrightarrow & \mathrm{SO}_4 \subset \mathrm{GL}(\mathbb{H}) \\ & \begin{pmatrix} za & 0 \\ 0 & zb \end{pmatrix} & \longmapsto & \boldsymbol{x} \mapsto zb\boldsymbol{x}(za)^{-1} = b\boldsymbol{x}a^{-1}, \end{array}$$

and

$$\begin{array}{rccl} \det\ : & \mathrm{Spin}_4^c & \longrightarrow & \mathrm{U}_1 \\ & \begin{pmatrix} za & 0 \\ 0 & zb \end{pmatrix} & \longmapsto & \det(za) = \det(zb) = z^2. \end{array}$$

An element of the complex (resp. conformal complex) spin group could be considered as an equivalence class $[a, z]$, $a \in \mathrm{Spin}_n$ (resp. $a \in \mathrm{CSpin_n}$), $z \in \mathrm{U}_1$, modulo $[a, z] = [-a, -z]$. The morphisms ξ and ζ induce double coverings

$$\begin{array}{rccl} \xi^c\ :\ \mathrm{Spin}_n^c & \longrightarrow & \mathrm{SO}_n \times \mathrm{U}_1 \\ {}[a, z] & \longmapsto & \left(\xi(a), z^2\right), \end{array}$$

$$
\begin{array}{rcl}
\zeta^c \,:\, \mathrm{CSpin}^{\mathrm{c}}_{\mathrm{n}} & \longrightarrow & \mathrm{CO}^+_{\mathrm{n}} \times \mathrm{U}_1 \\
[b, z] & \longmapsto & \left(\zeta(b), z^2\right) .
\end{array}
$$

We have the following commutative diagram of exact sequences relating ξ to ξ^c

$$
\begin{array}{ccccccccc}
\{1\} & \longrightarrow & \mathbb{Z}_2 & \longrightarrow & \mathrm{Spin}_n \times \mathrm{U}_1 & \xrightarrow{\xi \times \mathrm{Id}} & \mathrm{SO}_n \times \mathrm{U}_1 & \longrightarrow & \{1\} \\
 & & \wr\downarrow & & \downarrow\jmath & & \downarrow \mathrm{Id} \times \mathrm{sq} & & \\
\{1\} & \longrightarrow & \mathbb{Z}_2 & \longrightarrow & \mathrm{Spin}^c_n & \xrightarrow{\xi^c} & \mathrm{SO}_n \times \mathrm{U}_1 & \longrightarrow & \{1\}
\end{array}
$$

where

$$
\begin{array}{rcl}
\mathrm{sq} \,:\, \mathrm{U}_1 & \longrightarrow & \mathrm{U}_1 \\
z & \longmapsto & z^2 .
\end{array}
$$

There is also a similar commutative diagram corresponding to the conformal setup relating ζ to ζ^c. We shall use these exact sequences to get the explicit infinitesimal action of the morphisms ξ, ξ^c, ζ and ζ^c. The spin group being in the Clifford algebra $\mathrm{C}\ell_n$, its Lie algebra, $\mathfrak{spin}_n$, is a subalgebra of $\mathrm{C}\ell_n$, seen as a Lie algebra where $[a, b] := a \cdot b - b \cdot a$.

THEOREM 3.20. *With the obvious notations, the Lie algebras of the Lie groups involved in the real, complex and conformal setups satisfy the following*

$$
\begin{array}{rclrcl}
\mathfrak{spin}_n & \cong & \mathfrak{so}_n , & \xi_*(e_i \cdot e_j) & = & 2\, e_i \wedge e_j , \\
\mathfrak{spin}^c_n & \cong & \mathfrak{so}_n \oplus i\mathbb{R} , & \xi^c_* & = & \xi_* \oplus 2\,\mathrm{Id}_{i\mathbb{R}} , \\
\mathfrak{cspin}_n & \cong & \mathfrak{so}_n \oplus \mathbb{R} , & \zeta_* & = & \xi_* \oplus \mathrm{Id}_{\mathbb{R}} , \\
\mathfrak{cspin}^c_n & \cong & \mathfrak{so}_n \oplus \mathbb{R} \oplus i\mathbb{R} , & \zeta^c_* & = & \xi_* \oplus \mathrm{Id}_{\mathbb{R}} \oplus 2\,\mathrm{Id}_{i\mathbb{R}} ,
\end{array}
$$

where $e_i \wedge e_j \in \mathfrak{so}_n \subset \mathfrak{gl}_n$ *is defined by* $(e_i \wedge e_j)(e_k) = \delta_{ik} e_j - \delta_{jk} e_i$.

PROOF. Recall that $\Lambda^2\mathbb{R}^n$ is identified with a vector subspace of $\mathrm{C}\ell_n$ by $e_i \wedge e_j \mapsto e_i \cdot e_j$ for $1 \le i < j \le n$. This subspace lies in Spin_n, since $e_i \cdot e_j$ is the tangent vector at $t = 0$ to the curve lying in Spin_n and given by
(3.20)

$$
c(t) = (e_i \sin(t/2) - e_j \cos(t/2)) \cdot (e_i \sin(t/2) + e_j \cos(t/2)) = \cos t + e_i \cdot e_j \sin t .
$$

Since $\dim_{\mathbb{R}} \mathfrak{spin}_n = \dim_{\mathbb{R}} \mathfrak{so}_n = \dim_{\mathbb{R}} \Lambda^2\mathbb{R}^n$, we get $\Lambda^2\mathbb{R}^n = \mathfrak{spin}_n$. The other assertions could be obtained similarly.

Finally, for all i, j, k, $1 \le i < j \le n$ and $1 \le k \le n$ we have $\xi(c(t))(e_k) = c(t) \cdot e_k \cdot c(t)^{-1}$, hence

$$
\begin{aligned}
\xi_*(e_i \cdot e_j)(e_k) &= \left.\frac{d}{dt}\right|_{t=0} \xi(c(t))(e_k) \\
&= \left.\frac{d}{dt}\right|_{t=0} (\cos t + e_i \cdot e_j \sin t) \cdot e_k \cdot (\cos t - e_i \cdot e_j \sin t) \\
&= e_i \cdot e_j \cdot e_k - e_k \cdot e_i \cdot e_j = 2\delta_{ik}\, e_j - 2\delta_{jk}\, e_i \\
&= 2(e_i \wedge e_j)(e_k).
\end{aligned}
$$

The other relations are immediate consequences of the above exact sequences.

□

3.4. Spinor Representations

In this section we examine the complex representations of the Clifford algebras and those of the spin groups. Since every complex representation of $C\ell_n$ induces a complex representation of $\mathbb{C}\ell_n$ and reciprocally, we shall only consider the complex Clifford algebras. In Proposition 3.13 we gave an explicit isomorphism between the even dimensional complex Clifford algebra and a corresponding algebra of complex matrices. We shall now give an abstract proof of the isomorphisms (3.13).

THEOREM 3.21 (Classification of the complex Clifford algebras). (i) *There exists a canonical algebra isomorphism* $\mathbb{C}\ell_{n+2} \cong \mathbb{C}\ell_n \otimes \mathbb{C}\ell_2$.

(ii) *We have the following algebra isomorphisms*

$$\begin{array}{rcl} \mathbb{C}\ell_{2n+1} & \cong & \mathbb{C}(2^n) \oplus \mathbb{C}(2^n) \\ \mathbb{C}\ell_{2n+2} & \cong & \mathbb{C}(2^{n+1}) \end{array}$$

PROOF. (i) Denote by $\{e_1, \ldots, e_{n+2}\}$ the canonical basis of $\mathbb{C}^{n+2}$ and consider the identifications $\mathbb{C}^n = \operatorname{span}\{e_1, \ldots, e_n\}$, $\mathbb{C}^2 = \operatorname{span}\{e_{n+1}, e_{n+2}\}$. One can check that the linear map

$$\begin{array}{rcl} f : \mathbb{C}^{n+2} & \longrightarrow & \mathbb{C}\ell_n \otimes \mathbb{C}\ell_2 \\ e_k & \longmapsto & \begin{cases} e_k \otimes i e_{n+1} \cdot e_{n+2}, & \text{for } k \leq n \\ 1 \otimes e_k & \text{for } k > n \end{cases} \end{array}$$

satisfies (3.1). By Proposition 3.2, one can deduce that f could be extended to a morphism $\mathbb{C}\ell_{n+2} \to \mathbb{C}\ell_n \otimes \mathbb{C}\ell_2$. With the help of the standard bases (3.3) of the Clifford algebras, one shows that this is a surjective morphism and hence an isomorphism since $\dim_{\mathbb{C}} \mathbb{C}\ell_{n+2} = \dim_{\mathbb{C}} (\mathbb{C}\ell_n \otimes \mathbb{C}\ell_2)$.

(ii) By (i), knowing that $\mathbb{C}(k) \otimes \mathbb{C}(l) \cong \mathbb{C}(kl)$, it is sufficient to prove the assertion for $n = 0$. First observe that the $\mathbb{C}$–linear map

$$\begin{array}{rcl} \mathbb{C} & \longrightarrow & \mathbb{C} \oplus \mathbb{C} \\ e_1 & \mapsto & (i, i) \end{array}$$

satisfies property (3.1), hence it could be extended to an algebra homomorphism $\mathbb{C}\ell_1 \to \mathbb{C} \oplus \mathbb{C}$. It is clear that this homomorphism is surjective, hence an isomorphism since the vector spaces have the same dimension. Similarly, by the universal property, the linear map

$$\begin{array}{rcl} \mathbb{C}^2 & \longrightarrow & \mathbb{C}(2) \\ e_1 & \mapsto & \begin{pmatrix} 0 & 1 \\ -1 & 0 \end{pmatrix} \\ e_2 & \mapsto & \begin{pmatrix} 0 & i \\ i & 0 \end{pmatrix} \end{array}$$

could be extended to an algebra isomorphism $\mathbb{C}\ell_2 \to \mathbb{C}(2)$.

□

Since the only complex irreducible representation of the matrix algebra $\mathbb{C}(N)$ is the canonical representation on $\mathbb{C}^N$, we have

THEOREM 3.22. *The complex Clifford algebra $\mathbb{C}\ell_n$ has a unique irreducible representation*

$$\chi_n \ : \mathbb{C}\ell_n \to \mathrm{End}(\Sigma_n),$$

for n even, and two inequivalent irreducible representations

$$\chi_n^{\pm} \ : \mathbb{C}\ell_n \to \mathrm{End}(\Sigma_n),$$

for n odd, where Σ_n is a complex vector space of complex dimension $N := 2^{[\frac{n}{2}]}$, and $[\ ]$ stands for the integer part.

The only thing we still have to prove is that, for n odd, the two irreducible representations are inequivalent. This will be done by examining the action of the complex volume element. Recall that the complex volume element $\omega^{\mathbb{C}} := i^{[\frac{n+1}{2}]}\, e_1 \cdot \ldots \cdot e_n$, satisfies

$$(\omega^{\mathbb{C}})^2 = 1 \tag{3.21}$$

$$x \cdot \omega^{\mathbb{C}} = (-1)^{n-1}\omega^{\mathbb{C}} \cdot x, \qquad \forall x \in \mathbb{R}^n \subset \mathbb{C}\ell_n\,. \tag{3.22}$$

We now prove the following two propositions:

PROPOSITION 3.23. *For n odd, $\mathbb{C}l_n = \mathbb{C}l_n^+ \oplus \mathbb{C}l_n^-$, where $\mathbb{C}\ell_n^{\pm} = \pi^{\pm} \cdot \mathbb{C}\ell_n = \mathbb{C}\ell_n \cdot \pi^{\pm}$ and $\pi^{\pm} = \frac{1}{2}(1 \pm \omega^{\mathbb{C}})$. Moreover,*

$$\alpha(\mathbb{C}\ell_n^{\pm}) = \mathbb{C}\ell_n^{\mp}\,.$$

PROOF. Since $(\omega^{\mathbb{C}})^2 = 1$, one has:

$$\begin{aligned} \pi^+ + \pi^- &= 1\,, \\ (\pi^{\pm})^2 &= \pi^{\pm}\,, \\ \pi^- \cdot \pi^+ &= \pi^+ \cdot \pi^- = 0. \end{aligned}$$

Since n is odd, $\omega^{\mathbb{C}}$ and $\pi^{\pm}$ are central in $\mathbb{C}\ell_n$. It is then clear that $\mathbb{C}\ell_n^{\pm} = \pi^{\pm} \cdot \mathbb{C}l_n$ are two ideals of $\mathbb{C}\ell_n$ and $\mathbb{C}\ell_n = \mathbb{C}\ell_n^+ \oplus \mathbb{C}\ell_n^-$.

The volume element being odd, i.e., $\omega^{\mathbb{C}} \in \mathbb{C}\ell_n^1$, we have $\alpha(\pi^{\pm}) = \pi^{\mp}$, hence $\alpha(\mathbb{C}\ell_n^{\pm}) = \mathbb{C}\ell_n^{\mp}$ and the two subalgebras are isomorphic.

□

PROPOSITION 3.24. *For n odd, let χ_n be a complex irreducible representation of $\mathbb{C}\ell_n$. Then either $\chi_n(\omega^{\mathbb{C}}) = \mathrm{Id}$, or $\chi_n(\omega^{\mathbb{C}}) = -\mathrm{Id}$. The two possibilities occur and they are inequivalent.*

PROOF. Since $(\omega^{\mathbb{C}})^2 = 1$ and $\chi(\omega^{\mathbb{C}})^2 = \mathrm{Id}$, the vector space Σ_n could be written as the sum of the eigenspaces $\Sigma_n^{\pm}$ associated with the ±1-eigenvalues, i.e., $\Sigma_n = \Sigma_n^+ \oplus \Sigma_n^-$. The volume element $\omega^{\mathbb{C}}$ being central, the eigenspaces $\Sigma_n^{\pm}$ are $\mathbb{C}\ell_n$–invariant. The representation being irreducible, we conclude that $\Sigma_n = \Sigma_n^+$ or $\Sigma_n = \Sigma_n^-$. It is then clear that these two representations are inequivalent. By considering the action by left multiplication of $\mathbb{C}\ell_n$ on $\mathbb{C}\ell_n^{\pm}$ we see that the two possibilities occur.

□

We now define the graded tensor product in the category of $\mathbb{Z}_2$-graded algebras. Let $\mathcal{A}$ and $\mathcal{B}$ be two $\mathbb{Z}_2$-graded algebras and define $\mathcal{A}\widehat{\otimes}\mathcal{B}$ as the vector space $\mathcal{A}\otimes\mathcal{B}$ endowed with the product

$$(a\otimes b)\centerdot(a'\otimes b') := (-1)^{\deg(b)\deg(a')}(a\cdot a')\otimes(b\cdot b'),$$

whenever the elements a and b are of pure degrees. This is again a $\mathbb{Z}_2$-graded algebra with:

$$\begin{aligned}\left(\mathcal{A}\widehat{\otimes}\mathcal{B}\right)^0 &= \mathcal{A}^0\otimes\mathcal{B}^0+\mathcal{A}^1\otimes\mathcal{B}^1\\ \left(\mathcal{A}\widehat{\otimes}\mathcal{B}\right)^1 &= \mathcal{A}^0\otimes\mathcal{B}^1+\mathcal{A}^1\otimes\mathcal{B}^0\,.\end{aligned}$$

A $\mathbb{Z}_2$-graded $\mathcal{A}$-module is a module $\mathcal{M}:=\mathcal{M}^0\oplus\mathcal{M}^1$ such that for all $i,j\in\mathbb{Z}_2$, one has

$$\mathcal{A}^i\cdot\mathcal{M}^j\subset\mathcal{M}^{i+j}.$$

We now recall some algebraic facts which could be easily checked.

If $\mathcal{M}$ is a $\mathbb{Z}_2$-graded $\mathcal{A}$-module and $\mathcal{N}$ a $\mathbb{Z}_2$-graded $\mathcal{B}$-module, then $\mathcal{M}\otimes\mathcal{N}$ is a $(\mathcal{A}\widehat{\otimes}\mathcal{B})$-module with respect to the multiplication

$$(a\otimes b)\centerdot(\mu\otimes\nu) := (-1)^{\deg(b)\deg(\mu)}(a\cdot\mu)\otimes(b\cdot\nu),$$

for b and μ of pure degree.

If $\mathcal{M}$ is only an $\mathcal{A}$-module and $\mathcal{N}$ a $\mathbb{Z}_2$-graded $\mathcal{B}$-module, then $\mathcal{M}\otimes\mathcal{N}$ is a $(\mathcal{A}\widehat{\otimes}\mathcal{B})$-module with respect to the multiplication

$$(a\otimes b)\centerdot(\mu\otimes\nu) := (-1)^{\deg(a)\deg(\nu)}(a\cdot\mu)\otimes(b\cdot\nu),$$

for a and ν of pure degree.

Lemma 3.25. *There exists an algebra isomorphism $\mathbb{C}\ell_{m+n}\cong\mathbb{C}\ell_m\widehat{\otimes}\,\mathbb{C}\ell_n$.*

Proof. The universal property in Proposition 3.2 shows that the linear map

$$\begin{aligned} f\ :\ \mathbb{C}^{m+n} &\longrightarrow \mathbb{C}\ell_m\,\widehat{\otimes}\,\mathbb{C}\ell_n\\ (v,w) &\longmapsto v\otimes 1+1\otimes w\end{aligned}$$

could be extended to an algebra homomorphism $\widetilde{f}:\mathbb{C}\ell_{m+n}\to\mathbb{C}\ell_m\,\widehat{\otimes}\,\mathbb{C}\ell_n$ which is clearly surjective. As complex vector spaces, these two algebras are of the same dimension, hence $\widetilde{f}$ is an isomorphism.

□

Proposition 3.26. (i) *For n even, the restriction of χ_n to Spin_n (resp. $\mathbb{C}\ell^0_n$) splits into $\Sigma_n=\Sigma_n^+\oplus\Sigma_n^-$ where Σ_n^+ and Σ_n^- are two complex inequivalent irreducible representations of Spin_n (resp. $\mathbb{C}\ell^0_n$).*

(ii) *For n odd, the restrictions of $\chi_n^\pm$ to Spin_n (resp. $\mathbb{C}\ell^0_n$) are irreducible and equivalent.*

(iii) *For n even, for all $x\in\mathbb{R}\setminus\{0\}$, the linear maps $\chi_n^\pm(x):\Sigma_n^\pm\mapsto\Sigma_n^\mp$ are isomorphisms. Moreover, under the isomorphism $\mathbb{C}\ell^0_n\cong\mathbb{C}\ell_{n-1}$, the vector spaces $\Sigma_n^\pm$ correspond to the two inequivalent irreducible representations of $\mathbb{C}\ell_{n-1}$.*

Proof. Note that the complex subalgebra generated by $\mathrm{Spin}_n\subset\mathbb{C}\ell_n$ is the even part $\mathbb{C}\ell^0_n$ of $\mathbb{C}\ell_n$. Hence, two representations of $\mathbb{C}\ell^0_n$ are irreducible or equivalent if and only if it is the case for their restrictions to Spin_n. It is therefore sufficient to prove the assertions for $\mathbb{C}\ell^0_n$ instead of Spin_n.

(i) Let $n = 2m$. Since $\omega^{\mathbb{C}}$ commutes with $\mathbb{C}\ell_n^0$, the restriction of χ_n to $\mathbb{C}\ell_n^0 \cong \mathbb{C}\ell_{n-1}$ splits into $\Sigma_n = \Sigma_n^+ \oplus \Sigma_n^-$, where

$$\Sigma_n^{\pm} = \chi_n\left(\frac{1 \pm \omega^{\mathbb{C}}}{2}\right)(\Sigma_n). \tag{3.23}$$

By Theorem 3.22, it is sufficient to prove that Σ_n^+ and Σ_n^- are nontrivial vector spaces. The linear map χ_n being an isomorphism, it follows that $\chi_n(\omega^{\mathbb{C}})$ could not be equal to $\pm \mathrm{Id}_{\Sigma_n} = \chi_n(\pm 1)$.

(ii) Again, we make use of the isomorphism $\mathbb{C}\ell_n^0 \cong \mathbb{C}\ell_{n-1}$. For n odd, we know that $\alpha(\mathbb{C}\ell_n^{\pm}) = \mathbb{C}\ell_n^{\mp}$. Note that the even part $\mathbb{C}\ell_n^0$ sits diagonally in the decomposition $\mathbb{C}\ell_n = \mathbb{C}\ell_n^+ \oplus \mathbb{C}\ell_n^-$. In fact, since $\omega^{\mathbb{C}} \in \mathbb{C}\ell_n^1$, we have $\omega^{\mathbb{C}} : \mathbb{C}\ell_n^0 \to \mathbb{C}\ell_n^1$, hence $\mathbb{C}\ell_n^0 \cap \mathbb{C}\ell_n^{\pm} = \{0\}$. More precisely, we have

$$\mathbb{C}\ell_n^0 = \{u^{\pm} + \alpha(u^{\pm}) \mid u^{\pm} \in \mathbb{C}\ell_n^{\pm}\}.$$

Since the two inequivalent irreducible representations of $\mathbb{C}\ell_n$ are distinguished by the isomorphism α, by restriction to $\mathbb{C}\ell_n^0$, they become equivalent.

(iii) By (3.22) we have $x \cdot \omega^{\mathbb{C}} = -\omega^{\mathbb{C}} \cdot x$. Then,

$$\begin{aligned} \chi_n(x)\Sigma_n^{\pm} &= \chi_n(x)\chi_n\left(\frac{1 \pm \omega^{\mathbb{C}}}{2}\right)(\Sigma_n) \\ &= \chi_n\left(\frac{1 \mp \omega^{\mathbb{C}}}{2}\right)\chi_n(x)(\Sigma_n) \\ &\subset \Sigma_n^{\mp}. \end{aligned}$$

The linear map $\chi_n(x)$ is in fact an isomorphism since $(\chi_n(x))^2 = -\|x\|^2 \mathrm{Id}$.

For the last statement, it is sufficient to note that the isomorphism (3.7) maps the volume element $\omega_{n-1}^{\mathbb{C}} := i^{[\frac{n}{2}]} e_1 \cdot \ldots \cdot e_{n-1}$ of $\mathbb{C}\ell_{n-1}$ to the volume element $\omega_n^{\mathbb{C}} := i^{[\frac{n+1}{2}]} e_1 \cdot \ldots \cdot e_{n-1} \cdot e_n$ of $\mathbb{C}\ell_n$.

□

Definition 3.27 (Clifford multiplication). The map

$$\begin{aligned} \mathfrak{M} : \mathbb{C}\ell_n \otimes \Sigma_n &\longrightarrow \Sigma_n \\ \sigma \otimes \psi &\longmapsto \sigma \cdot \psi := \begin{cases} \chi_n(\sigma)(\psi), & \text{if } n \text{ is even} \\ \chi_n^+(\sigma)(\psi), & \text{if } n \text{ is odd}, \end{cases} \end{aligned}$$

is called Clifford multiplication of σ with ψ.

For n even, we define the *conjugation map*

$$\begin{aligned} \omega^{\mathbb{C}} \cdot \; : \Sigma_n = \Sigma_n^+ \oplus \Sigma_n^- &\longrightarrow \Sigma_n = \Sigma_n^+ \oplus \Sigma_n^- \\ \psi = \psi^+ + \psi^- &\longmapsto \overline{\psi} := \omega^{\mathbb{C}} \cdot \psi = \psi^+ - \psi^-. \end{aligned}$$

Lemma 3.28. (i) *If m and n are even, $(\chi_m, \Sigma_m) \otimes (\chi_n, \Sigma_n)$, endowed with the structure of a $(\mathbb{C}\ell_m \widehat{\otimes} \mathbb{C}\ell_n)$-module, is isomorphic to the $\mathbb{C}\ell_{m+n}$-module $(\chi_{m+n}, \Sigma_{m+n})$.*

(ii) *If m is even and n odd, then $(\chi_m, \Sigma_m) \otimes (\chi_n^+, \Sigma_n)$ is isomorphic to the $\mathbb{C}\ell_{m+n}$-module $(\chi_{m+n}^+, \Sigma_{m+n})$. Similarly, if m is odd and n is even, then $(\chi_m^+, \Sigma_m) \otimes (\chi_n, \Sigma_n)$ is isomorphic to the $\mathbb{C}\ell_{m+n}$-module $(\chi_{m+n}^+, \Sigma_{m+n})$.*

(iii) *If m and n are odd, then $(\chi_m^+, \Sigma_m) \otimes (\chi_n^+, \Sigma_n)$ is isomorphic to the $\mathbb{C}\ell_{m+n}^0$-module $(\chi_{m+n}, \Sigma_{m+n}^+)$.*

PROOF. (i) This is a straightforward consequence of Theorem 3.22, Lemma 3.25 and the fact that the two complex vector spaces Σ_{m+n} and $\Sigma_m \otimes \Sigma_n$ are of the same dimension.

(ii) Since $\dim_{\mathbb{C}} \Sigma_{m+n} = \dim_{\mathbb{C}} \Sigma_m \otimes \Sigma_n$, it is sufficient to examine the action of the complex volume element of $\mathbb{C}\ell_{m+n} \cong \mathbb{C}\ell_m \widehat{\otimes} \mathbb{C}\ell_n$ on $\Sigma_m \otimes \Sigma_n$. If m is even and n odd, the image of $\omega^{\mathbb{C}}_{m+n}$ in $\mathbb{C}\ell_m \widehat{\otimes} \mathbb{C}\ell_n$ by the above isomorphism is $\omega^{\mathbb{C}}_m \otimes \omega^{\mathbb{C}}_n$ and its action on $\Sigma_m \otimes \Sigma_n$ is given by

$$(\omega^{\mathbb{C}}_m \otimes \omega^{\mathbb{C}}_n)(\psi \otimes \varphi) = (\omega^{\mathbb{C}}_m \cdot \overline{\psi} \otimes \varphi) = \psi \otimes \varphi.$$

The other case is similar.

(iii) In this case $\Sigma_m \otimes \Sigma_n$ is no more a $(\mathbb{C}\ell_m \widehat{\otimes} \mathbb{C}\ell_n)$-module, since neither Σ_m, nor Σ_n are graded $\mathbb{C}\ell_m$ or $\mathbb{C}\ell_n$-modules. However, by Proposition 3.9 and Theorem 3.22, the space Σ_m is a graded $\mathbb{C}\ell_{m-1}$-module, which shows that $\Sigma_m \otimes \Sigma_n$ is a $\mathbb{C}\ell^0_{m+n}$-module, since

$$\mathbb{C}\ell_{m-1} \widehat{\otimes} \mathbb{C}\ell_n \cong \mathbb{C}\ell_{m+n-1} \cong \mathbb{C}\ell^0_{m+n}.$$

Noting that $\dim_{\mathbb{C}} \Sigma^+_{m+n} = \dim_{\mathbb{C}} \Sigma_m \otimes \Sigma_n$, it is sufficient to check that $\omega^{\mathbb{C}}_{m+n}$ acts, via the above isomorphism as the identity on $\Sigma_m \otimes \Sigma_n$. This is true because the image of $\omega^{\mathbb{C}}_{m+n}$ in $\mathbb{C}\ell_{n-1} \widehat{\otimes} \mathbb{C}\ell_m$ is, by this isomorphism, given by $\omega^{\mathbb{C}}_{m-1} \otimes \omega^{\mathbb{C}}_n$.

□

DEFINITION 3.29. The representation

$$\rho_n := \begin{cases} \chi_n\big|_{\mathrm{Spin}_n}, & \text{for } n \text{ even} \\ \chi^+_n\big|_{\mathrm{Spin}_n}, & \text{for } n \text{ odd}, \end{cases} \tag{3.24}$$

is called the *canonical complex spin representation* and is denoted by (ρ_n, Σ_n).

Note that since the algebra generated by the groups Spin_n and Spin^c_n in $\mathbb{C}\ell_n$ is exactly $\mathbb{C}\ell^0_n$, it is clear that the preceding facts are also true for the group Spin^c_n. In fact, observe that the complex representations of Spin_n are the same as that of the group Spin^c_n. Hence the restriction of χ_n (resp. χ^+_n) to Spin^c_n, also *denoted* by ρ^c_n, satisfies Proposition 3.26 too.

For each $k \in \mathbb{R}$ we consider the representation $\rho^{c,k}_n$ of CSpin_n (resp. CSpin^c_n) on Σ_n defined, for all $a \in \mathrm{Spin}_n$ (resp. $a \in \mathrm{Spin}^c_n$) and $t \in \mathbb{R}^*_+$, by

$$\rho^{c,k}_n((a,t)) = t^k \rho^c_n(a).$$

As before, one checks that these represntations are irreducible for n even and are the direct sum of two inequivalent irreducible representations, for n odd.

LEMMA 3.30. *The representation (ρ^c_n, Σ_n) of Spin^c_n is unitary, i.e., there exists a Hermitian scalar product on Σ_n denoted by $\langle \cdot, \cdot \rangle$, such that*

$$\langle a \cdot \psi, a \cdot \varphi \rangle = \langle \psi, \varphi \rangle, \qquad \forall a \in \mathrm{Spin}^c_n, \quad \forall \psi, \varphi \in \Sigma_n. \tag{3.25}$$

This Hermitian scalar product is unique up to a scalar if (3.25) is true for all unit vectors of $\mathbb{R}^n$.

PROOF. Let $\mathcal{G}$ be the multiplicative (finite) group generated by the canonical basis of $\mathbb{C}^n$ in $\mathbb{C}\ell_n$ and denote by $\langle \cdot, \cdot \rangle_0$ any Hermitian scalar product on Σ_n. Define $\langle \cdot, \cdot \rangle$ as

$$\langle \psi, \varphi \rangle := \sum_{g \in \mathcal{G}} \langle g \cdot \psi, g \cdot \varphi \rangle_0.$$

It is easy to check that Clifford multiplication by a unit vector is unitary w.r.t $\langle \cdot, \cdot \rangle$, hence (3.25) is satisfied. Uniqueness follows from the fact that the irreducibility of the action of $\mathcal{G}$ on Σ_n.

□

Oussama Hijazi, Institut Élie Cartan, Université Henri Poincaré, Nancy I, B.P. 239, 54506 Vandœuvre-Lès-Nancy Cedex, France.
E-mail address: hijazi@iecn.u-nancy.fr

CHAPTER 4

A Brief Introduction to Riemannian and Spinorial Geometries

JEAN-PIERRE BOURGUIGNON

ABSTRACT. Spin Geometry is undoubtedly an outgrowth of Riemannian Geometry, hence requires for its introduction that the basic ingredients of Differential Geometry be assimilated, with special emphasis on metric tools such as Riemannian and Lorentzian metrics, and objects directly connected to them such as covariant derivatives. This series of lectures aims to provide a brief introduction to these topics, keeping in mind the true target which is to give the basic tools for working in Spin Geometry.

Contents

4.1. Metrics from an Algebraic Point of View

In his famous inaugural lecture, Bernhard Riemann introduced almost all the basic ingredients that today make Riemannian Geometry a vast and extremely successful generalisation of surface theory, with many applications in Physics and other Natural Sciences.

His proposal is based on the use of scalar products which in local coordinates have general (i.e. variable) coefficients. The main invariant of this generalized geometry is the Riemann curvature tensor, that one now introduces through the canonical covariant derivative attached to a Riemannian metric.

This lecture is devoted to presenting the basic tools of Riemannian Geometry in a perspective which opens the way to the generalisation to Spinorial Geometry.

4.1.1. The Euclidean case. The creation of tools to measure space has been at the heart of Geometry from its very beginning. Several approaches have been used to discuss Geometry from the Ancient times on. The synthetic approach, which was so successfully developed by the Greeks, and in particular by Euclid, has now been replaced in the modern approach to Mathematics, in which linear structures play a central role, by a more analytical one. In this context *metrics* become *scalar products*, a change of terminology that we will adopt in this chapter before reverting to the less precise terminology of "metric" for the rest of the lectures.

DEFINITION 4.1. A Euclidean *scalar product* g on a vector space V is a bilinear form on V which is positive definite (hence non degenerate).

There are many ways in which a scalar product can be defined. We briefly review some of them.

Scalar products through orthonormal basis. A useful starting point is to realize that any basis $(f_i)_{1\leq i\leq n}$ on an n-dimensional vector space V can be declared *g-orthonormal* for a scalar product g, by stating that $g(f_i, f_j) = 0$ for $1 \leq i, j \leq n$, $i \neq j$ and $g(f_i, f_i) = 1$ for $1 \leq i \leq n$ and by defining the candidate scalar product for any vector $v = \sum_{i=1}^{n} v^i f_i$ by $g(v, v) = \sum_{i=1}^{n} (v^i)^2$ and for any pair of vectors by a bilinear extension. That g is a scalar product is then done by inspection. It is easy to see that any scalar product admits an orthonormal basis (hence many), e.g., through the orthonormalisation process applied to any linear basis.

Since any basis (f_i) of V determines a basis (φ^i) of the dual space V^*, the so-called *dual basis*, through $\varphi^i(f_j) = \delta^i_j$, where δ^i_j denotes the *Kronecker symbol* defined by $\delta^i_i = 1$ and $\delta^i_j = 0$ for $i \neq j$. It is traditional to denote a metric g in a general basis (f_i) through its components (g_{ij}) in this basis, so that $g = \sum_{ij}^{n} g_{ij}\, \varphi^i \otimes \varphi^j$. If (φ^i) is the dual basis of a g-orthonormal basis, one of course has $g_{ij} = \delta^j_i$.

A typical case of this construction is worthy of a special mention, namely that of $\mathbb{R}^n$, which inherits a natural metric from its canonical basis (e_i), where e_i is the n-uple $(0, ..., 0, 1, 0, ..., 0)$ where the only 1 is at the ith-place. It is traditional to denote the *standard Euclidean metric* by the letter e.

Scalar products induced by linear maps. Another useful method to define a scalar product h on a vector space V from a given scalar product g in a vector space W is via an injective linear map $\ell : V \longrightarrow W$ through the formula $h(v, v) = g(\ell(v), \ell(v))$. One then writes $h = \ell^* g$ and h is called the *pull-back* of g by h.

Two typical cases are of special interest: when $\dim V = \dim W$, and ℓ is then an isomorphism (a special case of which being of course when actually $V = W$), and when V is a vector subspace of W and ℓ the inclusion.

Objects naturally induced by metrics. It is of paramount importance to realize that a scalar product g on a vector space V, as any bilinear form, gives rise to a natural map $g^\sharp$ from V to V^* defining, for $v \in V$, $g^\sharp(v)$ as the linear form which to w associates $g(v, w)$. The non degeneracy of g implies that $g^\sharp$ is an isomorphism. This isomorphism, and its inverse, often called the *musical* isomorphisms because the operations of raising and lowering indices of tensors, are of common use in Euclidean geometry to "identify" contravariant and covariant objects. Later one we will try and avoid to use such a metric dependent identification since the dependence on the scalar product will be often of concern to us.

It is interesting to note that a basis (e_i) of V is g-orthonormal precisely when its dual basis coincides with its image through $g^\sharp$. The inverse isomorphism is called $g_\flat : V^* \longrightarrow V$ and can be defined as follows: for $\lambda \in V^*$, $g_\flat(\lambda)$ is the vector so that, for any $w \in V$, $g(g_\flat(\lambda), w) = \lambda(w)$.

In the same vein, i.e. playing with duality, it is important to point out that a scalar product on V gives naturally rise to a scalar product g^* on V^*, e.g., by declaring the dual basis of a g-orthonormal basis g^*-orthonormal, and verifying that this is consistent when one takes another g-orthonormal basis to start with. We will still use the notation g to denote this scalar product.

Later we will use freely that many tensorial spaces constructed on V inherit a natural scalar product from any scalar product g on V. A possible way is to declare orthonormal the basis induced on the tensorial space by the choice of a g-orthonormal basis on V. A case of special importance is that of the *volume elements* on V, namely the space $\Lambda^* V$ of n-forms on V if V is n-dimensional: for a g-orthonormal basis (f_i) with dual basis (φ^i), the n-form $\varphi^1 \wedge \cdots \wedge \varphi^n$ is a preferred volume element attached to the metric g defining an orientation of V making the basis (f_i) *positive*.

The space of scalar products of a vector space. It will of crucial importance later on to have a good understanding of the *space of scalar products* $\mathcal{M}et\,(V)$ of the vector space V. This can be achieved in several ways.

One of them can be described as follows. One starts with $\mathbb{R}^n$ with its canonical metric e and its standard basis (e_i) which is by definition e-orthonormal. Any other e-orthonormal basis can be obtained from (e_i) through application of an orthonormal matrix O (i.e. a matrix satisfying $O^t O = I$, where I denotes the identity matrix and O^t the transpose of O). Indeed, the scalar product h defined for an n-tuple of numbers (x^i) by $h((x^i),(x^i)) = e(O(x^i), O(x^i))$ actually coincides with e because of the identity satisfied by O, the adjoint of O admitting O^t as matrix in an orthonormal basis. Orthogonal matrices form the *orthogonal group* O_n, a subgroup of the general linear group Gl_n .

We now pass to a general vector space V of dimention n. The key point is to think of a basis (f_i) of V as an invertible linear map $f : \mathbb{R}^n \longrightarrow V$ by setting $f_i = f(e_i)$. The scalar product making (f_i) orthogonal is nothing but $(f^{-1})^* e$. The space $\mathcal{M}et\,(V)$ can then be described as the quotient of the space $\mathcal{I}nv(\mathbb{R}^n, V)$ of invertible linear maps from $\mathbb{R}^n$ to V by the action by composition on the right through orthonormal matrices, i.e.,

$$\mathcal{M}et\,(V) = Inv(\mathbb{R}^n, V)/\mathrm{O}_n \ .$$

In order to better understand the content of this statement, it is worth bringing into the picture the *polar decomposition theorem* which states that *"any matrix A can be decomposed into the product of an orthogonal matrix O and a positive definite one S"*. Indeed, it is clear that the scalar products induced by the maps whose matrices in the standard basis are OS, for $O \in \mathrm{O}_n$, coincide with the one induced by the map represented by the matrix S.

It is rather trivial to note that $\mathcal{M}et\,(V)$ *is a positive convex cone*, to mean that, for any $a \in \mathbb{R}^+$ and any $g \in \mathcal{M}et\,(V)$, $a\,g \in \mathcal{M}et\,(V)$, and for any g, $h \in \mathcal{M}et\,(V)$ and any $t \in [0,1]$, $t\,g + (1-t)\,h \in \mathcal{M}et\,(V)$.

4.1.2. The general case. The need to call upon non positive definite bilinear forms to describe the Geometry of Special Relativity has forced to broaden the Euclidean context, and to accept to consider "generalized" scalar products. The key new invariant is the *signature* of a non degenerate bilinear forms, i.e., the dimensions of the largest subspaces on which the form is respectively positive definite, and negative definite. Of course the non degeneracy implies that these two numbers add up to the dimension of the space.

Lorentzian scalar products. We owe to Hermann Minkowski and Henri Poincaré to have realized that the scheme proposed by Hendrik Lorentz to make the Michelson-Moreley experiments compatible with the basic assumptions on space and time was to endow the four-dimensional continuum formed by space-time with a metric of signature $(1,3)$ (or $(3,1)$, this a pure matter of convention). Such metrics are now called *Lorentzian* metrics. The standard Lorentzian metric l on $\mathbb{R}^n$ (which is often in this case denoted by $\mathbb{R}^{1,n-1}$) defined by $l(e_1, e_1) = -1$, $l(e_i, e_i) = 1$ for $2 \leq i \leq n$ and $l(e_i, e_j) = 0$ for $i \neq j$ is usually called the *Minkowski metric*.

Many of the constructions mentioned for Euclidean metrics carry over to generalized scalar products, such as pull back by invertible linear maps or induction on tensorial spaces of all types (although one must be careful that depending on the situation of the image of the map or the type of tensors, the induced bilinear map might change signature or be degenerate).

Spaces of generalized scalar products. The description of the space of Lorentzian metrics on a vector space V can be a priori done by mimicking what we did for Euclidean metrics, namely considering the quotient $\mathcal{I}nv(\mathbb{R}^n, V)$ by the *Lorentz group* $\mathrm{O}_{1,n-1}$, the group of automorphisms of the Minkowski metric. (Note that the name *Poincaré group* is given to the extension of the Lorentz group by translations, in the same way as the *Galileo group* is associated to the orthogonal group.)

It is worth pointing out here that the internal geometry of this space is more complicated. It is in particular not true that it is a convex cone (think of the path joining the Minkowski metric in 2 dimensions to its opposite which is another Lorentzian metric!).

This discussion can also be carried over to metrics with signatures $(p, n-p)$ $(p \leq 2)$ on an n-dimensional space V, and for these cases too the geometry of the space of generalized scalar products is more intricate than that of $\mathcal{M}et\,(V)$. Actually, a thorough description of their internal geometry is not readily available.

4.2. Manifolds and Riemannian Metrics

We refer to standard references such as [**GHL87**] or [**War71**] for thorough introductions to the (by now very classical) theory of *differential manifolds*. The purpose of this chapter is to provide a very biaised introduction, assuming the construction of the basic objects known to the reader. Emphasis will be put on the basic theorems and points of view which are important for the understanding of the subject matter of this course, namely *Spinorial Geometry*.

4.2.1. Differential manifolds and expressions in local coordinates. One of the key issues about the foundations of Modern Differential Geometry, which stresses the global aspects of the theory, is to properly understand how they relate to the more classical point of view stressing local coordinates patches and expressions of quantities of interest in local coordinates. The importance of being able to go swiftly from one to the other while working in Differential Geometry can never be underestimated.

Atlas, charts and coordinate systems. In this chapter (and as a matter of fact in this set of lectures) M will denote an n-dimensional differential manifold, i.e. a topological space which is endowed with an *atlas*, i.e. a family of maps (the *charts* of the atlas) $\varphi_i : U_i \longrightarrow \mathbb{R}^n$ which are homeomorphisms onto their images and defined over an open covering $(U_i)_{i\in I}$ of M with the conditions that, considering two charts φ_i and φ_j whose domain overlap, the map $\varphi_{ij} = \varphi_j \circ \varphi_i^{-1}$ from $\varphi(U_i \cap U_j) \subset \mathbb{R}^n$ to $\mathbb{R}^n$ is a diffeomorphism onto its image. (All throughout these lectures, without saying, differentiable will mean C^∞-differentiable.)

An equivalent way of speaking of the charts is to speak of the *local coordinate system* $(x^i)_{1\le i\le n}$ associated to a chart φ, simply by setting $x^i = \mathrm{pr}^i \circ \varphi$ where pr^i is simply the projection onto the ith-factor in $\mathbb{R}^n$. In a neighbourhood of a point p, this puts at our disposal the coordinate systems of charts whose domain contains p.

The tangent and cotangent bundles of a differential manifold. A key construction attaches naturally an important geometric object to M. It is its *tangent bundle* $\pi : TM \longrightarrow M$. To each point p of M, it attaches an n-dimensional vector space T_pM, which can be viewed equivalently as the equivalence classes of parametrized curves through p which have the same velocity vector at p when read in a chart (the definition of an atlas is precisely made such that this is independent of the chart), or the vector space dual to the space of *derivations at* p, namely the linear forms A on the space of (germs of) differentiable functions defined in a neighbourhood of p satisfying the Leibniz rule for the product of functions, namely $A(f_1.f_2) = f_2(p)\, A(f_1) + f_1(p)\, A(f_2)$.

Understanding in depth why these two definitions match one another is more or less the key to manifold theory. This shows that the algebra $C^\infty(M)$ of differentiable functions on M (or similar algebras defined only in subdomains) play an important role in Differential Geometry.

As mentioned just before, it directly leads to the notion of the *cotangent* bundle $\varpi : T^*M \longrightarrow M$, the collection of cotangent spaces T_p^*M at each point p. Typical elements of T_p^*M are *differentials* at p of functions, i.e. $df(p)$, among which one finds the differentials $dx^i(p)$ of the coordinate functions x^i. We note the fundamental formula

$$df(p) = \sum_{i=1}^{n} \left(\partial f/\partial x^i\right)(p)\, dx^i(p)\ .$$

4.2.2. Vector fields. An important notion is that of a *vector field*, i.e. a map X which to any point p in M associates a tangent vector $X(p)$ at p.

Natural vector fields associated to a chart. To each coordinate system (x^i) in the neighbourhood U of a point p are associated n natural vector fields $\partial/\partial x^i$, $1 \le i \le n$, which are characterized by the fact that, for $f \in C^\infty(U)$, $\partial/\partial x^i(f) = \partial f/\partial x^i$, where the right hand side is just the ordinary partial derivative expression when f is viewed as a function of the n variables x^i. (Remember, as mentioned before, that a tangent vector at p can be viewed as a linear form on the algebra $C^\infty(U)$.)

Moreover at each point $p \in U$ the n vector fields $\partial/\partial x^i$ form a basis of T_pM. As a result, any vector field X defined on U can be expressed as $X = \sum_{i=1}^n X^i\, \partial/'\partial x^i$ where the n components X^i are the local expression of X in the coordinate system (x^i). The differentiability of X is reflected in the differentiability of its compoments (X^i). This uses the module structure of the vector space $\mathcal{T}U$ of vector fields defined on the domain U over the algebra $\mathrm{C}^\infty(U)$.

In the same way as one assembled all tangent vectors to M in the tangent bundle $\pi : TM \longrightarrow M$, one can assemble all linear bases at points of M to form another bundle that we denote by $GlM \longrightarrow M$.

The Fundamental Theorem of Differential Geometry.

THEOREM 4.2. *Suppose that $\Lambda : \mathcal{T}M \longrightarrow \mathrm{C}^\infty(M)$ is linear with respect to multiplication by functions on $\mathcal{T}M$. Then Λ can be identified with a differential 1-form λ on M, i.e. a map from M to T^*M which is a section of the projection ϖ.*

A possible proof uses the following lemma:

LEMMA 4.3. *If X and Y are two vector fields which coincide in a neighbourhood of a point p in M, then*

$$(\Lambda(X))(p) = (\Lambda(Y))(p) \ .$$

The proof of the lemma goes as follows: If U is a neighbourhood of p, then one can construct a function $f \in \mathrm{C}^\infty(M)$ such that $f \equiv 1$ on an open neighbourhood U' of p such that $\overline{U'} \subset U$ and $f \equiv 0$ on the complement of U in M.

Then $\Lambda(f(X - Y)) = 0$ since $f(X - Y) \equiv 0$ on M. By evaluating at p, one has $0 = f(p)\,(\Lambda(X)(p) - \Lambda(Y)(p))$, hence the Lemma.

Let $X \in \mathcal{T}M$, and let U be the domain of definition of a system of coordinates (x^i). We write $X_{|U} = \sum_{i=1}^n X^i\, \partial/\partial x^i$. By evaluating $\Lambda(X)$ at $p \in U$, we obtain

$$(\Lambda(X))(p) = \sum_{i=1}^n X^i(p)\, \Lambda(\frac{\partial}{\partial x^i})(p) \ .$$

(Here Lemma 3 has been used to make sense of the right hand side when it is only defined in U.)

If we set $\lambda_i(p) = \Lambda(\partial/\partial x^i)(p)$, then one easily sees that the differential 1-form $\lambda = \sum_{i=1}^n \lambda_i\, dx^i$ does coincide with Λ in its action on $\mathrm{C}^\infty(M)$, as announced.

One also has a similar statement for higher order tensors. This makes this formulation the modern version of the old fashioned *tensoriality criteria*.

THEOREM 4.4. *Suppose that $B : \mathcal{T}M \times \mathcal{T}M \longrightarrow \mathrm{C}^\infty(M)$ is bilinear with respect to multiplication by functions on $\mathcal{T}M$. Then B can be identified with a covariant tensor field β of order 2 on M, i.e. a map from M to $T^*M \otimes T^*M$ which is a section of the projection $\varpi \otimes \varpi$.*

4.2.3. Encounter with Riemannian metrics. The central objects in this series of lectures are *Riemannian metrics*.

Definitions and coordinate expression.

DEFINITION 4.5. A *Riemannian metric* g is a field of scalar products on M, i.e. the data on each tangent space T_pM of a scalar product $g(p)$ whose coefficients are C^∞ in any coordinate local system (x^i),

$$g = \sum_{i,j}^{n} g_{ij}\, dx^i\, dx^j \ .$$

After what was said in Section 4.2.3, one can equivalently say that a Riemannian metric is a symmetric map $\mathcal{T}M \times \mathcal{T}M \longrightarrow C^\infty(M)$, which is bilinear with respect to the $C^\infty(M)$-module structure of $\mathcal{T}M$.

At a point p, orthonormal bases for the scalar product $g(p)$ will be denoted by $(O_g(M))_p$. Putting together all bases will form the bundle of g-orthonormal bases $O_gM \longrightarrow M$, which is actually a sub-bundle of $GlM \longrightarrow M$.

Euclidean metrics in general coordinate systems. On the vector space $\mathbb{R}^n$, a very special case is given by the data of a Riemannian metric e whose coefficients in a linear system of coordinates are constant. This is nothing but the classical notion of a Euclidean scalar product. (It is actually a deep, though elementary, theorem in Riemannian geometry to state that any metric which has constant coefficients in a coordinate system on a manifold is actually globally isometric to a Euclidean metric in an open set of a vector space.)

It is worth making the trivial remark that this constancy property is dependent on the system of coordinates that one uses. Indeed, if one uses polar coordinates (r,θ) in the plane instead of Cartesian (linear) coordinates (x^1, x^2), then the Euclidean metric $e = (dx^1)^2 + (dx^2)^2$ has the local expression $e = dr^2 + r^2\, d\theta^2$, in which the coefficients are not constant.

Existence of Riemannian metrics.

THEOREM 4.6. *On any manifold admitting a partition of unity, there exists at least one Riemannian metric, hence many.*

Idea of the proof: Let us fix a covering $(U_i)_{i\in I}$ by open sets corresponding to domains of charts φ_i. We take a partition of unity $(f_i)_{i\in I}$ subordinated to this covering. Therefore, over M, we have $\sum_{i=1}^n f_i \equiv 1$ (note that at each point the sum is actually a finite sum). Then we claim that $\sum_{i=1}^n f_i\, \varphi^* e_i$, where e denotes a Euclidean scalar product on $\mathbb{R}^n$, is a Riemannian metric on M. Indeed this follows readily from the construction, using the fact that each chart φ_i is a local diffeomorphism and the stability of scalar products by convex combinations.

4.3. Covariant Derivatives

Historically, Riemannian metrics were introduced at this level of generality, as the name shows, by Bernhard Riemann in his Habilitation defense in 1854.

They developed as a universal tool only after one more step had been taken, namely realizing that on a manifold endowed with a Riemannian metric, there is one privileged differential calculus for all tensor fields, something which was only clarified in

1888 by Grigorio Ricci-Curbastro, and finally widely available in the scientific community thanks to the article [**CRC01**] by Ricci-Curbastro and Tullio Levi-Civita, his student. Actually the decisive tool had been more or less interpreted some time earlier (in 1869) by Erwin-Bruno Christoffel in [**Chr70**], although it is quite clear that he did not fully envision the importance of the tool he was introducing.

To get the story straight, it is useful to first come back to vector fields, and to consider the consequences of the existence of two equivalent definitions for tangent vectors.

4.3.1. Vector fields as differential operators. Indeed, we said in 4.2.2 that a tangent vector could be viewed as a *derivation at a point*. If one pursues this point of view and extend it to vector fields, one gets the view that *a vector field on M is a derivation of the algebra* $C^*\infty(M)$, i.e. a linear map A which satisfies the *Leibnitz identity* which, for f_1, $f_2 \in C^\infty(M)$, states that

$$A(f_1.f_2) = f_1 \,.\, A(f_2) + f_2 \,.\, A(f_1) \;.$$

Equivalence between vector fields and derivations of the algebra of functions. To a vector field X one can attach the derivation ∂_X defined as $\partial_X f = df(X)$. The Leibnitz identity follows directly from the usual formula for the differential of the product of two functions. The key remark is then the following theorem

THEOREM 4.7. *The map ∂ which with a vector field X associates the derivation ∂_X is a one-to-one correspondence between vector fields on a manifold M and derivations of $C^\infty(M)$.*

The proof has two steps: first, to prove that the map $X \mapsto \partial_X$ is injective; second, to establish that it is onto.

Injectivity is easy by realising that the components (X^i) of a vector field X in the domain of a system of coordinates (x^i) are precisely given by $X^i = \partial_X(x^i)$. This implies that two vector fields X and Y who act in the same way on functions must have equal components in any system of coordinates, hence are actually identical.

Proving surjectivity can be obtained from a lemma, usually attributed to Jacques Hadamard, to the effect that, a coordinate system (x^i) being chosen in the neighbourhood of a point p and centered at p (i.e. such that $x^i(p) = 0$ for $1 \le i \le n$), *any C^∞ function f can be written locally around p as*

$$f = f(p) + \sum_{i=1}^{n} F_i\, x^i \;,$$

where the n functions F_i are C^∞.

Then, if A is a derivation of the algebra $C^\infty(M)$, using the fact that any derivation acts trivially on constant functions (an easy consequence of the Leibnitz identity applied to the square of the function identically equal to 1), one has $A(f)(p) = \sum_{i=1}^n F_i(p)\, A(x^i)(p)$. As a consequence, the derivation A acts in the same way as the vector field $X = \sum_{i=1}^n A(x^i)\, \partial/\partial x^i$ because one easily sees that $F_i(p) = \partial f/\partial x^i$. Therefore, we have proved surjectivity.

From now on, we systematically identify vector fields on M and derivations of the algebra $C^\infty(M)$.

The Lie bracket of vector fields. The first important consequence of this identification is the following definition.

DEFINITION 4.8. Let X and Y be two vector fields on M. The commutator $\partial_X \circ \partial_Y - \partial_Y \circ \partial_X$ of the two derivations ∂_X and ∂_Y is again a derivation of $C^\infty(M)$, hence defines a vector field that we denote by $[X, Y]$ and that is called the *Lie bracket* of X and Y.

It follows directly from the definition of the Lie bracket $[X, Y]$ that its components $[X, Y]^i$ in a coordinate system (x^i) are related to those of X and Y by the formula

$$[X, Y]^i = \sum_{j=1}^{n} \left(X^j \frac{\partial Y^i}{\partial x^j} - Y^j \frac{\partial X^i}{\partial x^j} \right) .$$

This shows very clearly that the value at p of the Lie bracket of X and Y does not depend only on $X(p)$ and $Y(p)$, but also on the value of the derivatives of their components at p. It therefore follows that the map $(X, Y) \mapsto [X, Y]$ is not C^∞-linear in any of its variables.

4.3.2. Definition of a Covariant Derivative. The need for a *covariant derivative* came from the fact that, contrary to the (after all very) easy road to derivations of the algebra $C^\infty(M)$ of functions, defining derivations of the spaces of tensor fields is more tortuous. Here is the critical definition:

DEFINITION 4.9. A *covariant derivative* is a map D from $\mathcal{T}M \times \mathcal{T}M \longrightarrow \mathcal{T}M$ (for X, $Y \in \mathcal{T}M$, one writes $D_X Y$) satisfying the following three axioms:

(CD 0) D is bilinear with respect to scalars acting on vector fields;

(CD 1) D is $C^\infty(M)$-linear in its first argument, namely, for any function $f \in C^\infty(M)$ and any vector fields X, $Y \in \mathcal{T}M$,

$$D_{f\,X} Y = f\, D_X Y .$$

(CD 2) D satisfies a Leibnitz condition with respect to the second argument, namely for any function $f \in C^\infty(M)$ and any vector fields X, $Y \in \mathcal{T}M$,

$$D_X(f\,Y) = f\,D_X Y + (\partial_X f)\,Y .$$

Local expression of a covariant derivative. Just applying successively axioms (CD 0), (CD 1) and (CD 2), one gets the local expression of the covariant derivative D in a system of local coordinates (x^i) in which X has components (X^i) and Y components (Y^i) as follows

$$D_X Y = \sum_{i,j,k=1}^{n} X^i \left(\frac{\partial Y^k}{\partial x^i} + Y^j \Gamma_{i\,j}^{\,k} \right) \frac{\partial}{\partial x^k} , \tag{4.1}$$

if one sets $D_{\partial/\partial x^i} \partial/\partial x^j = \Gamma_{i\,j}^{\,k}\, \partial/\partial x^k$. In order to determine D, one needs to know the $\Gamma_{i\,j}^{\,k}$, called the *Christoffel symbols* of D. Conversely, an explicit knowledge of D will allow to determine these symbols. To make sure one really understands the nature of covariant derivatives, it is useful to check how the Christoffel symbols change when changing coordinates.

REMARK 4.10. It is a tedious but useful exercice to determine the formula expressing the change in the Christoffel symbols when one goes from one local coordinate system to another. One then realizes that the $\Gamma_{i\,j}^{\,k}$ do not behave like a tensor field of type (1,2) as the notation of their indices seems to suggest. There is a contribution from the second order derivative of the coordinate change. This makes the linear changes of coordinates special from that point of view since for them this term drops. Actually there is a long story behind this, which connects to the long debated question of the proper definition of the *acceleration* in a general (i.e. non inertial) frame. It took the masterwork of Joseph-Louis de Lagrange who clarified this question through his work on the Euler-Lagrange equation in the Calculus of Variations.

Extending covariant derivatives to tensor fields. It is very useful to extend the covariant derivative from vector fields to general tensor fields. This can be done naturally provided one keeps in mind that the contraction of tensors (like evaluating a differential 1-form λ on a vector field X as $\lambda(X)$, and the like) is preserved under derivation. Concretely, this means that the definition of a covariant derivative of a differential 1-form λ should satisfy the following Leibnitz-type identity: for any vector fields X and Y,

$$\partial_X(\lambda(Y)) = \lambda(D_X Y) + (D_X\lambda)(Y) \ .$$

The previous "identity" can be turned around to become a definition of the covariant derivative of covariant objects by setting

$$(D_X\lambda)(Y) = \partial_X(\lambda(Y)) - \lambda(D_X Y) \ .$$

As usual it is important to give the expression in local coordinates for this extension. It follows again directly from the local expression of the covariant derivative of a vector field, and the identity above. If, in a system of coordinates (x^i), $\lambda = \sum_{i=1}^n \lambda_i \, dx^i$, then

$$D_X\lambda = \sum_{i,j,k=1}^{n} X^i \left(\frac{\partial \lambda_i}{\partial x^k} - \lambda_j \, \Gamma_{i\,k}^{\,j} \right) dx^k \ .$$

Extensions of the covariant derivative to tensor fields of higher orders are naturally obtained by imposing to the extended covariant derivative to behave like a derivation with respect to the tensor product, that is by requiring that, if t_1 and t_2 are two tensor fields, then $D_X(t_1 \otimes t_2) = D_X t_1 \otimes t_2 + t_1 \otimes D_X t_2$.

This allows us to define the covariant derivative of Riemannian metrics for example as covariant tensor fields of order 2.

The space of covariant derivatives. It is very easy to verify that, given any covariant derivative D and any tensor field C of type (2,1), the quantity $D+C$ acting on two vector fields X and Y as $D_X Y + C(X,Y)$ is again a covariant derivative. Conversely, the difference of two covariant derivatives is a tensor field of type (2,1), proving that the space of covariant derivatives is an affine space modeled after the space of tensor fields of type (2,1), hence an infinite dimensional space.

The torsion of a covariant derivative. Covariant derivatives have a few tensor fields attached to them. Here is the definition of one of them.

THEOREM 4.11. *Let D be a covariant derivative. The quantity $T^D(X,Y) = D_X Y - D_Y X - [X,Y]$ is a tensor field of type (2,1) attached to D called its torsion, which is skew-symmetric in its covariant arguments.*

The claim is a direct application of the Fundamental Theorem of Differential Geometry. The skew-symmetry follows directly from the definition.

In a local coordinate system (x^i), denoting by $\Gamma_{i\,j}^{\,k}$ the Christoffel symbols of D, one easily verifies that $T^D = \sum_{i,j,k=1}^n (\Gamma_{i\,j}^{\,k} - \Gamma_{j\,i}^{\,k})\, dx^i \otimes dx^j \otimes \partial/\partial x^k$.

Covariant derivatives with vanishing torsion are called *symmetric* or *torsion-free.* By adding $-T^D$ to any covariant derivative D , one obtains the symmetric covariant derivative attached to D.

4.3.3. The Levi-Civita Covariant Derivative. We are now ready to make the decisive step due to Ricci-Curbastro, namely to attach in a canonical way a covariant derivative to any Riemannian metric. Actually the construction goes through without any modification for any non-degenerate symmetric bilinear form, even if it is not positive definite. This is why this theorem is called the *Fundamental Theorem of Metric Geometry.*

THEOREM 4.12 (**Fundamental Theorem of Metric Geometry**). *Let g be a Riemannian metric. There is one and only one covariant derivative D, traditionally called the Levi-Civita*[1] *covariant derivative attached to g, verifying the following two conditions*

i) D is symmetric;

ii) D is metric, i.e. D satisfies, for any vector fields X, Y and Z, the identity

$$\partial_X(g(Y,Z)) = g(D_X Y, Z) + g(Y, D_X Z) \ .$$

The proof consists in proceeding first in a necessary fashion, namely one evaluates $(g(D_X Y, Z))$ by using repeatedly the two conditions put on the covariant derivative. Since g is non-degenerate, this is enough to determine $D_X Y$. If one does it right, i.e. if ones keep away from manipulations which would lead to circular reasoning, one ends up with a formula containing on the right hand side seven terms, one of them being the exact term we started with on the left hand side with the opposite sign. For the covariant derivative this gives the following necessary expression

$$2\,g(D_X Y, Z) = \partial_X(g(Y,Z)) + \partial_Y(g(X,Z)) - \partial_Z(g(X,Y)) \tag{4.2}$$

$$-g([X,Y],Z) + g(Y,[X,Z]) + g(X,[Y,Z]) \ . \tag{4.3}$$

To finish the proof, one then only needs to prove that the right hand side of the previous formula is indeed a covariant derivative and that it is indeed symmetric and metric.

Local expression of the Levi-Civita covariant derivative. We can use Formula (4.2) to get the Christoffel symbols of the Levi-Civita covariant derivative in a coordinate system (x^i) by giving to X, Y and Z some special values, typically

[1]As mentioned already in the short historical remark made at the beginning of this chapter, it is actually Ricci-Curbastro who made this fact clear, and therefore this covariant derivative should be called after his name.

$X = \partial/\partial x^i$, $Y = \partial/\partial x^j$, and $Z = \partial/\partial x^k$. Using the very definition of the Christoffel symbols, and the fact that the Lie brackets of the natural vector fields $\partial/\partial x^i$ all vanish (this is just the Schwarz Lemma on second derivatives), one gets the famous formula

$$\Gamma_{i\ j}^{\ k} = \frac{1}{2} g^{km} \left(\frac{\partial g_{im}}{\partial x^j} + \frac{\partial g_{jm}}{\partial x^i} - \frac{\partial g_{ij}}{\partial x^m} \right) , \tag{4.4}$$

where we recall that (g^{ij}) denotes the inverse matrix of the matrix (g_{ij}) of local coefficients of the metric g.

It is obvious from Formula (4.4) that the Christoffel symbols are symmetric in i and j, as they should since the Levi-Civita covariant derivative is symmetric.

The expression of the first derivative in the metric seems odd the first time one meets it. It does have some interesting internal symmetries, as one soon realizes. Its internal structure leads to the solution of the following interesting exercice.

EXERCISE 4.13. If, at p, $\Gamma_{i\ j}^{\ k}(p) = 0$ for $1 \leq i, j, k \leq n$, then $(\partial g_{ij}/\partial x^k)(p) = 0$ (the converse being obviously true by Formula (4.4)).

Since it is always possible by a geometric construction to construct a coordinate system at p for which $\Gamma_{i\ j}^{\ k}(p) = 0$, an important geometric consequence of this exercise is that the first order derivatives of the coefficients of the metric do not have any intrinsic meaning.

4.4. Riemannian Curvature Tensors

We now come to the main invariant attached to a Riemannian metric, namely its curvature tensor. It was the main point of Bernhard Riemann's inaugural article to notice that it measures the deviation for a Riemannian metric from being Euclidean.

4.4.1. The curvature tensor of a covariant derivative: definition.

THEOREM 4.14. *Let D be a covariant derivative. For vector fields X, Y and Z, the quantity*

$$R^D_{X,Y} Z = D_X(D_Y Z) - D_Y(D_X Z) - D_{[X,Y]} Z$$

is a tensor field of type $(3,1)$, called the curvature tensor of the covariant derivative, which is skew-symmetric in its first two arguments.

The fact that R^D is a tensor field follows by inspection from the Fundamental Theorem in Differential Geometry. The skew-symmetry is built into the definition.

Its local expression in a coordinate system (x^i) is obtained directly from the Christoffel symbols $\Gamma^k_{i\ j}$ of the covariant derivative. Namely, if one sets
$R^D = \sum_{i,j,k,l=1}^n R_{ijk}^{\ \ \ l}\, dx^i \otimes dx^j \otimes dx^k \otimes \partial/\partial x^l$, then

$$R_{ijk}^{\ \ \ l} = \frac{\partial \Gamma_{i\ k}^{\ l}}{\partial x^j} - \frac{\partial \Gamma_{j\ k}^{\ l}}{\partial x^i} + \sum_{m=1}^{n} \left(\Gamma_{i\ m}^{\ l} \Gamma_{j\ k}^{\ m} - \Gamma_{j\ m}^{\ l} \Gamma_{i\ k}^{\ m} \right) . \tag{4.5}$$

From Formula (4.5), it is clear that the curvature vanishes identically if the Christoffel symbols vanish in a system of coordinates. One recovers the fact that for the usual differential calculus in a vector space, the curvature tensor is identically zero.

4.4.2. The Riemann Curvature: Basic Properties. Specialising to the Levi Civita covariant derivative. When the covariant derivative D is the Levi Civita covariant derivative attached to a metric g, then it is traditional to call it the *Riemann(-Christoffel) curvature tensor*, and rightly so as suggested in the introductory paragraph. One will from time to time denote it by R^g to stress its dependence upon the metric.

Using Formula (4.4) relating the Christoffel symbols to the coefficients of the metric, one can derive an explicit formula for the Riemann curvature tensor as a complicated expression involving the local coefficients g_{ij} of the metric, almost linearly their partial derivatives of second order, and quadratically their partial derivatives of first order. Therefore, the Riemann curvature tensor is a (rather uggly looking) non-linear expression involving second order derivatives of the metric.

One can push one step further the remark made at the end of last section to the effect that, for any Euclidean metric, the curvature tensor vanishes. Indeed for such a metric one can find coordinates in which the coefficients of the metric are constant, and therefore their Christoffel symbols vanish. As promised, the curvature tensor is measuring the deviation of a Riemannian metric from being Euclidean.

The Riemann curvature extra symmetries

THEOREM 4.15. *The Riemann curvature tensor of a metric g satisfies two further symmetries:*

– *the first Bianchi identity which, for tangent vectors X, Y and Z, states that*

$$R_{X,Y}Z + R_{Y,Z}X + R_{Z,X}Y = 0 \ ;$$

– *another skew symmetry, namely for tangent vectors X, Y, Z and T,*

$$g(R_{X,Y}Z, T) = -g(R_{X,Y}T, Z) \ .$$

The first Bianchi identity follows directly from the Levi Civita covariant derivative being symmetric, while the other identity follows from its metric character.

One more symmetry is a consequence of the other three, namely for tangent vectors X, Y, Z and T, $g(R_{X,Y}Z,T) = g(R_{Z,T}X,Y)$. This is left as an (interesting) exercise. The previously mentioned skew symmetries and this symmetry allow one to consider the (4,0) version of the Riemann curvature tensor (i.e. $R(X,Y,Z,T) = g(R_{X,Y}Z,T)$) as an element of $S^2(\Lambda^2T^*M)$, and this turns out to be a useful remark. It is then interesting to reinterpret the first Bianchi identity as stating that (again the (4,0) version of) R lies in the kernel of the projector $b(R)(X,Y,Z,T) = \frac{1}{3}((R(X,Y,Z,T) + R(Y,Z,X,T) + R(Z,X,Y,T))$. One easily checks that the image of b is precisely the space Λ^4T^*M of exterior 4-forms. This allows us to determine the dimension of the space of Riemannian curvature tensors, namely $\frac{1}{12}n^2(n^2-1)$, since

$$\frac{1}{12}n^2(n^2-1) = \frac{1}{8}n(n-1)(n(n-1)+2) - \frac{1}{2\cdot 3\cdot 4}n(n-1)(n-2)(n-3) \ .$$

4.5. Principal bundles

It is only well in the XXth century that the proper geometric tools were formally available to discuss and work in Spin Geometry globally over a manifold. It does indeed require the language of principal and associated bundles to properly describe the problems encountered when one wants to define spinors over a manifold, and specify the piece of extra structure that is needed to make sense of them.

The purpose of this chapter is precisely to do that. Again, this is a piece of general knowledge for specialists of Differential Geometry that is developed here with the narrow focus of presenting what is necessary to work in Spin Geometry.

The key notion in Modern Geometry is that of bundles. We already encountered, and made use of, vector bundles in the form of tangent and cotangent bundles, and more generally tensor bundles.

In order to define spin structures, we need another approach, that of *principal bundles*.

4.5.1. Principal bundles. Definition and classification of principal bundles.

DEFINITION 4.16. Let P be a manifold on which a Lie group G acts freely on the right, i.e., for any $p \in P$ and $\gamma \in G$, $\{\gamma \in G \mid p.\gamma = p\} = \{e\}$.

The natural projection $\pi : P \longrightarrow P/G$ will be said to be a *principal G-bundle* if the following assumption[2] is met: each orbit $p.G$ admits a neighbourhood U in P/G such that, $\pi^{-1}(U) \approx U \times G$.

The weakness of this definition is that it is not very constructive, i.e., it requires that one already has at hand P and the action of G. Another more constructive approach is given by the following alternative definition.

DEFINITION 4.17. Let G be a Lie group, M a manifold and $(U_i)_{i\in I}$ an atlas of M. One supposes that for over each non empty intersection $U_i \cap U_j$ one has a map $\varphi_{ij} : U_i \cap U_j \longrightarrow G$, so that the cocycle condition $\varphi_{ij}(x)\,\varphi_{jk}(x) = \varphi_{ik}(x)$ holds for all $x \in U_i \cap U_j \cap U_k$, assumed to be non empty.

The link between the two definitions goes as follows: From the data presented in Definition 4.17, one can form a space P by identifying over $U_i \cap U_j$, the two copies of $(U_i \times U_j) \times G$ coming respectively from U_i and U_j through the map Φ_{ij} : $(U_i \times U_j) \times G \longrightarrow (U_i \times U_j) \times G$ defined as $\Phi_{ij}(x, \gamma) = (x, \varphi_{ij}(x)\gamma)$. The projection $\pi : P \longrightarrow M$ is given on each U_i by the projection onto the first factor. The *cocycle condition* states that, over $U_i \cap U_j \cap U_k$, if non empty, the two identifications coincide, ensuring that one has indeed defined a global object.

The action of the group on the right is given by the regular action of the group by multiplication on the right on the right factor in each local representation. Its freeness follows from the very definition.

The second presentation leads naturally to the classification of principal G-bundles, because it involves a cocycle condition. Equivalence classes of G-principal bundles modulo isomorphisms (i.e. identifying to the trivial bundle bundles defined by a coboundary, namely a collections of maps φ_{ij} such that $\varphi_{ij} = \varphi_i^{-1}.\varphi_j$ for maps $\varphi_i : U_i \longrightarrow G$) can be put into correspondence with the Cech cohomology group

[2]With some extra assumption on the action this assumption can actually be dropped.

$H^1(M, G)$. It should be noted that for G non commutative, higher cohomology groups cannot be defined.

Examples of principal bundles. Let M be a manifold. In 4.4.2 we already introduced the *linear frame bundle* $GlM \longrightarrow M$ by stating that, over every point x, Gl_xM is the set of linear bases of T_xM. As explained earlier, a basis can be viewed as an invertible map b between $\mathbb{R}^n$ and T_xM. Then the right action of $\ell \in \mathrm{Gl}_n(\mathbb{R})$ on b is just composition on the right in $\mathbb{R}^n$. This replaces the basis b by the basis $b \circ \ell$ which lies also in Gl_xM. It is clear that the action is free. We leave as an exercise to show the local triviality of the map $GlM \longrightarrow M$. (It can easily be obtained by working over an atlas defining the differential structure on M.) Therefore the linear frame bundle is a Gl_n-principal bundle.

If a metric g is given, then one can consider, as we did in 4.2.3, the bundle O_gM of g-orthonormal frames. It is an O_n-principal sub-bundle of the linear frame bundle $GlM \longrightarrow M$. Indeed a map from $\mathbb{R}^n$ to $\mathbb{R}^n$ maps an g-orthonormal basis to another one provided this map is a Euclidean isometry.

4.5.2. Associated bundles. Actually we have already met examples of associated bundles in the form of the tangent and the cotangent bundles of the manifold. The construction can be generalized as follows: if $\pi : P \longrightarrow M$ is a principal G-bundle, and $\rho : G \longrightarrow Gl(E)$ a linear representation of G, then, if we set $P \times_\rho E$ as the set of equivalence classes of pairs (p, v) for $p \in P$ and $v \in E$, where (p, v) and $(p.\gamma, \rho(\gamma^{-1})(v)$ are identified for $\gamma \in G$. It is easy to see that $P \times_\rho E$ naturally inherits a vector bundle structure with basis M and fibre the vector space E, and even a more refined structure induced by the group G, typically a scalar product if the representation is orthogonal.

In the examples given above, $G = \mathrm{Gl}(n)$ and $\rho = \mathrm{std}$, the standard representation of Gl_n on $\mathbb{R}^n$, so that $TM = Gl(M) \times_{\mathrm{std}} \mathbb{R}^n$, and $T^*M = Gl(M) \times_{\mathrm{std}^*} \mathbb{R}^n$, where std^* denotes the contragredient representation. This extends to tensor bundles on M by taking the appropriate tensor representations.

4.5.3. Connections on principal bundles. Let $P \longrightarrow M$ be a principal G-bundle. It will be necessary to refer to the Lie algebra $\mathbf{G}$ of G.

The key geometric situation to understand is the following. Through a point $p \in P$ passes the orbit $p.G$ of the group G. Its tangent space at p can be identified to $\mathbf{G}$ because of the freeness of the group action which ensures that any element $V \in \mathbf{G}$ is mapped injectively by the infinitesimal action to $\tilde{V} \in T_pP$.

Definition of a connection.

DEFINITION 4.18. A G-connection on a principal G-bundle $\pi : P \longrightarrow M$ is an element ω of $\Omega^1(P, \mathbf{G})$ such that

i) $\omega(\tilde{V}) = V$,
ii) $R_\gamma^*\omega = \mathrm{ad}\gamma^{-1}(\omega)$.

The two conditions put on a G-connection ensure that $\mathrm{Ker}(\omega_p)$ is complementary to $T_p(p.G)$. This space which is defined by the connection is called the *horizontal* space attached to it, and often denoted by H_p.

Conversely the data of a set of planes H_p transversal to the fibres of the projection $\pi : P \longrightarrow M$ and mapped onto one another by the G-action (i.e. $H_{p.\gamma} = R^*_\gamma(H_p)$) determines a G-connection. This more geometric approach is sometimes useful.

Comparaison with the covariant derivative approach. Let ψ be a local section of the vector bundle $P_\rho = P \times_\rho E$ associated via the representation ρ to the principal G-bundle $\pi : P \longrightarrow M$, i.e. $\psi : U \longrightarrow P$ so that, for $m \in M$ one can write $\psi(m) = [p(m), v(m)]$, where $v : U \longrightarrow E$ is an ordinary vector-valued function. Recall that $p(m)$ is a frame attached to each point, sometimes called in Élie Cartan's jargon "*a moving frame*". The local section ψ is a local *field* whose expression (v^α) in a moving frame (p_α) allows to write $\psi = \sum_\alpha v^\alpha\, p_\alpha$.

A connection ω being chosen on P, and a vector field X on M being fixed, one can associate with the local section ψ of P_ρ another local section

$$D_X\psi = [p, \partial_X v + \rho_*(\omega(dp(X)))(v)]$$

which coincides with the covariant derivative of ψ given by the set of Christoffel symbols $(\omega_i{}^\beta{}_\alpha)$.

Curvature of a connection in a principal bundle setting. In the principal bundle setting the curvature measures the non-integrability of the horizontal distribution. This is achieved through a Lie algebra-valued 2-form Ω, defined for X, $Y \in T_pP$ as

$$\Omega(X,Y) = d\omega(X,Y) + [\omega(X), \omega(Y)]\ .$$

Notice that the 2-form Ω vanishes on any vertical vector because in these directions ω reduces by definition to the Maurer-Cartan form of the group G. This shows that the curvature form Ω depends only on the value of the projections $d\pi(X)$ and $d\pi(Y)$ down to M.

Connecting the curvature expression in the two formalisms is simple since we already have related the covariant derivatives. Recall that the curvature $R_{X,Y}$ of the covariant derivative D is defined as $R_{X,Y} = [D_X, D_Y] - D_{[X,Y]}$ (where here $[\ ,\ ]$ denotes the commutator of two first order operators). The explicit relation can be formulated as follows: if ψ is a section of P_ρ given in the moving frame p by (v^α), then

$$\sum_\beta (p^*\Omega)^\alpha_\beta\, v^\beta = R(v^\alpha)$$

as an identity between Lie algebra-valued 2-forms on M.

4.6. Spin$_n$-bundles and spinorial structures

The case $G = \mathrm{Spin}_n$ is central to this set of notes. This is why we devote a special chapter to it.

We recall that the group Spin_n is a 2-fold cover of the group SO_n, i.e. its universal cover for $n \geq 3$. Fundamental to our considerations is the short exact sequence of groups

$$0 \longrightarrow \mathbb{Z}/2\mathbb{Z} \longrightarrow \mathrm{Spin}_n \longrightarrow \mathrm{SO}_n \longrightarrow 1\ . \tag{4.6}$$

4.6.1. Spin structures. Defining a *spin structure* globally over a manifold M took some time because of the need to develop the appropriate language to discuss global properties. An excellent reference for that purpose is the book [**LM89**].

There are several issues connected to this question that need to be addressed. First the whole question is purely topological. We will actually address it here with a background metric fixed.

Definition of a spin structure. For the sake of simplicity we will assume that the manifold M is oriented[3], so that we can reduce the structure group of the tangent bundle to the group Gl_n^+, leading to the principal Gl_n^+-bundle $Gl^+M \longrightarrow M$ of *positive* bases for this choice of an orientation of M.

If we fix a Riemannian metric g, we can therefore consider the SO_n-subbundle of the bundle of positive bases consisting of positive orthonormal frames, that we denote by $SO_gM \longrightarrow M$.

DEFINITION 4.19. A *spin structure* on a Riemannian manifold (M, g) is given by a principal Spin_n-bundle $\pi : P \longrightarrow M$ provided with an equivariant bundle map $s : P \longrightarrow SO_gM$ which is a double cover on each fibre with respect to the actions of Spin_n on the source and of SO_n on the target.

In we intend to do any spinorial geometry, to establish the existence and to classify spin structures are clearly key questions.

Obstruction to the existence of spin structures. Thanks to the fundamental exact sequence (4.6), we can deal with these questions by using the cohomology exact sequence

$$H^0(M, \mathbb{Z}/2\mathbb{Z}) \longrightarrow H^0(M, \mathrm{Spin}_n) \longrightarrow H^0(M, \mathrm{SO}_n) \longrightarrow$$

$$\longrightarrow H^1(M, \mathbb{Z}/2\mathbb{Z}) \longrightarrow H^1(M, \mathrm{Spin}_n) \longrightarrow H^1(M, \mathrm{SO}_n) \longrightarrow H^2(M, \mathbb{Z}/2\mathbb{Z}) \; .$$

Recall that the cohomology sets $H^1(M, \mathrm{SO}_n)$ and $H^1(M, \mathrm{Spin}_n)$ can be considered as representing the equivalence classes of principal G-bundles with respectively $G = \mathrm{SO}_n$ and $G = \mathrm{Spin}_n$. For the question of the existence of a principal Spin_n-bundle over the manifold M, one therefore has to see whether the equivalence class of the bundle of positive orthonormal bases $SO_gM \longrightarrow M$ is mapped to 0 in the exact sequence above. Going gack to the Cech approach of the cohomology one sees that this map is nothing but the *second Stiefel-Whitney* class of the bundle. Therefore

THEOREM 4.20. *An orientable manifold M admits a spin structure provided its second Stiefel-Whitney class $w_2(M) \in H^2(M, \mathbb{Z}/2\mathbb{Z})$ vanishes.*

One says that a manifold M is a *spin manifold* it it admits a spin structure, in other words from Theorem 4.18 if and only if $w_2(M) = 0$.

This condition does tell us that some manifolds do not admit spin structures. An important example is provided by even dimensional complex projective spaces. Indeed, one easily sees that for a complex manifold (i.e., a manifold for which one can find a coherent actual of complex-valued charts with holomorphic transition functions), the second Stiefel-Whitney class $w_2(M)$ is just the mod $\mathbb{Z}/2\mathbb{Z}$-reduction of

[3] if M is non-orientable, one has to consider the weaker notion of a *pin structure*.

the first Chern class $c_1(M) \in H^2(M, \mathbb{Z})$. Since even dimensional complex projective spaces have odd first Chern classes, they are abstructed.

To circumvent this difficulty it is useful to enlarge the notion of a spin structure to that of a spin$_c$ structure, i.e. one for which the structure group considered is the intermediate double cover of $S^1 \times \mathrm{SO}_n$ which is non trivial on both factors. The new notion mixes up, as the structure group shows, a line bundle (corresponding to the S^1 factor), and the tangent bundle (corresponding to the other factor). Obstructions to spin$_c$-structures can also be discussed in cohomological terms.

In this way it is possible to incorporate Kählerian metrics, i.e. Hermitian metrics whose almost complex structure is parallel with respect to the Levi-Civita connection of the metric, in the realm of Spinorial Geometry, something that would not be a priori possible in the strict context of spin structures.

Classification of spin structures. The exact sequence (4.6) also gives the answer to the question of classifying spin structures on a manifold. Indeed, the ambiguity of a spin structure, i.e. an equivalence class of principal Spin_n-bundles mapping onto the tangent bundle in $H^1(M, \mathrm{SO}_n)$, is precisely given by a class in $H^1(M, \mathbb{Z}/2\mathbb{Z})$, a set which can be identified with the set of double covers of M, an affine space over the field $\mathbb{Z}/2\mathbb{Z}$ acted upon by $H^1(M, \mathbb{Z}/2\mathbb{Z})$.

From this follows readily that a simply connected spin manifold M has exactly one spin structure.

One must be aware that Spin Geometry does deal with spin structures, that it to say that different spin structures on the same manifold may have different geometric properties.

4.6.2. Geometric Spinorial Objects. In this Section we review geometric objects that can be discussed on a spin manifold M and fix the terminology and notations concerning them that we use later.

Spinorial metrics. It is a bit emphatic to give a specific name and to have a specific notation for a pair consisting of a Riemannian metric g on M and a spin structure $\pi : P \longrightarrow M$. This is exactly what a *spinorial metric* is. We nevertheless do it.

There are instances though where it is indeed useful to have this short hand notation (see the Section 4.6.3). For such a pair we tend to use Greek letters, a typical choice being γ.

It is therefore natural to replace the notation P to denote the principal Spin_n-bundle defining γ by $\mathrm{Spin}_\gamma M$.

It is also worth recalling at this point that the space of complex spinors inherits a natural Hermitian metric from the Euclidean metric on the underlying vector space, the whole construction started with.

Spinor fields. A spinorial metric γ being fixed on M, we can make sense of *spinor fields* globally over M. They are the global sections of the associated vector bundle $\Sigma_\gamma M \longrightarrow M$ defined as $\Sigma_\gamma M = Spin_\gamma \times_{\sigma_n} \Sigma_n$, where Σ_n is the fundamental complex Spin_n-representation.

Indeed, it will be implicit later on that we deal with *complex* spinor fields. This is merely for the sake of simplicity, since the geometric tools we have introduced can equally deal with *real* spinor fields. As mentioned in the lectures on algebraic spinors, the main advantage of complex spin representations is their nice $\mathbb{Z}/2\mathbb{Z}$-periodicity. Real spin representations have an eight-fold periodicity which makes discussing them more cumbersome, but conceptually as easy.

Spinor fields will be denoted again by Greek letters, a typical choice being ψ. Recall that locally a spinor field can be represented by its components in Σ_n in a spinorial basis, i.e. a point ρ in $Spin_\gamma M$.

Spinorial connections. In Section 4.5.3 we discussed the way connections can be defined in principal G-bundles, and how they relate to the more traditional covariant derivative approach in vector bundles.

There are only a few comments that need to be made here to make sense of a *spinorial connection.* Indeed the data of a spinorial metric γ contains that of a Riemannian metric g, hence the availability of the Levi-Civita ∇^g on the tangent bundle, hence to a $\mathbf{G}$-valued connection 1-form ω on $SO_g M$.

Now two critical facts give rise to a *"Levi-Civita" spinorial connection*:

– first, the principal Spin$_n$-bundle $Spin_\gamma M$-bundle is, pointwise over M, a double cover of $SO_g M$ so that ω can be pulled back to a $\mathbf{G}$-valued 1-form $s^*(\omega)$ on $Spin_\gamma M$ (recall that $s : Spin_\gamma M \longrightarrow SO_g M$ is the double cover defining the spin structure);

– second, because Spin$_n$ is a double cover of SO$_n$, their Lie algebras are isomorphic, and even more precisely the tangent map at the identity to the natural map appearing in the fundamental exact sequence (4.6) is such an isomorphism that we denote by Φ.

The two properties together lead to the critical fact that $\tilde{\omega} = s^*(\omega) \circ \Phi$ is a Spin$_n$-connection on $Spin_\gamma M$, namely that $\tilde{\omega}$ enjoys the characteristic properties given in Definition 4.18.

In the rest of the notes, we systematically use this extended "Levi-Civita" connection attached to the spinorial metric γ. The covariant derivative that it induces on spinorial fields will be denoted by ∇^γ to recall that it involves both the Riemannian metric and taking lifts to the spinorial bundles, hence involves the spin structure. Explicitly, if $\psi = [\varphi, \sigma]$ and $X \in TM$, one has

$$\nabla^\gamma_X \psi = [\varphi, \partial_X \sigma + \tilde{\omega}(X)(\sigma)] \ .$$

Spinorial curvature. It is therefore natural that we call *spinorial curvature* of the spinorial metric γ the curvature of the covariant derivative ∇^γ, which is a tensor-spinor field $R^\gamma \in \Lambda^2 T^* M \otimes End(\Sigma_\gamma M)$, and also in the principal bundle formalism, a $\mathbf{Spin}_n$-valued 2-form on $Spin_\gamma M$.

Notice though that almost by definition of what spinors are $End(\Sigma_\gamma)$ is a bundle which is isomorphic, as an SO$_n$-representation bundle to the exterior algebra bundle, hence is actually a tensorial object. This is not a surprise to us since it must be determined by the Riemannian curvature tensor R^g.

4.6.3. Dependence of spinors on the metric. The very fact that we introduced spinors on a manifold M after a Riemannian metric g has been fixed shows that spinors are *metric-dependent*. Although this question is obviously central in Spinorial Geometry, it has not been given the attention it deserves for quite some time. In this Section we give the basic construction allowing to compare spinors for different metrics. From abstract nonsense it is known that spaces of spinors are isomorphic, but to deal with this situation geometrically one has to go further and make the isomorphisms somewhat explicit.

Algebraic preliminaries. In this subsection we will be working in a fixed vector space V. Let g and h be two Euclidean metrics on V. It will be useful to view them as points in the space $\mathcal{M}et\, V$ of Euclidean scalar products as introduced in Subsection 4.1.1.

The challenge is to find a natural process by which one can associate an h-orthonormal basis to a g-orthonormal one. To achieve this it will be of particular importance to understand the geometry of $\mathcal{I}nv(\mathbb{R}^n, V)$ viewed as a bundle over $\mathcal{M}et\, V$. Indeed the fibre of the projection from $\mathcal{I}nv(\mathbb{R}^n, V)$ to $\mathcal{M}et\, V$ is just the set of g-orthonormal bases, which is an orbit under the group O_n (recall that f being g-orthonormal exactly means that $(f^{-1})^* e = g$ where e denotes the natural Euclidean product in $\mathbb{R}^n$).

Therefore the question is: for every g-orthonormal basis f, viewed as a point in $\mathcal{I}nv(\mathbb{R}^n, V)$ lying above $g \in \mathcal{M}et\, V$, determine an h-orthonormal basis $\tilde{f}$ so that, if $f \circ \ell$ is another g-orthornormal basis of V for $\ell \in \mathrm{O}_n$, $\widetilde{f \circ \ell} = \tilde{f} \circ \ell$.

In fact one can even propose an a priori formula for that (this was first done by E. Binz and R. Pferschy): set $H_g = g^{-1}.h$; it is a linear map from V to V for which $h(v_1, v_2) = g(H_g(v_1), v_2)$ for any v_1, $v_2 \in V$, so that $\tilde{b} = (H_g)^{-1/2} \circ b$ does solve the problem, the equivariance being obvious. Notice that taking the square root of H_g does not pose any problem since it is a positive definitive map with respect to the metric g.

This would be enough if we were only interested in the question from a Euclidean point of view. We are actually interested in this from a spinorial point of view, and we must do more. To be in phase with the previous developments, from now on we stick to oriented maps and bases.

The bundle $\mathcal{I}nv(\mathbb{R}^n, V) \longrightarrow \mathcal{M}et\, V$ has a natural geometry that we will be able to use. For each $f \in Inv(\mathbb{R}^n, V)$, the tangent space at f can be identified with $End(\mathbb{R}^n, \mathbb{R}^n)$, i.e. with a space of matrices. The subspace made of the images of *symmetric* matrices is transversal to the fibres of the projection $\mathcal{I}nv(\mathbb{R}^n, V) \longrightarrow \mathcal{M}et\, V$, since the tangent spaces to the orbit are made of skew-symmetric matrices. Moreover it is equivariant under the action of O_n. The distribution of horizontal spaces gives rise to an O_n-connection on the bundle $\mathcal{I}nv(\mathbb{R}^n, V) \longrightarrow \mathcal{M}et\, V$.

Any curve c joining the points g and $h \in \mathcal{M}et\, V$, for example the segment $c(t) = (1-t)\, g + t\, h$, can be lifted horizontally to $\mathcal{I}nv(\mathbb{R}^n, V)$, i.e. through any point $b \in Inv(\mathbb{R}^n, V)$ there is a curve projecting to c whose tangent vector lies in the horizontal subspace at this point.

PROPOSITION 4.21. *The horizontal lift along the segment $t \mapsto (1-t)\, g + t\, h$ in the natural SO_n-bundle $\mathcal{I}nv(\mathbb{R}^n, V) \longrightarrow \mathcal{M}et\, V$ starting at the g-orthonormal basis f ends at the h-orthonormal basis $\tilde{f} = (H_g)^{-1/2} \circ f$. The transformation $b_g^h(f) = \tilde{f}$ is by its nature SO_n-equivariant.*

A word of caution is in order here, since the correspondence that was just defined is a bit peculiar. Indeed, if one takes three metrics g, h and k, it is not true in general that $b^k_h \circ b^h_g = b^k_g$. This holds only if the two endomorphisms H_g et K_g commute with one another. This stresses one more time the importance of the natural geometry of the space $\mathcal{M}et\, V$. The property just stated holds precisely when the three metrics belong to the same flat subspace of $\mathcal{M}et\, V$.

After this preparation, we can now turn our attention to the spinorial setting. We rely on the developments in Subsection 4.6.1, and use the fact that the natural map from the space $\tilde{B}$ of *spinorial bases* to $\mathcal{I}nv(\mathbb{R}^n, V)$ is a double cover. (It can be identified to the map from Spin_n to SO_n as soon as a base point has been chosen.) This allows us to lift the preceding construction to spinorial bases since every step can be followed, and a horizontal lift can be defined in the same way, giving rise to a transformation β mapping spinorial bases for a metric to spinorial bases for another.

Going to the geometric setting. We now work on a manifold M which is assumed to be spin. We consider two spinorial metrics γ and η and aim to comparing their bundles of spinors $\Sigma_\gamma M \longrightarrow M$ and $\Sigma_\eta M \longrightarrow M$. One way to do that is directly inspired by the associated bundle construction, namely one first compares the principal Spin_γ-bundles, and then one will be able to compare also the associated bundles, and in particular the spinor bundles. This shows that the comparison will be possible provided the two spinorial metrics refer to the same spin structure. If not, there is no way the local comparisons can be patched together globally.

The development of the previous subsection allows for the comparison to be made pointwise in M, but, since the method used is universal, the construction extends globally provided the spin structure is the same. One can then define a map β^η_γ mapping $Spin_\gamma M$ to $Spin_\eta M$. In this way, one can associate with any γ-spinor field $\psi = [\varphi, \sigma] \in \Sigma_\gamma M$ the η-spinor field $\beta^\eta_\gamma(\psi) = [\beta^\eta_\gamma(\varphi), \sigma] \in \Sigma_\eta M$.

Now consider three spinorial metrics γ, η and κ subordinated to the same spin structure but associated with Riemannian metrics g, h and k. The remark made in the Riemannian setting to the effect that, in general $b^k_h \circ b^h_g \neq b^k_g$ extends to the spinorial setting, and in general one does not have $\beta^\kappa_\eta \circ \beta^\eta_\gamma = \beta^\kappa_\gamma$.

There is one case which allows an easier descprition, namely when the two metrics are *conformally related*, i.e. $h = a^2\, g$. Then b^h_g and β^η_γ are just multiplication by a^{-1}.

4.7. Introduction to the Dirac Operator

After all the preparatory work done in the previous lectures, one is finally ready to introduce the Dirac operator, which is the major first order natural operator acting on spinor fields. We establish major identities concerning it.

We also introduce the twistor operator, which is the only other natural first order operator acting on spinor fields.

4.7.1. Definition of the Dirac Operator. Let M be an oriented spin manifold endowed with a spinorial metric γ. We denote the underlying Riemannian metric by g. We can consider the bundle of γ-spinors $\Sigma_\gamma M \longrightarrow M$.

Spinor fields, which are sections of this bundle, are acted upon by exterior forms (since $End(\Sigma)$ is identified as an SO_n-representation space with the Clifford algebra, hence with the exterior algebra) through Clifford multiplication.

DEFINITION 4.22. For any spinor field ψ on M, the *Dirac operator* $\mathcal{D}$ acting on ψ is defined as

$$\mathcal{D}\psi = \sum_{i=1}^{n} e_i.\nabla_{e_i}\psi \ ,$$

where (e_i) is a g-orthonormal basis, . denotes Clifford multiplication and ∇ the spinorial covariant derivative.

By definition $\mathcal{D}$ is a first order differential operator mapping spinor fields to spinor fields.

One easily verifies that the definition is independent of the orthonormal basis used. This is made apparent by the computation of its principal symbol Symb$\mathcal{D}$, which can be obtained for example by isolating the terms involving the derivatives of higher orders viewed as a map from cotangent directions ξ,

$$(\mathrm{Symb}\,\mathcal{D})(\xi)(\psi) = \xi.\psi \ .$$

From this formula, and a formula that we give later, it is clear that the principal symbol of $\mathcal{D}^2$ is $(\mathrm{Symb}\,\mathcal{D}^2)(\xi)(\psi) = -g(\xi,\xi)\,\psi$, hence that the Dirac operator is a square root of the Laplace-Beltrami operator. This was precisely the motivation of Paul-Adrien-Maurice Dirac who was looking for a square root of the wave operator, i.e. the natural operator for a Lorentzian metric. His basic discovery was that one could find such a square root provided one was willing to abandon the idea that wave functions are scalar. As a result, the wave functions of electrons, which were the physical objects he was interested in, had to be *spinor fields*.

To compute Symb $\mathcal{D}^2$, one needs an identity relating the spinorial covariant derivative and the Levi-Civita covariant derivative on tensor fields. The following formula holds for any vector field Y and spinor field ψ

$$\nabla_X(Y.\psi) = (\nabla_X Y).\psi + Y.(\nabla_X\psi) \ .$$

4.7.2. Main Properties of the Dirac Operator. Self-adjointness. When the manifold M is compact, one can consider integrals over M using the volume element v_g determined by the metric. Here is the fundamental property of the Dirac operator

THEOREM 4.23. *On a compact spin manifold M for any spinor fields ψ and ϕ on M, one has*

$$\int_M (\mathcal{D}\psi,\phi)\,v_g = \int_M (\psi,\mathcal{D}\phi)\,v_g$$

where (,) denotes the natural Hermitian product on the space of spinors, in other words the Dirac operator is self-adjoint.

The main ingredients in the proof of Theorem 4.23 are the following elementary identities involving two vector fields X and Y, and two spinor fields ψ and ϕ. The first one expresses the fact that the natural Hermitian product on spinor fields is parallel with respect to the spinorial metric

$$\partial_X((\psi,\phi)) = (\nabla_X\psi,\phi) + (\psi,\nabla_X\phi) \ ;$$

the second one expresses the fact that Clifford multiplication is parallel under the spinorial connection

$$D_X(Y.\psi) = (D_X Y).\psi + Y.D_X\psi \ .$$

They both make explicit some more relations tying together the Riemannian covariant derivative and the spinorial one.

It follows from Theorem 4.23 that the Dirac operator acting on spinor fields has index 0, its kernel and cokernel being identical. Spinor fields lying in the kernel of the Dirac operator are called *harmonic* spinors.

The even-dimensional case. In even dimensions spinors exhibit an extra structure that physicists call *chirality*. It has been presented in the part of the notes dealing with the algebraic side of the theory. It says that the space of spinors splits into the space of *positive* and *negative* spinors $\Sigma_{2m} = \Sigma^+_{2m} \oplus \Sigma^-_{2m}$. These spaces are the eigenspaces under the action of the Riemannian volume element viewed, since M is orientable, as a $2m$-form respectively for the eigenvalues +1 and -1. It is therefore easy to see that Clifford multiplication by a vector (or a 1-form since we identified them) maps positive spinors to negative spinors, and conversely.

It then follows directly from the definition of the Dirac operator $\mathcal{D}$ maps positive spinor fields to negative spinor fields, and conversely.

This allows for a very critical refinement of the Dirac operator by considering its restrictions $\mathcal{D}^+$ and $\mathcal{D}^-$ to positive and negative spinors respectively. It follows from Theorem 4.23 that one has $(\mathcal{D}^+)^* = \mathcal{D}^-$, and $(\mathcal{D}^-)^* = \mathcal{D}^+$. It is then natural to consider the index of $\mathcal{D}^+$ for example (the index of $\mathcal{D}^-$ is just its opposite) which can be expressed as follows

PROPOSITION 4.24. *If $\mathcal{H}^+$ denotes the space of positive harmonic spinors (resp. $\mathcal{H}^-$ the space of negative harmonic spinors), then*

$$\operatorname{ind}\mathcal{D}^+ = \dim\mathcal{H}^+ - \dim\mathcal{H}^- \ .$$

When one speaks of the index of the Dirac operator, one almost always means $\operatorname{ind}\mathcal{D}^+$, a convention we will use later in these notes.

4.7.3. The Lichnerowicz-Schrödinger Formula.

We already established that the leading terms of the square of the Dirac operator correspond to that of the Laplace-Beltrami operator. A key formula giving a full formula for the square of the Dirac operator in terms of geometric data is playing a very central role in the geometric use of Dirac operators.

The Formula. This formula was first discovered by Erwin Schrödinger in 1932 [**Sch32**], and rediscovered by André Lichnerowicz in 1963 [**Lic63**]. It follows from a direct calculation.

THEOREM 4.25 (**Lichnerowicz-Schrödinger Formula**). *For any spinor field ψ, the following identity holds*

$$\mathcal{D}^2\psi = \nabla^*\nabla\psi + \frac{1}{4}\operatorname{Scal_g}\psi \ ,$$

where ∇^* denotes the adjoint of the spinorial covariant derivative ∇, an d $Scal_g$ the scalar curvature of the metric g.

This formula is one of many so-called *Weitzenböck* formulas, which are formulas comparing natural second order differential operators on sections of geometric bundles. The most general case deals with the difference of two such operators, for which one can prove that it is a zero-th order operator, i.e. a tensor field. The universality of the operators ensures that this tensor must involve linearly the curvature of the Riemannian metric. In this formula, the a priori surprising fact is that the curvature appears only via the scalar curvature, i.e. its weakest form.

There are abstract reasons for this simplification which go as follows. We studied in Section 4.4.2 the Riemann curvature tensor and its symmetries. From a purely representation theoretic point of view, one has to decompose the space $\mathcal{R}V$ of (4 times covariant) algebraic curvature tensors on a Euclidean vector space V into irreducible components. The key decomposition is $S^2\Lambda^2V^* = \Lambda^4V^* \oplus \mathcal{R}V$. The representations that occur in $\mathcal{R}V$ are a trivial representation, corresponding to the scalar curvature, for $n \geq 3$ a representation isomorphic to traceless symmetric tensors, corresponding to the traceless part of the Ricci curvature, and for $n \geq 4$ a third representation corresponding to the Weyl conformal curvature (its Young diagramme has four squares arranged as a square). For a formula involving an action on spinors, whose space of endomorphisms is isomorphic as an SO_n-representation space to a full exterior algebra, only representations occuring both in the space of curvatures and acting faithfully on spinors can occur. This leaves the trivial representation as only possibility, and explains why the curvature contribution to the formula is only through the scalar curvature.

The preceding algebraic facts are hidden in the direct computation through the way the symmetries of the curvature tensor and the action of forms of different types on spinors interact. Of course the direct calculation comes up with a more precise information, namely the exact value of the coefficient in front of the scalar curvature, in particular its sign (see the next Section).

4.7.4. Geometric Consequences of the Formula. The formula has played a very important role in the linkage between geometrical and topological properties of a compact manifold[4].

Indeed, thanks to the Atiyah-Singer Index Theorem, the index of the Dirac operator can be expressed using topological invariants of the spin manifold M. On the other hand, one easily derives from the Lichnerowicz-Schrödinger Formula that a harmonic spinor ψ satisfies the following identity

$$\int_M \left((\nabla\psi, \nabla\psi) + \frac{1}{4}\,\mathrm{Scal}_g\,(\psi, \psi) \right) v_g = 0\;,$$

and this was the motivation of A. Lichnerowicz in his 1963 article to derive the Formula.

This shows that, if (M, g) has positive scalar curvature, it cannot admit a non trivial harmonic spinor. On the other hand the topology of M may force such spinor fields to exist if the index of $\mathcal{D}^+$ does not vanish. Hence, through these considerations involving the Dirac operator the fact that on spin manifolds there can be topological obstructions to the existence of metrics with positive scalar curvature.

[4]Note that extensions to non-compact manifolds with metrics having a suitable behaviour at infinity have also been considered (see in particular the work of M. Gromov and H.B. Lawson).

These considerations can be considerably sophisticated. This made spin geometry a key ingredient in discussing altogether the question of existence of metrics with positive curvature on manifolds. (See in particular the work of M. Gromov and H.B. Lawson [**GL80a, GL80b, GL83**], S. Stolz [**Sto92, Sto95**], J. Rosenberg and S. Stolz [**RS95, RS01**], and many others.)

4.8. Introduction to the Twistor Operator

On the space of spinor fields there is another universally defined first order operator, the so-called *twistor* operator, which has been introduced by R. Penrose. It plays an important role in the context of supersymmetries.

4.8.1. Definition of the Twistor Operator. The algebraic fact that gives rise to the twistor operator is the decomposition $T^*M \otimes \Sigma M = \Sigma^{1/2} \otimes \Sigma^{3/2}$, where $\Sigma^{3/2}$ is the irreducible representation of the group Spin_n corresponding to what physicists call particles of spin 3/2. The projection onto the first factor of this decomposition represents the principal symbol of the Dirac operator, the projection onto the second factor, which is the complement to the identity of the other projection, represents the symbol of the twistor operator.

DEFINITION 4.26. For any spinor field ψ on M, the *twistor operator* $\mathcal{P}$ acting on ψ is defined as

$$\mathcal{P}_X\psi = \nabla_X\psi + \frac{1}{n}\,X.\mathcal{D}\psi\ .$$

4.8.2. A Brief Look at Possible Uses. Spinor fields that belong to the kernel of the twistor operator are called *spinor-twistors*. The equation defining them is overdetermined. Among them eigenspinors for the Dirac operators are called *Killing spinors*, and have been used initially by R. Penrose to give a full set of conserved quantities for the geodesic flow in special spaces. Theoretical physicists have been using them to generate *supersymmetries*.

Several important results to estimate the lowest eigenvalue of the Dirac operator on compact spin manifolds have been obtained thanks to the use of the twistor operator after a seminal result by T. Friedrich. Killing spinors, or sophistications of them when the supersymmetries are more involved, appear in the limiting cases of these estimates.

JEAN-PIERRE BOURGUIGNON, INSTITUT DES HAUTES ÉTUDES SCIENTIFIQUES, 91440 BURES-SUR-YVETTE, FRANCE.
E-mail address: jpb@ihes.fr

CHAPTER 5

Holonomy and Special Geometries

ROBERT L. BRYANT

ABSTRACT. Basic notions of fiber bundles, connections, curvature, and holonomy are reviewed. The classification by Berger of the Riemannian holonomies is then discussed and then each of the cases is taken up in turn, with a discussion of the local and global geometries associated to the Riemannian metrics with these holonomies. Particular attention is paid to the space of parallel forms in each case, as these are the key to understanding the exceptional cases in dimensions 7 and 8.

Contents

Introduction

According to the Oxford English Dictionary, it was Heinrich Hertz[1] who introduced the words *holonomic* and *nonholonomic* to describe a property of velocity constraints in mechanical systems.

In the language of differential geometry, a *state space* is a smooth manifold M and a set of *velocity constraints* is represented by a subset $\Sigma \subset TM$, that should be thought of as the allowable velocities of differentiable curves in M. Such a set $\Sigma \subset TM$ of velocity constraints is said to be *holonomic* if it forces a curve in M to stay in a proper subspace. For example, the condition $\mathbf{p} \cdot d\mathbf{p} = 0$ for a vector particle $\mathbf{p} \in \mathbb{R}^n$ forces $\mathbf{p}$ to have constant length, while the constraint $\mathbf{p} \wedge d\mathbf{p} = 0$ forces $\mathbf{p}$ to move on a line. These are holonomic constraints.

Nonholonomic constraints, on the other hand, imply no such 'finite' constraints. A classical example is that of a ball rolling on a table without slipping or twisting. The state space is $B = \mathrm{SO}(3) \times \mathbb{R}^2$, where the $\mathrm{SO}(3)$ records the orientation of the

[1]Hertz died in 1894 and his book *Principien der Mechanik* was published in 1895. An English translation appeared in 1899 and it is this translation that the OED uses to date the terms under discussion.

ball and the $\mathbb{R}^2$ records its contact point on the plane. The rolling constraint is expressed as the set of differential equations

$$\alpha = a^{-1}\,da + a^{-1}\begin{pmatrix} 0 & 0 & -dx \\ 0 & 0 & -dy \\ dx & dy & 0 \end{pmatrix} a = 0 \tag{5.1}$$

for a curve $\big(a(t); x(t), y(t)\big)$ in B. The curves in B satisfying this constraint are those tangent to the 2-plane field $D = \operatorname{Ker}\alpha$, which is transverse to the fibers of the projection $\mathrm{SO}(3) \times \mathbb{R}^2 \to \mathbb{R}^2$. It is not difficult to show that any two points of B can be joined by a curve tangent to D. I.e., one can roll a ball from any position to any other by choosing an appropriate rolling path $\big(x(t), y(t)\big)$.

Such constraints and their geometry have long been of considerable interest in the calculus of variations and control theory. For recent results, see the foundational work on Carnot-Carathéodory geometries by Gromov [**Gro96**].

A generalization of this idea, namely *holonomy*, was introduced into differential geometry by Schouten [**Sch18**] and Cartan [**Car25, Car26**] and has come to play an important role in differential geometry.

Of course, these notes can only serve as an overview and introduction to the subject. For details and further information, the reader is referred to the standard works on the subject, i.e., [**KN69b, Hel78a, Sal89, Joy00**].

5.1. Bundles and Connections

5.1.1. Principal Bundles and Connections. Since I will only need to discuss bundles over manifolds, I will adopt a definition that is well-adapted to to this case.

DEFINITION 5.1 (Principal H-bundles). Given a Lie group H and a smooth manifold M, a *principal right H-bundle* over M is a smooth manifold B endowed with both a smooth, surjective submersion $\pi : B \to M$ and a free, right H-action $R : B \times H \to B$ such that the fibers of π are the orbits of the H-action R.

Two principal right H-bundles (B_1, π_1, R_1) and (B_2, π_2, R_2) over M are said to be *equivalent* if there exists a diffeomorphism $f : B_1 \to B_2$ such that $\pi_1 = \pi_2 \circ f$ and $f\big(R_1(u,h)\big) = R_2(u,h)$.

REMARK 5.2 (Notation). For $u \in B$ and $h \in H$, it is customary to write $u{\cdot}h$ for $R(u,h)$ and I will adopt this practice.

REMARK 5.3 (Trivial Bundles). An obvious example is to let $B = M{\times}H$, with $\pi : B \to M$ being the projection on the first factor and $R : (M{\times}H) \times H \to M{\times}H$ being defined by group multiplication in the obvious way. This is said to be the *trivial* bundle over M.

More generally, a principal right H-bundle (B, π, R) over M is trivial (i.e., equivalent to the trivial bundle) if there is a smooth map $\sigma : M \to B$ that satisfies $\pi\big(\sigma(m)\big) = m$ for all $m \in M$. The mapping σ is said to be a *section*.

REMARK 5.4 (Induced bundles and local triviality). If $\big(B, \pi, R\big)$ is a principal right H-bundle over M and $N \subset M$ is any submanifold, then, by the implicit function theorem, $B_N = \pi^{-1}(N)$ is a smooth submanifold of B that inherits the structure of a principal right H-bundle over N.

More generally, if $f : N \to M$ is any smooth mapping, then the submanifold $f^*(B) = \{(n, u) \mid f(n) = \pi(u)\} \subset N \times B$ is a principal right H-bundle over N.

Note that the implicit function theorem implies that each point $m \in M$ has an open neighborhood $U \subset M$ such that B_U is a trivial H-bundle over U.

5.1.1.1. *Examples.* Of course, there are many examples of principal H-bundles. Here are a few of the most common:

EXAMPLE 5.5 (Homogeneous bundles). If H is a closed subgroup of a Lie group G, then the left coset space G/H carries a unique smooth structure such that the coset projection $\pi : G \to G/H$ is a smooth submersion [**Hel78a**]. Letting H act on G on the right via group multiplication makes G into a principal right H-bundle over $M = G/H$.

More generally, if $A_1 \subset A_2 \subset A_3$ is a flag of Lie groups, with A_i closed in A_{i+1} and with A_1 a normal subgroup of A_2, then the manifold $B = A_3/A_1$ is naturally a principal right H-bundle over $M = A_3/A_2$ where $H = A_2/A_1$. One simply defines $\pi(aA_1) = aA_2$ and the right action $R : B \times H \to B$ is defined by $R(aA_1, hA_1) = (ah)A_1$. I leave to the reader the task of checking that π is a submersion and that R is both well-defined and a free right H-action whose orbits are the fibers of π.

Note that A_1 acts on the left on both B and M and that the projection π and the right action R commute with this action, in the appropriate sense. In fact, this construction gives the most general case of a 'homogeneous principal right H-bundle. This general construction provides a large class of important examples, as will be seen.

EXAMPLE 5.6 (Coframe bundles). If $E \to M$ is a smooth (real) vector bundle of rank r, let $\mathsf{F}(E)$ denote the set of pairs (m, u) where $m \in M$ and where $u : E_m \to \mathbb{R}^r$ is a linear isomorphism. (One says that u is a *coframe* of E_m.)

One can regard $\mathsf{F}(E)$ as an open subset of the vector bundle $\mathrm{Hom}(E, \mathbb{R}^r) = E^* \otimes \mathbb{R}^r$, so it is naturally endowed with a smooth structure for which the basepoint projection $\pi : \mathsf{F}(E) \to M$ defined by $\pi(m, u) = m$ is a surjective submersion. Let $H = \mathrm{GL}(r, \mathbb{R})$ act on $\mathsf{F}(E)$ on the right via the action $R\big((m, u), h\big) = \big(m, h^{-1} \circ u\big)$. Then the mappings π and R make $\mathsf{F}(E)$ into a principal right $\mathrm{GL}(r, \mathbb{R})$-bundle over M that is called the *coframe bundle* of E.

More generally, if $E \to M$ is a vector bundle over M that is endowed with some extra structure, one can often encode this extra structure into an associated principal coframe bundle. For example, if $E \to M$ is an Hermitian r-plane bundle, i.e., if E is endowed with a complex structure J and a compatible inner product $\langle, \rangle$, then one can consider the bundle $\mathsf{U}(E)$ consisting of the pairs (m, u) where m lies in M and $u : E_m \to \mathbb{C}^r$ is both complex linear and isometric (where $\mathbb{C}^r$ is endowed with its standard inner product and complex structure). Then $\mathsf{U}(E)$ is naturally a principal right $\mathrm{U}(r)$-bundle over M.

5.1.1.2. *Associated (vector) bundles.* The process of constructing principal H-bundles from vector bundles (possibly endowed with extra structure) can be reversed. If (B, π, R) is a principal right H-bundle over M and $\rho : H \to \mathrm{GL}(V)$ is a representation of H as a set of linear transformations of some vector space V (where the ground 'field' can be $\mathbb{R}$, $\mathbb{C}$, or $\mathbb{H}$), then one can construct an *associated vector bundle* $E = B \times_\rho V = \big(B \times V\big)/\sim$, where one sets $(u, x) \sim \big(u\cdot h, \rho(h^{-1})x\big)$ and denotes the corresponding equivalence class of (u, x) by $[u, x]$. (To save notational clutter, the notation $[u, x]$ supresses the dependence on the representation ρ.)

The associated bundle inherits those properties from V that are invariant under the action of the elements $\rho(h)$ for $h \in H$. Thus, for example, if $\rho(H)$ preserves an inner product (resp., orientation, complex structure, etc.) on V, then E inherits a corresponding inner product (resp., orientation, complex structure, etc.)

If the representation ρ is faithful, one can then naturally interpret B as an H-subbundle of the $\mathrm{GL}(V)$-coframe bundle $\mathsf{F}(E)$.

This is useful because it is frequently important to be able to treat sections of the bundle E as functions in some sense. One can do this by interpreting a section $\sigma : M \to E$ as the mapping $s : B \to V$ that satisfies

$$\sigma\bigl(\pi(u)\bigr) = \bigl[u, s(u)\bigr]. \tag{5.2}$$

It is easy to see that

$$s(u{\cdot}h) = \rho(h^{-1})(u) \tag{5.3}$$

and, conversely, any function $s : B \to V$ that satisfies the equivariance (5.3) defines a section σ of E via (5.2).

Of course, this construction can be generalized considerably: If H acts smoothly on the left on a smooth manifold Y, then one can define $B \times_H Y$ as the quotient of the manifold $B \times Y$ by the free right action $(b, y){\cdot}h = \bigl(b{\cdot}h,\, h^{-1}{\cdot}y\bigr)$. The well-defined projection $\tilde{\pi} : B \times_H Y \to M$ defined by the rule $\tilde{\pi}\bigl([b, y]\bigr) = \pi(b)$ is then a submersion of $B \times_H Y$ onto M whose typical fiber is diffeomorphic to Y.

5.1.1.3. *Extension and Retraction.* There is an important special case of this associated bundle construction that yields new principal bundles, but with different groups: If H is a subgroup of a Lie group G, one can 'extend' a principal right H-bundle B over M to a principal right G-bundle B' over M in a natural way:

Set $B' = \bigl(B \times G\bigr)/\!\sim$ where one sets $(u, g) \sim \bigl(u{\cdot}h, h^{-1}g\bigr)$ and denotes the equivalence class of (u, g) by $[u, g]$. It is easy to see that G acts naturally on the right on B' via $[u, g]{\cdot}g' = [u, gg']$ and that the quotient B' carries a unique smooth structure for which $B \times G \to B'$ is a smooth submersion. Then the natural projection $\pi' : B' \to M$ defined by $\pi'\bigl([u, g]\bigr) = \pi(u)$ completes the data needed to make B' into a principal right G-bundle over M. One says that B' is the *extension* of B to a G-bundle.

It is not generally true that every principal right G-bundle over M is the extension of a principal right H-bundle, but this is true when H contains a maximal compact subgroup of G (as follows from a partition of unity argument). In general, a principal right G-bundle B' over M is an extension of a principal right H-bundle B over M if and only if B can be found as a submanifold of B' that is invariant under the action of H. In this case, one says that the submanifold $B \subset B'$ is a *retraction* (or sometimes *reduction*) of the bundle B'.

5.1.1.4. *Connections.* Even though every principal right H-bundle $B \to M$ is locally trivial, it is not canonically trivial, i.e., there is no canonical way to identify the fibers B_x and B_y for points $x \neq y$ in M. Even in the case in which the bundle B is globally trivial, so that there exists a section $\sigma : M \to B$, this section is far from unique. Any other section σ' can be written in the form $\sigma'(m) = \sigma(m){\cdot}f(m)$ where $f : M \to H$ is any smooth function.

The notion of a *connection* was introduced to, at least partly, remedy this situation.

DEFINITION 5.7. A *connection* on a principal right H-bundle B over M is a (linear) subbundle $L \subset TB$ that is invariant under the right action R (i.e., $(R_h)'(L_u) = L_{u\cdot h}$ for all $u \in B$ and $h \in H$ and that is complementary to the 'vertical subbundle' $\operatorname{Ker}\pi' \subset TB$, i.e., so that $L \cap \operatorname{Ker}\pi' = \{0\}$ and $TB = L \oplus \operatorname{Ker}\pi'$).

Every principal right H-bundle B has a connection, in fact, many of them [**KN69b**].

EXAMPLE 5.8 (Homogeneous connections). Consider the case where H is a closed subgroup of a Lie group G and let $\mathfrak{h} \subset \mathfrak{g} = T_eG$ denote the Lie algebras of the corresponding groups. The representation $\mathrm{Ad} : H \to \mathrm{GL}(\mathfrak{g})$ then preserves the subspace $\mathfrak{h} \subset \mathfrak{g}$, but need not preserve a complementary subspace $\mathfrak{m} \subset \mathfrak{g}$ to $\mathfrak{h}$ and, if it does, such an $\mathrm{Ad}(H)$-invariant complement need not be unique. However, if there is such a complement $\mathfrak{m}$ (and one always exists if H is compact, or, more generally, reductive), then there is a unique left-invariant subbundle $L \subset TG$ that satisfies $L_e = \mathfrak{m}$ and this defines a connection on G (considered as a principal right H-bundle over $M = G/H$) that is invariant under the left action of G. Conversely, every such G-invariant connection on G over G/H comes from such an invariant complement $\mathfrak{m} \subset \mathfrak{g}$ to $\mathfrak{h}$.

I leave as an exercise for the reader the project of working out a similar description of the general homogeneous connection for the homogeneous principal H-bundles as described in Example 5.5.

5.1.1.5. *Connection 1-forms.* An equivalent formulation is in terms of *connection 1-forms.* Let $\mathfrak{h} = T_eH$ denote the Lie algebra of H. For any $v \in \mathfrak{h}$, there is a corresponding homomorphism $\mathbb{R} \to H$ denoted by $t \mapsto e^{tv}$. This homomorphism induces a unique vector field Y_v on B whose associated flow Φ_v satisfies $\Phi_v(t,u) = u\cdot e^{tv}$. The mapping $\iota_u : \mathfrak{h} \to T_uB$ defined by $\iota_u(v) = Y_v(u)$ is a linear isomorphism of $\mathfrak{h}$ with $\operatorname{Ker}\pi'(u) \subset T_uB$. Linear algebra now shows that, for each connection $L \subset TB$, there is a unique *connection 1-form* θ, i.e., an $\mathfrak{h}$-valued 1-form θ on B with the two properties that, first, $\theta(Y_v) = v$ for all $v \in \mathfrak{h}$, and, second, that $\theta(L) = 0$.

The invariance of L under the right action of H translates into the condition that θ have the equivariance

$$R_h{}^*(\theta) = \mathrm{Ad}(h^{-1})(\theta) \tag{5.4}$$

for all $h \in H$. Conversely, an $\mathfrak{h}$-valued 1-form θ on B that satisfies $\theta(Y_v) = v$ for all $v \in \mathfrak{h}$ and (5.4) for all $h \in H$ defines a unique connection L by simply setting $L = \operatorname{Ker}\theta$.

5.1.1.6. *L-horizontal curves and vector fields.* Let $L \subset TB$ be a connection. By definition, $\pi'(u) : L_u \to M_{\pi(u)}$ is an isomorphism for all $u \in B$. A piecewise differentiable curve $u : [a,b] \to B$ is said to be *L-horizontal* if $u'(t)$ lies in $L_{u(t)}$ for all $t \in [a,b]$ for which $u'(t)$ exists. Because of the H-invariance of L, if $u : [a,b] \to B$ is L-horizonal, then so is $R_h \circ u : [a,b] \to B$ for all $h \in H$.

Standard results in ordinary differential equations then imply that, for every piecewise differentiable curve $c : [a,b] \to M$ and every $u_0 \in B_{c(0)}$, there is a unique L-horizontal curve $u : [a,b] \to B$ that satisfies $u(a) = u_0$ and that lifts c through π in the sense that $\pi\bigl(u(t)\bigr) = c(t)$ for all $t \in [a,b]$. Because of the H-invariance of L, the curve $c : [a,b] \to M$ induces a diffeomorphism $P_c(t) : B_{c(a)} \to B_{c(t)}$ for all $t \in [a,b]$ that satisfies $P_c(t)(u_0) = u(t)$ and $P_c(t)\bigl(u\cdot h\bigr) = P_c(t)\bigl(u\bigr)\cdot h$.

In general, the map $P_c(t)$ depends on the curve c and not just on the endpoints $c(a)$ and $c(t)$.

One can also use L to lift vector fields on M in the following sense: For each vector field X on M, there is a unique vector field X^L on B that satisfies $\pi'(u)\big(X^L(u)\big) = X\big(\pi(u)\big)$ and $X^L(u) \in L_u$ for all $u \in B$. The H-invariance of L implies that X^L is invariant under the action of H.

This process of lifting vector fields allows one to define an important invariant of a connection: For vector fields X and Y on M, consider the vector field on B defined by the formula

$$Z = [X^L, Y^L] - [X, Y]^L. \tag{5.5}$$

Now, $Z(u)$ depends only on $X\big(\pi(u)\big)$ and $Y\big(\pi(u)\big)$, is linear in each factor, and takes values in $\operatorname{Ker}\pi'(u) \simeq \mathfrak{h}$. In particular, there is a well-defined map $K(u) : \Lambda^2(T_{\pi(u)}M) \to \mathfrak{h}$ that satisfies the equivariance

$$K(u{\cdot}h)(x, y) = \operatorname{Ad}(h^{-1})\big(K(u)(x, y)\big). \tag{5.6}$$

This is known as the *curvature* of the connection L.

5.1.2. Holonomy. It was Cartan who introduced the notion of the *holonomy group* of a connection and this has turned out to be a central notion in differential geometry.

DEFINITION 5.9 (Holonomy bundle and group). Let (B, π, R) be a principal right H-bundle over a connected manifold M and let $L \subset TB$ be a connection on B.

For each $u \in B$, the *holonomy bundle* of L through u is the set $B^L(u)$ consisting of those points v for which there exists an L-horizontal curve $f : [a, b] \to B$ that satisfies $f(a) = u$ and $f(b) = v$. The *holonomy* of L at u is the set $\operatorname{Hol}^L(u) \subset H$ consisting of those $h \in H$ for which $u{\cdot}h$ lies in $B^L(u)$.

5.1.2.1. *Holonomy group structure.* The set $\operatorname{Hol}^L(u)$ is actually a subgroup of H. This can be seen as follows: First, since the constant path $f : [0, 1] \to B$ defined by $f(t) = u$ is L-horizontal, it follows that the identity element e lies in $\operatorname{Hol}^L(u)$. Second, if h lies in $\operatorname{Hol}^L(u)$, then, by definition, there exists an L-horizontal curve $f : [0, 1] \to B$ such that $f(0) = u$ while $f(1) = u{\cdot}h$. The mapping $g : [0, 1] \to B$ defined by $g(t) = g(1{-}t) \cdot h^{-1}$ is then L-horizontal and joins u to $u{\cdot}h^{-1}$, so $\operatorname{Hol}^L(u)$ is closed under inverses. Finally, if h_1 and h_2 lie in $\operatorname{Hol}^L(u)$, then let $f_i : [0, 1] \to B$ be L-horizontal curves that satisfy $f_i(0) = u$ and $f_i(1) = u{\cdot}h_i$. Define a new L-horizontal curve $g : [0, 1] \to B$ by setting $g(t) = f_1(2t)$ for $t \in [0, \frac{1}{2}]$ while $g(t) = f_2(2t{-}1) \cdot h_1$ for $t \in [\frac{1}{2}, 1]$. Then $g(0) = u$ and $g(1) = (u{\cdot}h_2){\cdot}h_1 = u{\cdot}(h_2 h_1)$, showing that $\operatorname{Hol}^L(u)$ is closed under multiplication.

5.1.2.2. *Holonomy properties.* As the reader can verify, the group $\operatorname{Hol}^L(u)$ has the following properties:

$$\operatorname{Hol}^L(u{\cdot}h) = h^{-1}\operatorname{Hol}^L(u)h \tag{5.7}$$

and

$$\operatorname{Hol}^L(v) = \operatorname{Hol}^L(u) \qquad \text{for all } v \in B^L(u). \tag{5.8}$$

In particular, the connection L determines a conjugacy class of subgroups of H. Strictly speaking, the *holonomy* of L is this conjugacy class of subgroups, but it is common to speak of the holonomy of L as being a particular subgroup in this conjugacy class.

What is not so obvious is that $\mathrm{Hol}^L(u)$ is actually a (not necessarily closed) Lie subgroup of H. In fact, Cartan asserted this without giving what would now be considered to be an adequate proof. It was eventually proved (to modern standards) by Borel and Lichnerowicz:

THEOREM 5.10 (Borel-Lichnerowicz [**BL52**]). *The group* $\mathrm{Hol}^L(u) \subset H$ *is a Lie subgroup, connected if* M *is simply connected. Moreover,* $B(u) \subset B$ *is a submanifold of* B *and a principal right* $\mathrm{Hol}^L(u)$*-bundle over* M.

For the proof, see [**KN69b**]. The main idea is to reduce to the case in which M is simply connected, in which case one easily sees that $\mathrm{Hol}^L(u)$ is a path-connected subgroup of the Lie group H and this implies directly that it is a Lie subgroup of H [**Hel78a**].

A further important result relates the curvature of the connection L to its holonomy. To state this result, recall the definition (5.6) of the curvature map $K^L(u) : \Lambda^2(T_{\pi(u)}M) \to \mathfrak{h}$. Let $\mathfrak{k}^L(u) \subset \mathfrak{h}$ denote the image of this map.

THEOREM 5.11 (Ambrose-Singer [**AS53**]). *The Lie algebra of* $\mathrm{Hol}^L(u) \subset H$ *is the smallest subalgebra of* $\mathfrak{h}$ *that contains the subspaces* $\mathfrak{k}^L(v) \subset \mathfrak{h}$ *for all* $v \in B^L(u)$.

Again, for the proof, see [**KN69b**].

Note, however, a curious difficulty posed by Theorem 5.11: One cannot directly determine the smallest subalgebra of $\mathfrak{h}$ that contains the subspaces $\mathfrak{k}^L(v) \subset \mathfrak{h}$ for all $v \in B^L(u)$ without determining $B^L(u)$, which will, of course, determine $\mathrm{Hol}^L(u)$ directly. In fact, the main uses of Theorem 5.11 are theoretical, in which this result is used to provide a 'lower bound' for the holonomy group rather than to compute it directly, as will be seen.

5.2. Riemannian Holonomy

Undoubtedly, one of the most important uses, so far, of holonomy has been its application to the 'classification' of special geometries that can be compatible with Riemannian geometry.

5.2.1. Orthonormal coframe bundle. Let (M, g) be a Riemannian manifold. For simplicity, in these notes, I am going to assume that M is both connected and simply connected (though there are certainly interesting effects to consider when M has a non-trivial fundamental group). In this case, M is orientable, so I will go ahead and assume that M has been oriented as well.

Using the metric g and orientation, there is a natural $\mathrm{SO}(n)$-bundle defined over M. Namely, let $\pi : F \to M$ denote the set of orientation preserving, isometric isomorphisms $u : T_xM \to \mathbb{R}^n$, let $\pi(u) = x$, and let $\mathrm{SO}(n)$ act on F on the right by $u{\cdot}h = h^{-1}{\circ}u$. (The reader should recognize this as a special case of the construction in Example 5.6.) Then F is the bundle of oriented, orthonormal coframes of (M, g).

5.2.2. The Levi-Civita connection. Now, as is well-known, a Riemannian metric has a distinguished connection on its tangent bundle, i.e., the Levi-Civita connection. From now on, this will be understood to be the connection on F unless it is otherwise explicitly stated.

The resulting holonomy group through $u \in F$ is known as the *holonomy* at u of the Riemannian metric g and so will be denoted as $\mathrm{Hol}^g(u) \subset \mathrm{SO}(n)$. Because M has been assumed to be simply connected, $\mathrm{Hol}^g(u)$ is connected. Note that the equivariance property (5.7) implies that there is a well-defined subgroup $\mathrm{Hol}^g(x) \subset \mathrm{SO}(T_xM)$ that satisfies $\mathrm{Hol}^g(x) = u^{-1}\mathrm{Hol}(u)u$ for any $u : T_xM \to \mathbb{R}^n$ in F_x. Moreover, because M is connected, for any two points x and y in M, there is an isometric linear map $\ell : T_xM \to T_yM$ that conjugates the two subgroups $\mathrm{Hol}^g(x) \subset \mathrm{SO}(T_xM)$ and $\mathrm{Hol}^g(y) \subset \mathrm{SO}(T_yM)$.

A special feature of F is that, for $u \in F$ with $\pi(u) = x$, the isomorphism $u : T_x \to \mathbb{R}^n$ allows one to interpret the curvature, as defined via (5.6), as a linear map

$$K_x : \Lambda^2(T_xM) \to \Lambda^2(T_xM) \simeq \mathfrak{so}(T_xM) \tag{5.9}$$

and the *first Bianchi identity* is expressed as the relation

$$K_x(u,v)w + K_x(v,w)u + K_x(w,u)v = 0 \qquad \text{for all } u,v,w \in T_xM. \tag{5.10}$$

Let $\mathfrak{h}(u) \subset \mathfrak{so}(n)$ denote the Lie algebra of $\mathrm{Hol}^g(u)$.

5.2.3. Reducibly acting holonomy. If $\mathrm{Hol}^g(u) \subset \mathrm{SO}(n)$ preserves a subspace $V \subset \mathbb{R}^n$ then it must also preserve its orthogonal $V^\perp$. In particular, the representation of $\mathrm{Hol}^g(u)$ on $\mathbb{R}^n$ is completely reducible, i.e., there is an orthogonal direct sum decomposition

$$\mathbb{R}^n = V_1 \oplus V_2 \oplus \cdots \oplus V_k \tag{5.11}$$

where each of the V_i is irreducible as a representation of $\mathrm{Hol}^g(u)$.

Now, a consequence of the first Bianchi identity (5.10) is that, if one sets $\mathfrak{h}_i(u) = \mathfrak{h}(u) \cap \mathfrak{so}(V_i)$, then one has a direct sum decomposition

$$\mathfrak{h}(u) = \mathfrak{h}_1(u) \oplus \mathfrak{h}_2(u) \oplus \cdots \oplus \mathfrak{h}_k(u). \tag{5.12}$$

This observation is due to Cartan [**Car25**], who then went on to observe that, in such a case, the point $x = \pi(u) \in M$ has an neighborhood U that has a coordinate system

$$x = (x_1, \dots, x_k) : U \to V_1 \oplus V_2 \oplus \cdots \oplus V_k \simeq \mathbb{R}^n \tag{5.13}$$

where each $x_i : U \to V_k$ is a submersion and for which there exist Riemannian metrics g_i on V_i such that, on U one has

$$g = (x_1)^*(g_1) + \cdots + (x_k)^*(g_k). \tag{5.14}$$

In other words, the metric g is locally a product metric.

Two further refinements can be made of this observation:

First, G. de Rham proved that if (M, g) is also complete (and, as usual, M is connected and simply connected), then the above product decomposition can be globalized:

THEOREM 5.12 (de Rham Splitting). *If (M,g) is a complete, simply-connected Riemannian manifold then (M,g) can be written as a Riemannian product*

$$(M,g) = (M_1,g_1) \times (M_2,g_2) \times \cdots \times (M_k,g_k) \tag{5.15}$$

where (M_i,g_i) is a simply-connected Riemannian n_i-manifold whose Riemannian holonomy is an irreducibly acting subgroup of $\mathrm{SO}(n_i)$.

Second, while the relation (5.12) shows that $\mathrm{Hol}^g(u)$ is a direct product of groups $H_i \subset \mathrm{SO}(V_i)$, it is not immediately clear (without the assumption that the metric g is complete) that each of the groups H_i is the holonomy of some Riemannian metric on a manifold of dimension $\dim V_i$. Nevertheless, this can be shown to be true. Consequently, in order to classify the possible holonomy groups of Riemannian metrics (of simply connected manifolds), it suffices to classify the possible holonomy groups of Riemannian metrics on n-manifolds whose holonomy groups act irreducibly on $\mathbb{R}^n$.

5.2.4. Berger's Classification. I now want to describe the classification of the possible Riemannian holonomy groups.

5.2.4.1. *Locally symmetric metrics.* There is a class of metrics for which the holonomy is not difficult to compute and that plays an important role in differential geometry: the locally symmetric metrics.

DEFINITION 5.13 (Locally symmetric metrics). A connected Riemannian manifold (M,g) is said to be *locally symmetric* if each point $x \in M$ has an open neighborhood U on which there exists a g-isometric involution $\iota_x : U \to U$ that fixes x and satisfies $(\iota_x)'(v) = -v$ for all $v \in T_xM$.

It is not difficult to show that a locally symmetric (M,g) is locally homogeneous, i.e., that for any two points x and y in M there are open neighborhoods U of x and V of y and a g-isometry $\phi : U \to V$ that satisfies $\phi(x) = y$. Moreover, any locally symmetric (M,g) is locally isometric to a unique, simply-connected, complete locally symmetric space $(\overline{M},\overline{g})$ and that, $(\overline{M},\overline{g})$ is *globally symmetric* in the sense that, for each $x \in \overline{M}$, the involution ι_x can be taken to be defined on the entire manifold $\overline{M}$.

If (M,g) is a simply-connected, complete, locally symmetric Riemannian manifold, then, when one factors it as a product using the de Rham splitting theorem, each of the factors (M_i,g_i) is a simply-connected, complete, locally symmetric Riemannian manifold that cannot be further decomposed as a product. Thus, to classify the simply-connected, complete locally symmetric Riemannian manifolds, it suffices to classify the irreducible ones. Of course, if the dimension of (M,g) is 1, then (M,g) is isometric to the real line with its standard metric, so one can further assume that $\dim M > 1$.

Finally, in this case, one can write $M = G/K$ where G is the identity component of $\mathrm{Isom}(M,g)$ and $K \subset G$ is the subgroup of G that fixes a point, say $m_0 \in M$. It can then be shown that the holonomy of (M,g) is isomorphic to K as embedded into $\mathfrak{so}(\mathfrak{m})$ (where $\mathfrak{m} = \mathfrak{g}/\mathfrak{k}$) via the adjoint homomorphism.

Thus, the local classification of the locally symmetric spaces is reduced to a classification problem involving Lie groups and Lie algebras. This is a long and beautiful story that I cannot do justice to here. The reader can consult [**Hel78a**] or [**KN69b**] for details. The main point is that the locally symmetric case is well-understood and that these metrics have been classified by (fairly sophisticated) algebraic means.

5.2.4.2. *Non locally symmetric metrics.* In Marcel Berger's thesis [**Ber55**], he showed that a subgroup $H \subset \mathrm{SO}(n)$ that acts irreducibly on $\mathbb{R}^n$ and can be the holonomy of a Riemannian n-manifold (M, g) that is not locally symmetric has to be conjugate to one of the groups on the following list: $\mathrm{SO}(n)$, $\mathrm{U}(\frac{1}{2}n)$, $\mathrm{SU}(\frac{1}{2}n)$, $\mathrm{Sp}(\frac{1}{4}n)\cdot\mathrm{Sp}(1)$, $\mathrm{Sp}(\frac{1}{4}n)$, G_2 (if $n = 7$), $\mathrm{Spin}(7)$ (if $n = 8$), or $\mathrm{Spin}(9)$ (if $n = 16$).

5.2.4.3. *Berger's method.* Berger's method can be described as follows: If $H \subset \mathrm{SO}(n)$ is to be the holonomy of a Riemannian metric (M, g), then one can endow the bundle $\mathsf{F}(g)$ of oriented orthonormal coframes for the metric g with its Levi-Civita connection and find a holonomy subbundle $B(u) \subset \mathsf{F}(g)$ whose structure group is H. The associated Levi-Civita connection 1-form θ on F then pulls back to $B(u)$ to take values in the Lie algebra $\mathfrak{h} \subset \mathrm{SO}(n)$. Now, if one looks at the first Bianchi identity (5.10) applied to the curvature of θ pulled back to $B(u)$, one finds that it has to take values in a certain subspace $K(\mathfrak{h}) \subset \mathfrak{h}$ that, in fact, must be an ideal in $\mathfrak{h}$. One can describe this subspace as follows: For any Lie subalgebra $\mathfrak{h} \subset \mathfrak{so}(n)$, set

$$R(\mathfrak{h}) = \{k : \Lambda^2(\mathbb{R}^n) \to \mathfrak{h} \mid k(x,y)z + k(y,z)x + k(z,x)y = 0\} \tag{5.16}$$

and then define the subspace

$$K(\mathfrak{h}) = \{k(x,y) \mid k \in R(\mathfrak{h}),\ x, y \in \mathbb{R}^n\} \subseteq \mathfrak{h}. \tag{5.17}$$

Now, the classification of the connected subgroups of $\mathrm{SO}(n)$ that act irreducibly on $\mathbb{R}^n$ is known (this is essentially the list of irreducible faithful representations of compact Lie groups), and, for each subgroup $H \subset \mathrm{SO}(n)$ on this list, one can check whether or not $K(\mathfrak{h})$ is a proper subalgebra of $\mathfrak{h}$. If $K(\mathfrak{h})$ is a proper subalgebra of $\mathfrak{h}$, then by Theorem 5.11, the holonomy of the connection pulled back to $B(u)$ must be a proper subgroup of H, so H itself cannot be holonomy. In this way, one is reduced to considering the subgroups $H \subset \mathrm{SO}(n)$ that act irreducibly on $\mathbb{R}^n$ and for which $K(\mathfrak{h}) = \mathfrak{h}$. This latter list of subgroups is much, much smaller than the full list of subgroups $H \subset \mathrm{SO}(n)$ that act irreducibly on $\mathbb{R}^n$.

For the subgroups $H \subset \mathrm{SO}(n)$ that act irreducibly on $\mathbb{R}^n$ and satisfy $K(\mathfrak{h}) = \mathfrak{h}$, there is a further test one can apply: Using the second Bianchi identity, one can define a subspace $R^1(\mathfrak{h}) \subset R(\mathfrak{h}) \otimes \mathbb{R}^n$ with the property that the covariant derivative of the curvature tensor of the connection form on $B(u)$ must take values in this space $R^1(\mathfrak{h})$, which can also be computed via representation theory. If $H \subset \mathrm{SO}(n)$ satisfies $R^1(\mathfrak{h}) = (0)$, then any metric with holonomy H has to have covariant constant curvature, which, it turns out, is equivalent to being locally symmetric.

Thus, Berger went through the list of $H \subset \mathrm{SO}(n)$ that act irreducibly on $\mathbb{R}^n$ and that satisfy $K(\mathfrak{h}) = \mathfrak{h}$ and computed the corresponding $R^1(\mathfrak{h})$. He found that this latter vector space was nearly always zero and that, in fact, any subgroup on this list for which $R^1(\mathfrak{h})$ was not zero had to lie on the list as given above.

5.2.4.4. *Another approach and an elimination.* That was where matters stood for some time. Berger's proof was regarded by some as unsatisfactory because it relied so heavily on a case-by-case analysis of the irreducible representations of compact Lie groups.

It was eventually noticed that the groups listed above are subgroups that act transitively on the unit sphere in $\mathbb{R}^n$. In a ground-breaking 1962 paper, Simons [**Sim62**] gave a direct proof that any subgroup $H \subset \mathrm{SO}(n)$ that acts irreducibly on $\mathbb{R}^n$ and that can be the holonomy of a Riemannian n-manifold (M, g) that is not locally symmetric must act transitively on the unit sphere in $\mathbb{R}^n$.

However a few years later, Alekseevski [**Ale68**] and Brown and Gray [**BG72**] (independently) noticed that, for $H = \mathrm{Spin}(9) \subset \mathrm{SO}(16)$, one actually has $R^1(\mathfrak{h}) = (0)$, so that the group $\mathrm{Spin}(9) \subset \mathrm{SO}(16)$ never belonged on Berger's list in the first place[2].

5.2.4.5. *Existence.* Berger's and Simons' arguments do not show that there *are* metrics with these holonomies, just that these were the only possibilities for locally irreducible metrics that were not locally symmetric.

Some of these holonomies were known to exist already when the classification was proved. I will just give some examples now, saving the more extensive discussion for the next section.

For example, it is easy to show that the generic metric on an (orientable) n-manifold has holonomy $\mathrm{SO}(n)$.

The condition of (M^n, g) having holonomy in $H = \mathrm{U}(\frac{1}{2}n)$ is equivalent to the manifold supporting a complex structure J that is orthogonal and parallel with respect to g, i.e., that (M, g, J) is a Kähler manifold. Of course, many of these had been known for quite some time because of their appearance in the study of complex manifolds and algebraic geometry.

Calabi constructed an example of a metric on the canonical bundle of $\mathbb{CP}^{n-1}$ whose holonomy is $H = \mathrm{SU}(\frac{1}{2}n)$ and an example of a metric on the cotangent bundle of $\mathbb{CP}^{n/2}$ whose holonomy is $H = \mathrm{Sp}(\frac{1}{4}n)$.

Alekseevski constructed examples of left-invariant metrics on solvmanifolds of dimension $n = 4m$ whose holonomy is $H = \mathrm{Sp}(\frac{1}{4}n){\cdot}\mathrm{Sp}(1)$.

Finally, in 1987, Bryant constructed examples of metrics in dimension 7 (resp. 8) with holonomy G_2 (resp. Spin(7)). (Abstract existence proofs had been completed in 1984, but explicit examples were not available until 3 years later.)

Thus, all of the groups that remained on Berger's original list (after the elimination of $\mathrm{Spin}(9) \subset \mathrm{SO}(16)$) are known to occur as holonomy groups of Riemannian manifolds. These holonomies and their commonly accepted names are summarized in Table 5.1.

5.2.5. The holonomy principle. The reader may be wondering just why the holonomy group of a metric says anything important about it. The answer is that the holonomy of a Riemannian manifold (M, g) governs what sorts of algebraic structures can be defined on M that are as compatible as possible with the metric g.

For example, suppose that $\sigma \in \Gamma\big(\otimes^k(T^*M)\big)$ is a covariant tensor field of degree k on M. Its covariant derivative with respect to the Levi-Civita connection is denoted by $\nabla\sigma \in \Gamma\big(\otimes^{k+1}(T^*M)\big)$ and, by definition, vanishes if and only if σ is parallel with respect to the Levi-Civita connection.

Since the bundle $\otimes^p(T^*M)$ is the vector bundle associated to F via the p-th tensor power of the standard representation of $\mathrm{SO}(n)$ on $\mathbb{R}^n$, it follows that σ is represented

[2] In fact, it is interesting to look at Berger's original paper: He had done the calculations needed to show that Spin(9) satisfies $R^1(\mathfrak{h}) = (0)$, but had not realized it.

TABLE 5.1. The irreducible nonsymmetric Riemannian holonomies

Manifold Dimension	Holonomy Group	Geometry
n	$\mathrm{SO}(n)$	Generic
$n=2m$	$\mathrm{U}(m)$	Kähler
$n=2m$	$\mathrm{SU}(m)$	Calabi-Yau
$n=4m$	$\mathrm{Sp}(m)\cdot\mathrm{Sp}(1)$	quaternion-Kähler
$n=4m$	$\mathrm{Sp}(m)$	HyperKähler
$n=7$	G_2	Associative
$n=8$	$\mathrm{Spin}(7)$	Cayley

by a function $s : F \to \otimes^k(\mathbb{R}^n)$ that is equivariant with respect to the action of $\mathrm{SO}(n)$. The condition of being covariant constant is then seen to be equivalent to the condition that s be constant on each holonomy bundle $F(u) \subset F$. By equivariance, it then follows that $s(u) \in \otimes^k(\mathbb{R}^n)$ must be fixed by $H = \mathrm{Hol}^g(u)$.

Conversely, if $s_0 \in \otimes^k(\mathbb{R}^n)$ is fixed by a subgroup $H \subset \mathrm{SO}(n)$ and (M, g) is a Riemannian manifold whose holonomy is conjugate to a subgroup of H, then, by choosing a $u \in F$ such that $\mathrm{Hol}^g(u) \subset H$, one can identify $\otimes^k(T^*M)$ as the associated bundle of $F(u)$ defined by the k-th tensor power of the induced representation of $\mathrm{Hol}^g(u)$ on $\mathbb{R}^n$ and hence construct the section $\sigma \in \Gamma\big(\otimes^k(T^*M)\big)$ that corresponds to the constant function $s_0 : F(u) \to \otimes^k(\mathbb{R}^n)$. This resulting section σ will then be covariant constant with respect to the metric g.

This correspondence between the elements in the representations of $\mathrm{SO}(n)$ that are fixed by a subgroup $H \subset \mathrm{SO}(n)$ and the covariant constant tensor fields on Riemannian manifolds whose holonomy is conjugate to a subgroup of H is known as the *holonomy principle*. In particular, the holonomy of a metric determines what sorts of parallel tensor fields the metric supports.

Given the classification of the possible holonomy groups of metrics that are not locally symmetric, there are three particularly important applications of the holonomy principle:

5.2.5.1. *(Almost) complex structures.* Three of the possible holonomy groups, $\mathrm{U}(\frac{1}{2}n)$, $\mathrm{SU}(\frac{1}{2}n)$, and $\mathrm{Sp}(\frac{1}{4}n)$, fix (at least one) orthogonal complex structure on $\mathbb{R}^n$. Consequently, a Riemannian manifold (M, g) with such a holonomy group supports a g-parallel, orthogonal almost complex structure J. Since such a J is g-parallel, i.e., $\nabla^g J = 0$, it follows that, in geodesic normal coordinates about a point $x \in M$, the coefficients of J are constant to first order, i.e., that, centered on any point of M, there are coordinates $x = (x^i)$ such that, when J is written in the form

$$J = a^i_j(x)\,\frac{\partial}{\partial x^i} \otimes dx^j, \tag{5.18}$$

the coefficients satisfy $a^i_{j,k}(0) = 0$, where

$$a^i_{j,k} = \frac{\partial a^i_j}{\partial x^k}. \tag{5.19}$$

In particular, since the coordinate expression of the Nijnhuis tensor of J is linear in the functions $a^i_{j,k}$, it follows that $\nabla^g J = 0$ implies that the Nijnhuis tensor of J must vanish, which is the condition that the almost complex structure J is actually be complex (i.e., that there exist local J-holomorphic coordinate systems on M).

5.2.5.2. *Differential forms.* Similar considerations apply when one considers the space of parallel exterior differential forms that a Riemannian manifold (M, g) supports. If the holonomy of g is conjugate to a subgroup of $H \subset \mathrm{SO}(n)$, then each of the elements of $\Lambda^*(\mathbb{R}^n)^H$, the ring of H-invariant differential forms on $\mathbb{R}^n$, gives rise to a g-parallel differential form.

By an argument similar to the one above showing that every g-parallel almost complex structure must be integrable, one can show that any g-parallel differential form on M is necessarily both closed and co-closed.

It is an interesting fact that, for each of the subgroups $H \subset \mathrm{SO}(n)$ listed in Table 5.1, the ring $\Lambda^*(\mathbb{R}^n)^H$ actually characterizes H, i.e., H is actually equal to the subgroup of $\mathrm{SO}(n)$ that fixes the elements of $\Lambda^*(\mathbb{R}^n)^H$. For each of the groups listed in Table 5.1, generators (and their degrees) of $\Lambda^*(\mathbb{R}^n)^H$ are listed in Table 5.2.

An even more remarkable fact emerges when the individual entries are examined: Except for the cases $H = \mathrm{SO}(n)$ and $H = \mathrm{U}(\frac{1}{2}n)$ and $H = \mathrm{Sp}(1){\cdot}\mathrm{Sp}(1)$, (i.e., the first two lines in the table and the case $m = 1$ in the quaternion-Kähler case), the elements of $\Lambda^*(\mathbb{R}^n)^H$ serve to characterize H as a subgroup of $\mathrm{GL}(n, \mathbb{R})$ (rather than just as a subgroup of $\mathrm{SO}(n)$). This observation leads to the following theorem, which is a combination of several different results:

THEOREM 5.14. *Let $H \subset \mathrm{SO}(n)$ be one of the groups $\mathrm{SU}(\frac{1}{2}n)$, $\mathrm{Sp}(\frac{1}{4}n){\cdot}\mathrm{Sp}(1)$ ($n > 8$), $\mathrm{Sp}(\frac{1}{4}n)$, G_2 ($n = 7$), or $\mathrm{Spin}(7)$ ($n = 8$) and let $R = \Lambda^*(\mathbb{R}^n)^H$ be the ring of H-invariant differential forms. Suppose that M^n is an n-manifold endowed with a ring $\mathcal{R} \subset \Omega^*(M)$ of closed differential forms with the property that, for each $x \in M$, there exists an isomorphism $u : T_xM \to \mathbb{R}^n$ such that $\mathcal{R}_x = u^*(R)$. Then there exists a Riemannian metric g on M whose holonomy is conjugate to a subgroup of H and with respect to which each of the forms in $\mathcal{R}$ is parallel.*

REMARK 5.15. As already remarked, this theorem is a combination of results individually proved for each of the groups H mentioned in the theorem. For example, the results for $H = \mathrm{G}_2$ or $\mathrm{Spin}(7)$ are due to a combination of results of Bryant and Fernandez and Gray (see [**Bry87**] for an account). The result for $H = \mathrm{SU}(\frac{1}{2}n)$ or $\mathrm{Sp}(\frac{1}{4}n)$ is due to Hitchin. For the history of the result for $H = \mathrm{Sp}(\frac{1}{4}n){\cdot}\mathrm{Sp}(1)$ ($n > 8$), see [**Sal89**]. (The fact that this latter result requires $n > 8$, i.e., that the theorem fails in this case for $n = 8$, has been known for quite some time, but I am not sure to whom it should be attributed.)

The main use of Theorem 5.14 is that it allows one to construct examples of metrics with these holonomies via exterior differential systems techniques and to analyze their local generality. For more details on this approach, see [**Bry96**].

5.2.5.3. *Spinors.* Finally, there is the relation with parallel spinor fields. Each of the groups $\mathrm{SU}(\frac{1}{2}n)$, $\mathrm{Sp}(\frac{1}{4}n)$, G_2 ($n = 7$), or $\mathrm{Spin}(7)$ ($n = 8$) can be characterized as the stabilizer of one or more spinors in a spin (or semi-spin) representation of $\mathrm{SO}(n)$ and hence metrics with these holonomies are characterized by the parallel spinor fields that they admit.

TABLE 5.2. Invariant Exterior Forms

Manifold Dimension	Holonomy Group	Exterior forms (generators)
n	$\mathrm{SO}(n)$	$1 \in \Lambda^0,\ *1 \in \Lambda^n$
$n = 2m$	$\mathrm{U}(m)$	$1 \in \Lambda^0,\ \omega \in \Lambda^2$
$n = 2m$	$\mathrm{SU}(m)$	$1 \in \Lambda^0,\ \omega \in \Lambda^2,\ \phi, \psi \in \Lambda^m$
$n = 4m$	$\mathrm{Sp}(m){\cdot}\mathrm{Sp}(1)$	$1 \in \Lambda^0,\ \Phi \in \Lambda^4$
$n = 4m$	$\mathrm{Sp}(m)$	$1 \in \Lambda^0,\ \omega_1, \omega_2, \omega_3 \in \Lambda^2$
$n = 7$	G_2	$1 \in \Lambda^0,\ \phi \in \Lambda^3,\ *\phi \in \Lambda^4$
$n = 8$	$\mathrm{Spin}(7)$	$1 \in \Lambda^0,\ \Phi \in \Lambda^4$

Because such a Riemannian manifold admits parallel spinor fields, it follows from Lichnerowicz' formula that such a metric metrics must have vanishing scalar curvature. In fact, more is true: Riemannian manifolds with these holonomies must have vanishing Ricci tensor.

5.3. Special Riemannian Geometries

In this final section, I will make some short remarks about interesting features characteristic of each of these geometries. These remarks are intended merely to give the reader some sense of the overall geometric features and cannot even begin an exhaustive treatment. After all, these geometries are currently the focus of very active research and an extensive treatment of even one of these cases would be very long indeed. The references mentioned in the introduction constitute a very good place to start to read more about these geometries.

5.3.1. Kähler and Calabi-Yau geometries. As has already been mentioned, a Riemannian manifold (M^{2m}, g) has holonomy conjugate to a subgroup of $\mathrm{U}(m)$ if and only if it admits a g-parallel, orthogonal almost complex structure J that must, in fact, be integrable. The combined structure (M, g, J) defines what is known as a *Kähler* manifold.

Given g, specifying J is equivalent to specifying the 2-form ω defined by

$$\omega(x, y) = g(Jx, y) \tag{5.20}$$

This 2-form is known as the *Kähler form* of the structure. Any two of (g, J, ω) determine the third via (5.20).

5.3.1.1. *Local analysis for* $H = \mathrm{U}(\frac{1}{2}n)$. In fact, this relationship gives one a way locally to generate all Riemannian metrics whose holonomy is conjugate to a subgroup of $\mathrm{U}(\frac{1}{2}n)$: Since J is integrable, setting $n = 2m$, one can find J-holomorphic local coordinate charts $z = (z^a)$ and it is not difficult to show that, in such coordinates, one can always find a local function (called the *Kähler potential*) h such that

$$\omega = \frac{\sqrt{-1}}{2}\frac{\partial^2 h}{\partial z^a \partial \overline{z}^b}\, dz^a \wedge d\overline{z}^b \qquad \text{and} \qquad g = \frac{\partial^2 h}{\partial z^a \partial \overline{z}^b}\, dz^a \circ d\overline{z}^b \tag{5.21}$$

Conversely, if h is a function on a domain in $\mathbb{C}^m$ such that the quadratic-form g defined by (5.21) is positive definite, then g is a metric for which the standard complex structure on $\mathbb{C}^m$ is orthogonal and parallel. (Of course, the 2-form ω is also g-parallel.)

It follows that, in a certain sense, the local Kähler metrics depend on one 'arbitrary' function of $n = 2m$ real variables.

5.3.1.2. *Global examples.* Global examples are easy to come by: Any complex submanifold of a Kähler manifold is Kähler when it is given the induced metric. In particular, since the (locally symmetric) $\mathrm{SU}(n+1)$-invariant metric on $\mathbb{CP}^n = \mathrm{SU}(n+1)/S\big(\mathrm{U}(1)\mathrm{U}(n)\big)$ is Kähler, every nonsingular algebraic variety inherits a Kähler metric.

Kähler geometry is of enormous importance in algebraic geometry and complex analysis, and these examples only scratch the surface of a vast and beautiful theory.

5.3.1.3. *Local analysis for* $H = \mathrm{SU}(\frac{1}{2}n)$. If the holonomy of (M, g) is a subgroup of $\mathrm{SU}(\frac{1}{2}n)$, then, in addition to a parallel complex structure J and $(1,1)$-form ω, the Riemannian manifold supports a nonzero parallel $(n,0)$-form, i.e., a complex volume form $\Psi = \phi + \sqrt{-1}\,\psi$. Because this form must be closed, it follows that this volume form is actually holomorphic. Moreover, one can then choose local J-holomorphic coordinates so that

$$\Psi = dz^1 \wedge dz^2 \wedge \cdots \wedge dz^m. \tag{5.22}$$

The fact that Ψ is parallel with respect to g then is equivalent to having the Kähler potential h satisfy the complex Monge-Ampère equation

$$\det\left(\frac{\partial^2 h}{\partial z^a \partial \overline{z}^b}\right) = C \tag{5.23}$$

for some constant c.

This gives a complete local description of the metrics on n-dimensional domains whose holonomy lies in $\mathrm{SU}(\frac{1}{2}n)$. One finds that, in Cartan's sense of 'generality', the 'general' local metric with holonomy in $\mathrm{SU}(\frac{1}{2}n)$ depends on two functions of $n-1$ variables. For more details on this, see [**Bry96**].

5.3.1.4. *Global examples.* On the global front, it is a famous theorem of Yau that a conjecture of Calabi is true:

THEOREM 5.16. *If (M, J) is a compact complex manifold that has a nonvanishing holomorphic volume form Ψ and that admits a Kähler metric g_0 with associated Kähler 2-form ω_0, then there exists a unique metric g_1 on M so that (M, g_1, J) is Kähler, the associated Kähler form ω_1 belongs to the same cohomology class as ω_0, and Ψ is g_1-parallel.*

It is for this reason that metrics with holonomy in $\mathrm{SU}(\frac{1}{2}n)$ are nowadays referred to as *Calabi-Yau metrics.*

For Calabi-Yau metrics, there are, in addition to the complex submanifolds that play a role in their modern analysis, other special types of submanifolds. For example, a very important class are the m-dimensional submanifolds $L^m \subset M^{2m}$ that are Lagrangian with respect to the 2-form ω and on which Ψ pulls back to be a real m-form (i.e., the imaginary part of Ψ pulls back to zero on L). These submanifolds are, by an argument due to Harvey and Lawson [**HL82**], calibrated submanifolds and hence are homologically mass-minimizing. These submanifolds are now known as *special Lagrangian submanifolds* and currently play a very important role in string theory.

5.3.2. Hyperkähler and Quaternion Kähler. Recall that an almost quaternion structure on a manifold M^n is a triple (J_1, J_2, J_3) where each J_i is an almost complex structure on M and the three almost complex structure anticommute ($J_1J_2 = -J_2J_1$, etc.) and satisfy $J_1J_2 = J_3$ (and hence, $J_2J_3 = J_1$, etc.). In particular, it follows that $J_\lambda = \lambda_1 J_1 + \lambda_2 J_2 + \lambda_3 J_3$ is an almost complex structure whenever ${\lambda_1}^2 + {\lambda_2}^2 + {\lambda_3}^2 = 1$. Such an almost quaternion structure is said to be *integrable* if each J_λ is integrable, which is equivalent to the condition that any two of the complex structures in $\{J_1, J_2, J_3\}$ be integrable. The reader should note that this notion of integrability does *not* imply that there are local coordinates in which each of the J_i has constant coefficients.

A Riemannian manifold (M^n, g) where $n = 4m$ has holonomy conjugate to a subgroup of $\mathrm{Sp}(m) \subset \mathrm{SO}(n)$ if and only if it admits a g-parallel, orthogonal almost quaternion structure (J_1, J_2, J_3). Of course, since each J_i is g-parallel, it follows that each J_i is integrable and hence the almost quaternion structure (J_1, J_2, J_3) is integrable and, moreover, that the nondegenerate 2-form ω_i associated to the Kähler structure (M, g, J_i) is g-parallel and hence closed and co-closed for each i.

5.3.2.1. *Definition via 2-forms.* The three 2-forms $(\omega_1, \omega_2, \omega_3)$ alone are sufficient to recover the metric g and the almost quaternion structure (J_1, J_2, J_3). Indeed, one has

$$
\begin{aligned}
\omega_1(J_2x, y) &= \omega_3(x, y),\\
\omega_2(J_3x, y) &= \omega_1(x, y),\\
\omega_3(J_1x, y) &= \omega_2(x, y),
\end{aligned}
\tag{5.24}
$$

and these equations serve to determine (J_1, J_2, J_3) while g is then defined by

$$
g(x, y) = -\omega_1(J_1x, y) = -\omega_2(J_2x, y) = -\omega_3(J_3x, y).
\tag{5.25}
$$

Moreover, given a triple $(\omega_1, \omega_2, \omega_3)$ of closed nondegenerate 2-forms on M for which the maps J_i defined by (5.24) form an almost quaternion structure and the quadratic form g then defined by (5.25) is a positive definite quadratic form on M, one finds that each J_i is g-parallel. (As already mentioned, this observation is due to Hitchin.)

5.3.2.2. *Local analysis for $H = \mathrm{Sp}(\frac{1}{4}n)$.* Although the analysis is somewhat more complicated, it is not difficult to show that, locally, these metrics depend on $\frac{1}{2}n$ functions of $\frac{1}{2}n + 1$ variables. Again, see [**Bry96**] for details. There is also a twistor description of these local metrics that is due to Hitchin. See [**Bes87**] for this description.

5.3.2.3. *Compact construction for $H = \mathrm{Sp}(\frac{1}{4}n)$.* The construction of compact examples can be accomplished with the aid of complex geometry, Berger's classification of holonomy, and Theorem 5.16 because of the following observation: Using the definitions, one sees that

$$
\omega_2(J_1x, y) + \sqrt{-1}\,\omega_3(J_1x, y) = \sqrt{-1}\big(\omega_2(x, y) + \sqrt{-1}\,\omega_3(x, y)\big),
\tag{5.26}
$$

so it follows that the $\mathbb{C}$-valued 2-form $\Omega = \omega_2 + \sqrt{-1}\omega_3$ is actually of type $(2,0)$ with respect to the complex structure J_1. Moreover, it is visibly nondegenerate and hence defines a holomorphic symplectic structure on the complex manifold (M, J_1).

Since Ω^m is then a non-vanishing holomorphic volume form, it follows that (M, g, J_1) is a Calabi-Yau manifold.

Conversely, suppose that one starts with a compact complex manifold (M, J_1) that admits both a holomorphic symplectic structure $\Omega_0 \in \Omega^{2,0}(M)$ and a Kähler 2-form $\omega_0 \in \Omega^{1,1}_+(M)$. (Several methods of constructing simply connected examples of such manifolds are now known, building on the original constructions of Futaki and Beauville.) Then, by Theorem 5.16, there exists a unique Kähler 2-form $\omega_1 \in \Omega^{1,1}_+(M)$ that lies in the same cohomology class as ω_0 and for which ${\Omega_0}^m$ is g-parallel for the metric g associated to (M, J_1, ω_1). Now, it is not difficult to show that Ω_0 itself must be parallel with respect to g. (This uses the fact that g must be Ricci-flat since its holonomy lies in $\mathrm{SU}(2m)$ and then a Weitzenböck formula that shows that, because Ω_0 is closed and holomorphic, it must be parallel.) It then easily follows, from the classification of holonomies, that the holonomy of g must be conjugate to a subgroup of the form

$$\mathrm{Sp}(m_1) \times \mathrm{Sp}(m_2) \times \cdots \times \mathrm{Sp}(m_k) \subset \mathrm{SU}(2m) \tag{5.27}$$

where $m_1 + \cdots + m_k \leq m$. If one now chooses M so that it is not a product (again, this extra condition can be satisfied by examples constructed using algebro-geometric techniques), then the holonomy is forced to be $\mathrm{Sp}(m) \subset \mathrm{SU}(2m)$, as desired.

5.3.2.4. $\mathrm{Sp}(\frac{1}{4}n)\cdot\mathrm{Sp}(1)$-*structures.* The three 2-forms $(\omega_1, \omega_2, \omega_3)$ suffice to determine the underlying $\mathrm{Sp}(\frac{1}{4}n)$-structure on an n-manifold (where $n = 4m$). On the other hand, the 4-form Ω defined by

$$6\,\Omega = {\omega_1}^2 + {\omega_2}^2 + {\omega_3}^2 \tag{5.28}$$

has a significantly larger stabilizer group, since it is stabilized not only by $\mathrm{Sp}(\frac{1}{4}n)$, the orthogonal quaternion linear transformations, but by scalar multiplication by unit quaternions as well, i.e., the commutator of $\mathrm{Sp}(\frac{1}{4}n)$ in $\mathrm{SO}(n)$.

In fact, it is not difficult to prove that, when $m > 1$, the group $\mathrm{Sp}(m)\cdot\mathrm{Sp}(1) \subset \mathrm{SO}(4m)$ is defined (as a subgroup of $\mathrm{GL}(4m, \mathbb{R})$) as the stabilizer of a 4-form $\Omega_0 \in \Lambda^4(\mathbb{R}^{4m})$. Thus, one can define a $\mathrm{Sp}(m)\cdot\mathrm{Sp}(1)$-structure on a $4m$-manifold by specifying a 4-form Ω of the same algebraic type at each point as Ω_0. One could say that the pair (M, Ω) is an *almost quaternion-Kähler* manifold. As was mentioned in Theorem 5.14, when $m > 2$, the condition that this $\mathrm{Sp}(m)\cdot\mathrm{Sp}(1)$-structure underly a metric g with respect to which Ω is parallel is simply that $d\Omega = 0$. (When $m = 2$, this is not adequate and further conditions must be imposed.)

There are some very well-known examples of such metrics: The *Wolf* spaces. These are the quaternion-Kähler symmetric spaces of compact type and there is one such symmetric space associated to each of the simple Lie groups. The best known are, of course, $\mathbb{HP}^n = \mathrm{Sp}(m{+}1)/\big(\mathrm{Sp}(m){\times}\mathrm{Sp}(1)\big)$, i.e., the quaternionic projective spaces, but the other classical ones are $\mathrm{SO}(n{+}4)/\big(\mathrm{SO}(n){\times}\mathrm{SO}(4)\big)$, the Grassmannian of oriented 4-planes in $\mathbb{R}^{n+4}$ and $\mathrm{SU}(n{+}2)/S\big(\mathrm{U}(n){\times}\mathrm{U}(2)\big)$, the Grassmannian of complex 2-planes in $\mathbb{C}^{n+2}$. The first exceptional case $\mathrm{G}_2/\mathrm{SO}(4)$ plays an important role in calibrated geometry in dimension 7.

Assume from now on that (M^{4m}, g) is a quaternion-Kähler manifold whose holonomy acts irreducibly. One can show without difficulty that g must be Einstein. If the scalar curvature of g is zero, then, in fact, the holonomy lies in $\mathrm{Sp}(m)$ and so the metric is actually hyperKähler. If the scalar curvature is nonzero, then either g

is locally symmetric (and hence is locally isometric to some Wolf space or its noncompact dual) or else, by Berger's classification, it must have holonomy conjugate to $\mathrm{Sp}(m)\times\mathrm{Sp}(1)$.

5.3.2.5. *Local generality.* By using methods similar to those for studying local hyperKähler structures, one can show that the 'general' quaternion-Kähler metric depends on $2m$ functions of $2m+1$ variables (the same generality as for local hyperKähler metrics). Thus, there are many such metrics, locally.

5.3.2.6. *Global constructions.* Aside from the Wolf spaces (and their products), no complete examples of quaternion-Kähler manifolds with positive curvature are known and there is a conjecture that all such manifolds are products of symmetric spaces. This conjecture has been verified in the low dimensions.

On the other hand, quaternion-Kähler orbifolds of positive scalar curvature have been constructed by a sort of reduction process and many examples are now known.

For negative scalar curvature, Alekseevski has constructed examples on certain solvmanifolds that are homogeneous under the action of a Lie group.

5.3.3. G_2 geometry. This is the first of the two possible *exceptional* holonomy groups. For many years, this was just a curiosity (in fact metrics with this holonomy were not known to exist, even locally, until 1984), but it became more important after Havey and Lawson did their groundbreaking work on calibrated geometries [**HL82**]. Even then, interest in such manifolds languished for a while until the work of Joyce (see below) and the appearance of G_2-geometry in certain extensions of string theory, which stimulates interest in this geometry among theoretical physicists these days.

One way of defining G_2-geometry is in terms of a metric in dimension 7 that admits a parallel spinor field. However, in this brief discussion, I will concentrate on its definition via 3-forms.

DEFINITION 5.17. The group $G_2 \subset GL(7,\mathbb{R})$ is the group of linear transformations that preserve the 3-form ϕ_0 on $\mathbb{R}^7$ defined by

$$\phi_0 = dx^{123} - dx^{145} - dx^{167} - dx^{246} - dx^{275} - dx^{347} - dx^{356} \tag{5.29}$$

where dx^{ijk} is shorthand for $dx^i \wedge dx^j \wedge dx^k$.

It can be shown that G_2 is a connected and simply-connected Lie group of dimension 14 that lies in SO(7). It follows that the orbit of ϕ_0 under the action of $GL(7,\mathbb{R})$ has dimension $49 - 14 = 35 = \dim\Lambda^3(\mathbb{R}^7)$. In particular, this orbit is open in $\Lambda^3(\mathbb{R}^7)$, so that ϕ_0 is what is nowadays called 'stable', i.e., it is linearly equivalent to any small perturbation of itself (just as a positive definite quadratic form is linearly equivalent to any small perturbation of itself).

A G_2-structure on a 7-manifold M can now be defined by specifying a 3-form ϕ on M such that, at every point $x \in M$, there exists a linear isomorphism $u : T_xM \to \mathbb{R}^7$ satisfying $\phi_x = u^*(\phi_0)$. By the above remark, if ϕ is such a form on M, then any sufficiently small perturbation of ϕ also defines a G_2-structure on M.

Moreover, because G_2 is a subgroup of SO(7), it follows that a G_2-structure on M defines an underlying metric g_ϕ and orientation, which in turn, determines a Hodge duality operator $*_\phi$.

The fundamental theorem in the subject is a combination of work by Fernandez and Gray with that of Bryant:

THEOREM 5.18. *If ϕ is a* G_2*-defining* 3*-form on* M^7, *then ϕ is parallel with respect to the metric g_ϕ if and only if* $d\phi = d(*_\phi \phi) = 0$.

Consequently, in this case, the holonomy of g_ϕ will be a subgroup of G_2. By the classification of holonomy, if the holonomy of g_ϕ acts irreducibly (i.e., if g_ϕ is not locally a product metric), then the holonomy must be all of G_2.

5.3.3.1. *Local generality.* This result provides the essential step needed to determine the generality of such metrics locally. By an exterior differential systems analysis [**Bry87**], one shows that the local metrics with holonomy G_2 depend on six functions of six variables up to local diffeomorphism.

5.3.3.2. *Global existence.* In a stunning development, Joyce proved that compact manifolds with holonomy G_2 exist and, in fact, constructed many series of such manifolds. This is a long and beautiful story that I do not have time to go into here, but the reader can learn all about it (and the beautiful geometry of such manifolds) by consulting the fundamental reference [**Joy00**].

5.3.4. Spin(7)**.** This is the second of the two possible *exceptional* holonomy groups. Again, metrics of with this holonomy remained mysterious until their existence was proved and then became more interesting when Joyce proved that there were compact examples. The significance of these metrics for mathematical physics still remains unclear.

One way of defining Spin(7)-geometry is in terms of a metric in dimension 8 that admits a parallel spinor field. However, in this brief discussion, I will concentrate on its definition via 4-forms.

DEFINITION 5.19. The group $\mathrm{Spin}(7) \subset \mathrm{GL}(8,\mathbb{R})$ is the group of linear transformations that preserve the 4-form Φ_0 on $\mathbb{R}^8 = \mathbb{R}\oplus\mathbb{R}^7$ defined by

$$\Phi_0 = dx^0 \wedge \phi_0 + *(\phi_0) \tag{5.30}$$

where ϕ_0 is as defined in (5.29).

It can be shown that Spin(7) is a connected and simply-connected Lie group of dimension 21 that lies in SO(8) and is the double cover of SO(7). The orbit of Φ_0 under the action of $\mathrm{GL}(8,\mathbb{R})$ has dimension $64-21 = 43 < \dim \Lambda^4(\mathbb{R}^8)$, so this orbit is not open in $\Lambda^4(\mathbb{R}^8)$. (In fact, for dimension reasons, $\Lambda^4(\mathbb{R}^8)$ clearly does not have any open $\mathrm{GL}(8,\mathbb{R})$-orbits.) It is not difficult to show that Φ_0 and its multiples are the only nontrivial forms (other than the ones of degrees 0 or 8) on $\mathbb{R}^8$ that are left invariant by Spin(7).

A Spin(7)-structure on an 8-manifold M can now be defined by specifying a 4-form Φ on M such that, at every point $x \in M$, there exists a linear isomorphism $u : T_xM \to \mathbb{R}^8$ satisfying $\Phi_x = u^*(\Phi_0)$.

Again, because Spin(7) is a subgroup of SO(8), it follows that a Spin(7)-structure on M defines an underlying metric g_Φ.

The fundamental theorem in the subject is essentially due to Fernandez:

THEOREM 5.20. *If Φ is a* Spin(7)*-defining* 4*-form on* M^8, *then Φ is parallel with respect to the metric g_Φ if and only if* $d\Phi = 0$.

Consequently, in this case, the holonomy of g_Φ will be a subgroup of Spin(7). By the classification of holonomy, if the holonomy of g_Φ acts irreducibly (i.e., if g_Φ is not locally a product metric), then the holonomy must be one of Sp(2), SU(4), or Spin(7). The former two are distinguished by the fact that metrics with these holonomies admit parallel 2-forms. Thus, such a closed form Φ whose underlying metric g_Φ is not locally a product and does not admit nontrivial parallel 2-forms necessarily has Spin(7) as its holonomy group.

5.3.4.1. *Local generality.* This result provides the essential step needed to determine the generality of such metrics locally. By an exterior differential systems analysis [**Bry87**], one shows that the local metrics with holonomy Spin(7) depend on twelve functions of seven variables, up to local diffeomorphism.

5.3.4.2. *Global existence.* Joyce has also proved that compact manifolds with holonomy Spin(7) exist and, in fact, has constructed many series of such manifolds. Again, the reader is referred to [**Joy00**] for further information.

Robert L. Bryant, Mathematics Department, Duke University, PO Box 90320, Durham, NC 27708-0320, USA.
E-mail address: `bryant@math.duke.edu`

CHAPTER 6

A Visit to Representation Theory

Robert J. Stanton

Abstract. A sketch of the elementary algebraic properties of Lie groups and Lie algebras is presented. The structure theory of complex simple Lie algebras is illustrated, then used as a model of the Cartan theory of the weight space structure of an irreducible module for a compact simple Lie group. To understand the geometric applications in other lectures, a summary of the fundamentals of representation theory of compact simple Lie groups is presented, leading to the statement of the Bott-Borel-Weil theorem. Ample references for an alternative approach using Dirac induction are provided in the bibliography.

Contents

6.1. Introduction

A Lie group is a topological space endowed with two compatible structures: (i) it is a real analytic manifold; (ii) it is a group. Compatibility requires the group operations to be analytic maps. In the nineteenth century Sophus Lie introduced these objects as local transformations of space, a goal of his being to use them as analogs of Galois groups in order to extend to differential equations the Galois theory of algebraic equations. Since then their rich mixture of topological and algebraic structure has made them a staple for diverse mathematical subjects ranging from analytic number theory through harmonic analysis to differential geometry, as well as to mathematical physics.

Each of the two distinct structures (group and manifold) is of equal importance to Lie theory and their interaction the source of the beauty of the subject. On the algebraic side one obtains the usual constructions from group theory: subgroups, quotient groups and coset spaces; specialized structures like abelian, nilpotent, solvable and simple; representation theory and characters. On the topological side, besides common properties like connectedness and compactness, one has tangent and cotangent bundles; differential forms and integration; vector bundles and differential operators on sections of vector bundles; connections and differential geometry. Invoking the compatibility of the two structures it happens that the most interesting and useful constructs are those which interact in a simple way. For example, the differential operators on coset spaces of Lie groups about which one can say the most are those which commute with translations from the group.

Of course the subject is interesting only to the extent that there are sufficiently many examples occurring in several settings. Fortunately, this is indeed the case.

In this series of lectures I will present a sample of topics mostly about compact reductive groups. In keeping with the theme of the Summer School I will aim for those parts of the subject that will provide the necessary background for the other lectures on Dirac operators and related topics; in addition, material that will present the background to delve into the extensive literature on differential operators on homogeneous spaces. Probably I will just barely accomplish this much, and in doing this I will omit many important topics. The references provide a good jump-off point for those who want to explore further these subjects.

6.2. Elementary Lie Theory

6.2.1. Lie groups and Lie algebras. I begin with a sketch of elementary Lie theory. Much more can be found in the books [**Bou71, Bou72, Bou68, How92, Hus94, Kna96, Wal88, Wal92, Zel73, DK00, Ros02**] listed in the Bibliography.

DEFINITION 6.1. A Lie group is a Hausdorff topological space endowed with the structure of a group and with the structure of a real analytic manifold such that multiplication and inversion are analytic maps.

An important source of examples of Lie groups is that of matrix groups. Let V be a finite dimensional vector space over $\mathbb{R}$ or $\mathbb{C}$. Set

$$\mathrm{GL}(V) = \{A \in \mathrm{End}\,(V) | A \text{ is invertible}\}.$$

From the determinant characterization of $\mathrm{GL}(V)$ it is immediate that $\mathrm{GL}(V)$ is open in $\mathrm{End}\,(V)$. So $\mathrm{GL}(V)$ can be given the analytic structure from the underlying Euclidean space $\mathrm{End}\,(V)$. It is easy to see that the group operations are analytic maps with respect to this analytic structure.

DEFINITION 6.2. A matrix group is a closed subgroup of $\mathrm{GL}(V)$, V some finite dimensional vector space.

We will see later that any closed subgroup of a Lie group is a Lie group so that matrix groups are further examples of Lie groups.

EXAMPLE 6.3. $\mathrm{SL}(V) := \{A \in \mathrm{GL}(V) | \det A = 1\}$ is closed.

EXAMPLE 6.4. Let J be a non-degenerate bilinear form on V.

If J is symmetric, set

$$O(V;J) = \{A \in \mathrm{GL}(V) | J(Av, Aw) = J(v,w) \text{ for all } v, w \in V\},$$

and call $O(V;J)$ the orthogonal group associated to the form J, while $SO(V;J) = O(V;J) \cap \mathrm{SL}(V)$ is the special orthogonal group.

If J is skew-symmetric, set

$$\mathrm{Sp}(V;J) = \{A \in \mathrm{GL}(V) | J(Av, Aw) = J(v, w)\},$$

and call $\mathrm{Sp}(V;J)$ the symplectic group associated to the form J.

$O(V;J)$, $SO(V;J)$ and $\mathrm{Sp}(V;J)$ are closed in $\mathrm{GL}(V)$, so are matrix groups.

Now let G be a Lie group and f a function on G. For any $y \in G$ define left and right translation by

$$l_y f(x) = f(yx)$$
$$r_y f(x) = f(xy).$$

Then l_y and r_y preserve the class of analytic functions.

DEFINITION 6.5. If X is an analytic vector field on G then we say X is left (resp. right) invariant if we have on analytic functions for all $y \in G$,

$$l_y \circ X = X \circ l_y$$
$$(\text{resp.} \quad r_y \circ X = X \circ r_y).$$

Since G acts transitively on itself by left or right translation, any left invariant (resp. right invariant) vector field is completely determined by what it does at the group identity, e. In other words, the set of left (resp. right) invariant vector fields is canonically isomorphic to the tangent space to G at e. Evidently as a manifold G is then orientable.

DEFINITION 6.6. Set $\mathfrak{g} = \{\text{left invariant vector fields on } G\}$ and call $\mathfrak{g}$ the Lie algebra of G. Then $\mathfrak{g}$ is a vector space over $\mathbb{R}$ of dimension that of G.

DEFINITION 6.7. A Lie algebra is a vector space $\mathfrak{a}$ over a field k together with a k-bilinear map

$$[,] : \mathfrak{a} \times \mathfrak{a} \longrightarrow \mathfrak{a}$$

satisfying

(skew symmetry) (i) $[x, y] = -[y, x]$

(Jacobi identity) (ii) $[x, [y, z]] + [y, [z, x]] + [z, [x, y]] = 0.$

If $\mathfrak{g}$ is the Lie algebra of a Lie group G, then defining $[X, Y] = X \circ Y - Y \circ X$ one sees that $\mathfrak{g}$ is a Lie algebra over $\mathbb{R}$.

A Lie group G, being a manifold, has a simply connected covering space $\widetilde{G}$; but G also being a group, one can show that $\widetilde{G}$ is also a group. Since any covering space of G is a quotient of $\widetilde{G}$ by the action of the discrete group of covering transformations, one can show that any covering space of G is also a Lie group. Any cover of G will have a neighborhood of e locally diffeomorphic with one in G. Hence one can identify the tangent space at the identity of G and the tangent space at the identity of any cover of G. Thus we see that the Lie algebras of G and any cover of G are isomorphic. Summarizing one has

PROPOSITION 6.8. *Let G be a Lie group with Lie algebra $\mathfrak{g}$. Then any covering space of G is a Lie group with Lie algebra naturally isomorphic to $\mathfrak{g}$. In particular, the simply connected covering space $\widetilde{G}$ of G is a Lie group with Lie algebra isomorphic to $\mathfrak{g}$.*

EXAMPLE 6.9. Let J be the usual symmetric positive definite bilinear form on $\mathbb{R}^n$. Then $SO(n, \mathbb{R}) := SO(\mathbb{R}^n; J)$ is called the special orthogonal group. The group $Spin_n$ defined in the other lectures is the simply connected covering group of $SO(n, \mathbb{R})$.

When $G \subseteq \mathrm{GL}(V)$ is a matrix group, the computation of $\mathfrak{g}$ is relatively easy. The starting point is the exponential function defined by the usual power series:

$$\exp A = \sum_{n=0}^{\infty} \frac{A^n}{n!}, \quad A \in \mathrm{End}\,(V).$$

THEOREM 6.10. exp *is well defined from* $\mathrm{End}\,(V)$ *to* $\mathrm{GL}(V)$ *and is an analytic diffeomorphism of a neighborhood of zero in* $\mathrm{End}\,(V)$ *with a neighborhood of* e *in* $\mathrm{GL}(V)$. *Moreover*

$$\det\ \exp(A) = e^{\mathrm{Trace}(A)}.$$

Next let $G \subseteq \mathrm{GL}(V)$ be a matrix group and denote the pre-image of G under the exponential map by $\exp^{-1}(G)$, i.e. $A \in \exp^{-1}(G)$ if and only if $\exp A \in G$. Using that G is a group as well as properties of exp one has for $A \in \exp^{-1}(G)$ and $n \in \mathbb{Z}$, $nA \in \exp^{-1}(G)$, i.e. $\exp nA = (\exp A)^n \in G$. More generally, set

$$\mathfrak{g} = \{A \in \mathrm{End}\,(V) | \exp tA \in G\ \forall t \in \mathbb{R}\}.$$

PROPOSITION 6.11. $\mathfrak{g}$ *is the Lie algebra of* $G \subseteq \mathrm{GL}(V)$.

EXERCISE 6.12. The Lie algebra of $O(V; J)$ consists of all A in $\mathrm{End}\,(V)$ that are skew-symmetric with respect to J.

6.2.2. Homomorphisms.

DEFINITION 6.13. Let G, H be Lie groups and $\varphi : G \to H$ a group homomorphism. One says that φ is a Lie group homomorphism if φ is an analytic map. Similarly, homomorphisms between Lie algebras are linear maps that preserve the Lie bracket, $[\,,\,]$.

Let G, H be Lie groups and $\varphi : G \to H$ a Lie group homomorphism. Since $\varphi(e_G) = e_H$, the differential of φ at e_G, $d\varphi_e$, is a linear map from $\mathfrak{g}$ to $\mathfrak{h}$. It is not difficult to check that $d\varphi_e$ preserves the Lie bracket and so is a homomorphism of Lie algebras. Thus to each pair of Lie groups with a homomorphism between them there is a homomorphism between their Lie algebras. The collection of Lie groups and Lie group homomorphisms forms the category of Lie groups; similarly for finite dimensional Lie algebras and their homomorphisms there is the category of (finite dimensional) Lie algebras. From the above discussion one can show that the assignment to a Lie group of its Lie algebra is a functor from the category of Lie groups to the category of Lie algebras.

In the converse direction one can ask whether given Lie algebras $\mathfrak{g}$ and $\mathfrak{h}$ of Lie groups G and H, and a Lie algebra homomorphism from $\mathfrak{g}$ to $\mathfrak{h}$, there is a Lie group homomorphism from G to H? A topological argument shows this is true at least for G simply connected.

In general, there is a useful local relationship between φ and $d\varphi$ which is a consequence of the analyticity of φ.

PROPOSITION 6.14. *Let $\varphi : G \to H$ be a Lie group homomorphism. Then the following diagram is commutative where defined:*

$$\begin{array}{ccc} \mathfrak{g} & \xrightarrow{d\varphi} & \mathfrak{h} \\ \exp_G \downarrow & & \downarrow \exp_H \\ G & \xrightarrow{\varphi} & H \end{array}$$

A Lie algebra homomorphism fundamental to the subject comes from considering the automorphism given by conjugation. Let G be a Lie group and let $g \in G$. Set $i_g(x) = gxg^{-1}$. Then i_g is an automorphism of G. As just described, there is an associated Lie algebra automorphism in GL($\mathfrak{g}$) usually denoted Ad(g).

The map $g \to$ Ad (g) is itself a homomorphism of the Lie group G to the Lie group GL($\mathfrak{g}$), so there is Lie algebra homomorphism from $\mathfrak{g}$ to End ($\mathfrak{g}$). This is commonly denoted $ad(\cdot)$. Using the exponential map one can compute the effect of ad(X) on an element Y to be $ad(X)(Y) = [X, Y]$. From Proposition 6.14 we then obtain the useful formula

$$\mathrm{Ad}(\exp X)(Y) = e^{\mathrm{ad} X}(Y).$$

Generalizing on the above example of the homomorphism Ad one can ask for other homomorphisms of a Lie group G into GL(V), V a finite dimensional vector space. These will be the focus of attention in a later section of the paper but for now we introduce only the definition.

DEFINITION 6.15. Let G be a Lie group. A representation of G is a homomorphism

$$\tau : G \to GL(V)$$

for some vector space V. The representation is called faithful if the kernel of τ is trivial. For a Lie algebra one makes a similar definition with the target space being End (V). [Warning: When V is infinite dimensional some care is needed about the topology on $GL(V)$ so that τ is *analytic*.]

THEOREM 6.16 (Ado). *Let $\mathfrak{g}$ be a finite dimensional Lie algebra over $\mathbb{R}$ or $\mathbb{C}$. Then $\mathfrak{g}$ has a faithful finite dimensional representation.*

Hence no generality is lost in Lie algebra theory by considering only matrix Lie algebras. Moreover, since $\tau(\mathfrak{g})$ will be a subalgebra of End (V), there is (from 1.15) a matrix subgroup of $GL(V)$, say G, with Lie algebra $\tau(\mathfrak{g})$. Thus the functor from the category of Lie groups to the category of finite dimensional Lie algebras is surjective, and the fiber over any object $\mathfrak{g}$ consists of all covering groups obtained from $\widetilde{G}$. Were $\widetilde{G}$ to be a matrix group, then the theory of Lie groups would be that of matrix groups. However, there are the following examples.

EXAMPLE 6.17. Let $SL(2, \mathbb{R})$ be the group of 2x2 real matrices with determinant $= 1$. Let $\widetilde{SL}(2, \mathbb{R})$ be its simply connected covering space. Then $\widetilde{SL}(2, \mathbb{R})$ has no faithful finite dimensional representation.

EXAMPLE 6.18. Let J be the usual symplectic form on $\mathbb{R}^{2n}$ and set $Sp(n, \mathbb{R}) = Sp(\mathbb{R}^{2n}; J)$. Then $Sp(n, \mathbb{R})$ has a double cover called the metaplectic group, $Mp(n, \mathbb{R})$, which has no faithful finite dimensional representation.

6.2.3. Algebraic constructions. The next several results concern the modifications of some natural constructions in the category of groups to ones in the category of Lie groups, and the extent to which the functor maps them to similar objects in the category of Lie algebras. Of course, sometimes the definition of a structure in Lie algebra theory is made exactly for consistency of this functor.

PROPOSITION 6.19. *If G_1 and G_2 are Lie groups then so is $G_1 \times G_2$ and the Lie algebra of $G_1 \times G_2$ is isomorphic to $\mathfrak{g}_1 \oplus \mathfrak{g}_2$.*

PROOF. This is easy and follows from the definitions. □

DEFINITION 6.20. Let G be a Lie group. A subgroup H of G is a Lie subgroup if H is a Lie group and the inclusion of H into G is an immersion. Similarly, a vector subspace $\mathfrak{b}$ of a Lie algebra $\mathfrak{a}$ is a subalgebra if $[\mathfrak{b}, \mathfrak{b}] \subseteq \mathfrak{b}$.

PROPOSITION 6.21. *(i) If H is a Lie subgroup of a Lie group G then the Lie algebra $\mathfrak{h}$ of H is a Lie subalgebra of $\mathfrak{g}$ the Lie algebra of G.*

(ii) If $\mathfrak{h} \subseteq \mathfrak{g}$ is a Lie subalgebra where $\mathfrak{g}$ is the Lie algebra of the Lie group G, then there is a unique connected Lie subgroup H of G having Lie algebra $\mathfrak{h}$.

PROOF. (i) Let $i : H \to G$ be the inclusion map. Since $i(e) = e$, the differential $di_e : T_e(H) \to T_e(G)$. Clearly, di_e is injective, so $di_e(\mathfrak{h})$ can be identified with $\mathfrak{h}$ as vector spaces. It remains to show that di_e is a Lie algebra homomorphism. This is easy viewing elements of $\mathfrak{h}$ as invariant vector fields.

(ii) is a little difficult and is, essentially, a special case of the Frobenius integrability theorem. □

The existence of a compatible topological structure on an abstract subgroup H of a Lie group G is a more subtle problem. The following important result gives a condition for this to happen.

THEOREM 6.22. *If H is a subgroup of a Lie group G and closed in the subspace topology, then H is a Lie subgroup and in a unique way.*

This is one of the more difficult results in elementary Lie theory. The proof uses the extension of the exponential map to Lie groups that are not matrix groups.

The next example shows that there are Lie subgroups that are not closed.

EXAMPLE 6.23. Set

$$G = \left\{ \begin{pmatrix} \cos\theta_1 & \sin\theta_1 & & \\ -\sin\theta_1 & \cos\theta_1 & & \\ & & \cos\theta_2 & \sin\theta_2 \\ & & -\sin\theta_2 & \cos\theta_2 \end{pmatrix} \right\}$$

and for an irrational number x set

$$H = \left\{ \begin{pmatrix} \cos\theta_1 & \sin\theta_1 & & \\ -\sin\theta_1 & \cos\theta_1 & & \\ & & \cos x\theta_1 & \sin x\theta_1 \\ & & -\sin x\theta_1 & \cos x\theta_1 \end{pmatrix} \right\}$$

Thus H is a Lie subgroup of G but one can check that H is not closed in G (the irrational flow on the torus).

DEFINITION 6.24. Let $\mathfrak{a}$ be a Lie algebra and $\mathfrak{b} \subseteq \mathfrak{a}$ a Lie subalgebra. Then $\mathfrak{b}$ is called an ideal if $[\mathfrak{a}, \mathfrak{b}] \subseteq \mathfrak{b}$. It follows that on the quotient vector space $\mathfrak{a}/\mathfrak{b}$ one has a natural Lie algebra structure.

PROPOSITION 6.25. *(i) If H is a normal Lie subgroup of G, then the Lie algebra $\mathfrak{h}$ of H is an ideal in the Lie algebra $\mathfrak{g}$ of G.*

(ii) If $\mathfrak{h}$ is an ideal in $\mathfrak{g}$, the Lie algebra of G, then the corresponding connected subgroup H of G is normal.

The existence of quotient groups requires some care. The following lemma points out the problem.

LEMMA 6.26. *G/H satisfies the T_1 separation axiom if and only if H is closed.*

Given a closed normal Lie subgroup H of a Lie group G one can show that the quotient group has the structure of a Lie group. As the Lie algebra $\mathfrak{h}$ will be an ideal in $\mathfrak{g}$ there is a quotient Lie algebra $\mathfrak{g}/\mathfrak{h}$. There is the expected correspondence between the Lie algebra of the quotient group and the quotient Lie algebra. Thus the functor that assigns to a Lie group its Lie algebra preserves short exact sequences.

EXAMPLE 6.27. Let G and H be as in the preceding example. Since G is abelian, H is a normal Lie subgroup of G. So while G/H is a group it is not a Lie group.

More generally, one tries to construct a manifold structure on a quotient topological space G/H with H a Lie subgroup but not normal. For closed subgroups H of a Lie group G there is in fact a natural analytic manifold structure on the quotient space G/H. Indeed, such are a primary source of examples from Lie theory appearing in topology.

EXAMPLE 6.28. Let J be the usual inner product on $\mathbb{R}^n$ and $SO(\mathbb{R}^n)$ the elements of $O(\mathbb{R}^n; J)$ of determinant $=1$. Then $SO(\mathbb{R}^n)/SO(\mathbb{R}^m) \times SO(\mathbb{R}^{n-m})$ is analytically diffeomorphic to the Grassmannian of m-planes in $\mathbb{R}^n$.

EXAMPLE 6.29. Other familiar coset examples are spheres, S^n, and projective spaces over $\mathbb{R}$, $\mathbb{C}$, or $\mathbb{H}$.

Modulo some technicalities there is a consistent relationship between the group theoretic properties of a Lie group G and analogous properties of its Lie algebra $\mathfrak{g}$. For example,

DEFINITION 6.30. Let $\mathfrak{a}$ be a Lie algebra. Set $\mathfrak{a}^{(0)} = \mathfrak{a}$ (resp. $\mathfrak{a}^0 = \mathfrak{a}$). For $i \geq 1$ set

$$\mathfrak{a}^{(i)} = [\mathfrak{a}^{(i-1)}, \mathfrak{a}^{(i-1)}]$$

$$(\text{resp. } \mathfrak{a}^i = [\mathfrak{a}, \mathfrak{a}^{i-1}]).$$

$\mathfrak{a}$ is called abelian if $\mathfrak{a}^{(1)} = 0 = \mathfrak{a}^1$. $\mathfrak{a}$ is called solvable if $\mathfrak{a}^{(i)} = 0$ for i large (resp. nilpotent if $\mathfrak{a}^i = 0$ for i large).

This definition resembles the definition of nilpotent (resp. solvable) group and not surprisingly one has

PROPOSITION 6.31. *G is a connected nilpotent (resp. solvable) Lie group if and only if its Lie algebra $\mathfrak{g}$ is nilpotent (resp. solvable).*

DEFINITION 6.32. $\mathfrak{g}$ is called a simple Lie algebra if $\mathfrak{g}$ is non-abelian and has no non-trivial ideal. $\mathfrak{g}$ is called semisimple if it is a direct sum of simple ideals.

EXAMPLE 6.33. Let $\mathrm{sl}(V) = \{A \in \mathrm{End}\,(V) | \mathrm{Trace}\ A = 0\}$. Then $\mathrm{sl}(V)$ is a simple Lie algebra for $k = \mathbb{R}$ or $\mathbb{C}$.

PROPOSITION 6.34. *Let G be a connected Lie group with Lie algebra $\mathfrak{g}$. Then $\mathfrak{g}$ is semi-simple if and only if G has no non-discrete, closed normal subgroup.*

Now we have the building blocks for a general connected Lie group. So let G be connected with Lie algebra $\mathfrak{g}$. It is easy to see that $\mathfrak{g}$ contains a maximal solvable ideal, which we denote by $\mathfrak{r}$ and call the solvable radical of $\mathfrak{g}$. Then either $\mathfrak{g} = \mathfrak{r}$ or $\mathfrak{g}/\mathfrak{r}$ is a non-zero Lie algebra which must be semisimple. One shows that there exists a (non-unique) choice of subalgebra $\mathfrak{s} \subseteq \mathfrak{g}$ with $\mathfrak{s}$ isomorphic to $\mathfrak{g}/\mathfrak{r}$. As $\mathfrak{r}$ is an ideal, then there is a corresponding connected closed normal subgroup R of G. Similarly, as $\mathfrak{s}$ is a subalgebra there is a corresponding connected subgroup S of G. Building on these constructions one proves the fundamental

THEOREM 6.35 (Levi-Malcev Theorem). *Let G be a connected Lie group with Lie algebra $\mathfrak{g}$. Then $\mathfrak{g}$ is the semidirect product of its solvable radical with a semisimple subalgebra. Moreover, $G = R \rtimes S$.*

Because of this result, as well as other reasons, the theory of Lie groups diverges into results about solvable Lie groups and results about semi-simple Lie groups. There is a small technical variation of semisimple that occurs frequently and naturally enough that it must be given. I present a simplified version of the usual definition.

DEFINITION 6.36. Let G be a connected Lie group with Lie algebra $\mathfrak{g}$. Then G is called reductive if $\mathfrak{g}$ is the direct sum of its center and a semisimple ideal.

EXAMPLE 6.37. $\mathrm{SL}(V)$ is semisimple while $\mathrm{GL}(V)$ is reductive.

6.2.4. Killing form.

There is a natural symmetric bilinear form on any finite dimensional Lie algebra. This will provide an invariant Riemannian metric on compact Lie groups and various coset spaces. Let G be a Lie group with Lie algebra $\mathfrak{g}$. For X and Y in $\mathfrak{g}$ define

$$B(X, Y) = \mathrm{Trace}(\mathrm{ad}\ X \circ \mathrm{ad}\ Y).$$

This symmetric bilinear form B is called the Killing form and we have Cartan's Criterion

PROPOSITION 6.38. *A finite dimensional Lie algebra $\mathfrak{g}$ over $\mathbb{R}$ is semisimple if and only if $B(\cdot,\cdot)$ is non-degenerate.*

For any X, Y, Z in $\mathfrak{g}$ one has

$$\begin{aligned} B([X,Y],Z) &= \mathrm{Tr}(\mathrm{ad}[X,Y] \circ \mathrm{ad}\ Z) \\ &= \mathrm{Tr}((\mathrm{ad}\ X \circ \mathrm{ad}\ Y - \mathrm{ad}\ Y \circ \mathrm{ad}\ X) \circ \mathrm{ad}\ Z) \\ &= \mathrm{Tr}(\mathrm{ad}\ Y \circ \mathrm{ad}\ Z \circ \mathrm{ad}\ X) - \mathrm{Tr}(\mathrm{ad}\ Y \circ \mathrm{ad}\ X \circ \mathrm{ad}\ Z) \\ &= -\,\mathrm{Tr}(\mathrm{ad}\ Y, \mathrm{ad}\ [X,Z]) \\ &= -\,B(Y,[X,Z]). \end{aligned}$$

Thus $\mathfrak{g}$ acts on itself by $\mathrm{ad}(\cdot)$ through skew-symmetric transformations relative to $B(\cdot,\cdot)$. Since ad is the derivative of the homomorphism Ad we have that G maps into $O(\mathfrak{g}; B)$ by means of Ad. We say then that the form B is invariant. In particular, for $\mathfrak{g}$ semisimple we have a non-degenerate inner product on the tangent space to G at e. When G is also compact, this is a (negative) definite inner product.

6.2.5. Integration. The final topic from elementary Lie theory that will be needed subsequently is integration on Lie groups. Let G be a connected Lie group with Lie algebra $\mathfrak{g}$. Recall that $\mathfrak{g}$ can be viewed as the left invariant vector fields on G or equivalently as the tangent space to G at e, $T_e(G)$. Let $T_e(G)^*$ be the dual space. Then, in a similar way, the left invariant differential 1-forms on G can be identified with $T_e(G)^*$. Let $\omega_1, \ldots, \omega_n$ be a basis for $T_e(G)^*$ and extend the ω_i to left invariant 1-forms on G. Set $\omega = \omega_1 \wedge \cdots \wedge \omega_n$. Then ω is a nowhere vanishing top degree form on G. Assuming, as we may, that $\omega > 0$ then from ω and the usual partition of unity construction we obtain the (left) Haar integral for G. Its defining property is that for any y in G and compactly supported continuous function f,

$$\int_G \ell_y \circ f(x)\omega = \int_G f(x)\omega,$$

or $\ell_y^*\omega = \omega$.

If G is also compact then multiplying ω by a non-zero constant we obtain a normalized Haar integral satisfying

$$\int_G \omega = 1.$$

Continuing with G compact, let $y \in G$ and consider $r_y^*\omega$. Since left and right multiplication commute, for any $x \in G$, $\ell_x^* \circ r_y^*\omega = r_y^* \circ \ell_x^*\omega = r_y^*\omega$. Thus $r_y^*\omega$ is another left invariant top degree form on G. Since ω is normalized to have total mass 1 it is easy to see that we must have $r_y^*\omega = \omega$. Thus for compact connected Lie groups the normalized Haar integral is bi-invariant.

6.2.6. Analyticity. To conclude this introduction to Lie theory we want to clarify the use of analytic manifolds. By means of the exponential function and a coordinate system provided by it one can show that a group with a differentiable structure whose group operations are smooth maps is a Lie group. For this reason most texts require only smoothness not analyticity. The 5th Problem on Hilbert's famous list was to determine whether a locally Euclidean Hausdorff topological group has the structure of an analytic manifold with respect to which it became a Lie group. This much harder question was resolved affirmatively in the 1950's by Gleason and Montgomery-Zippin. A useful consequence of this is that continuous homomorphisms between Lie groups are automatically analytic.

6.3. Compact Reductive Groups: Representation Theory

6.3.1. Classical example: S^1. The most elementary connected Lie group is Euclidean space $\mathbb{R}^n$ or its quotient by the integer lattice, $\mathbb{Z}^n$, the n-torus $S^1 \mathrm{x} \cdots \mathrm{x} S^1$. As a realization of the circle ($n = 1$) we can take $S^1 = \{e^{i\theta} \in \mathbb{C}\}$, or, as a quotient of $\mathbb{R}$ by $\mathbb{Z}$, we can take $S^1 = \{0 \leq x \leq 2\pi | 0 \sim 2\pi\}$. As S^1 is the zero set of the function $x^2 + y^2 - 1$ it is a compact analytic manifold. From the obvious isomorphism between the two realizations, $\theta \to e^{i\theta}$, it is clear that S^1 is a compact connected Lie group. Functions on S^1 can be identified with periodic functions on $\mathbb{R}^1$, and the translation invariant Lebesgue measure on $\mathbb{R}^1$ (given by the 1-form "dx") descends to give a translation invariant measure $d\theta$ on S^1. The normalized Haar measure is $d\theta/2\pi$.

For abelian harmonic analysis there are many routes one could take. Here, first we sketch the "representation theory" approach. Afterwords, we describe an alternate route using differential operators.

DEFINITION 6.39. A character of S^1 is a continuous homomorphism to $\{z \in \mathbb{C} | \, |z| = 1\}$. In particular from Definition 6.15, it is a representation of S^1 on $\mathbb{C}$.

Certainly $e^{i\theta} \mapsto e^{in\theta}$ is a character for any $n \in \mathbb{Z}$. It is an easy exercise to see that any character of S^1 must be of this form. Also, for n, m distinct integers, the characters $e^{in\theta}, e^{im\theta}$ are different. So for the set of distinct characters on S^1 we have

$$\hat{S}^1 = \{\text{characters}\} \leftrightarrow \mathbb{Z}.$$

For f a continuous function on S^1, one can define a function $\hat{f}$ on $\hat{S}^1$ by

$$\hat{f}(n) = \frac{1}{2\pi} \int_0^{2\pi} f(\theta) e^{-in\theta} d\theta.$$

Harmonic analysis on S^1 is concerned with the information content of the dual objects f and $\hat{f}$. Two fundamental results in harmonic analysis are the Plancherel theorem and the inversion formula. Their familiar statements are

(i) $\|f\|_{L^2(S^1, d\theta/2\pi)} = \|\hat{f}\|_{\ell^2(\hat{S}^1)}$

(ii) for suitable f,

$$f(\theta) = \sum_{\hat{S}^1} \hat{f}(n) e^{in\theta}.$$

A more modern formulation of the above results involves the determination of the Fourier transform of the Dirac delta distribution supported at the identity element, δ. Define an involution on $C^\infty(S^1)$ by $f^*(\theta) = \overline{f(-\theta)}$, and a convolution on S^1 by

$$f * g(\theta) = \frac{1}{2\pi} \int_0^{2\pi} f(\varphi) g(\theta - \varphi) d\varphi.$$

Then the Plancherel theorem reformulates to

(i) $\langle \delta, f * f^* \rangle_{S^1} = \langle 1, |\hat{f}|^2 \rangle_{\hat{S}^1}$,

while inversion becomes

(ii) $\langle \ell_{\theta *} \delta, f \rangle_{S^1} = \langle 1, \widehat{\ell_\theta^* f} \rangle_{\hat{S}^1}$.

Every approach to harmonic analysis requires a parametrization of the characters of S^1, i.e. $\hat{S}^1$, as well as some analytic properties of them. We shall see that for non-abelian groups there are several different approaches to determining the parametrization. That there might be one involving the Dirac operator is suggested by the following.

It is easy to see that the differential operators on $\mathbb{R}^1$ commuting with translation are the constant coefficient ones. Appealing to the other lectures at the Summer School, it is evident that the Dirac operator for S^1 is $D = \frac{1}{i} \frac{d}{d\theta}$ and the spinors are just the complex numbers. The general results (here easily proven) give that D is essentially self-adjoint, with L^2 eigenfunctions $\{\varphi_n | n \in \text{Spec } D\}$ forming a complete orthogonal basis for $L^2(S^1)$. Since D is constant coefficient, it commutes with translation by elements of S^1. Thus S^1 acts on each eigenspace of D. Computation of Spec D shows it to be $\mathbb{Z}$ with eigenvalues of multiplicity one, and φ_n proportional to $e^{in\theta}$.

Determination of the complete spectrum of a Dirac operator is, in general, a difficult problem. Hence we will not follow this approach to determine $\hat{G}$, although with sufficient background we could. Rather, in this section I shall sketch the extension from S^1 to a compact reductive group G of the earlier discussion. A character of S^1 will be replaced with an (equivalence class of) irreducible unitary representation of G. Our goal in this section is to give a parametrization of them, i.e. describe $\hat{G}$. This will require a little more structure theory of semisimple Lie algebras.

6.3.2. Representation theory. We have seen that a connected reductive group G is the direct product of a central subgroup with a semisimple group. If G is compact then the central subgroup is a torus, essentially already discussed. So we may assume that G is a connected, compact, semisimple Lie group with Lie algebra $\mathfrak{g}$.

DEFINITION 6.40. Let V be a vector space over $\mathbb{C}$ and let $GL(V)$ denote the set of invertible linear operators given a suitable topology. A representation of G on V is a continuous homomorphism $\tau : G \to \mathrm{GL}(V)$. If V has an Hermitian inner product and $\tau(G) \subseteq U(V)$, the unitary operators on V, then the representation is called unitary.

REMARK 6.41. A more refined question to ask for finite dimensional V is when V has an $\mathbb{R}$-form $V_{\mathbb{R}}$ with a symmetric inner product J so that $\tau(G) \subseteq O(V_{\mathbb{R}}; J)$. There is a similar question for symplectic forms. Answers to these questions can be found in [**Bou68**].

It is necessary to define an equivalence relation on representations that is analogous to a change of basis. Two representations (τ_1, V_1) and (τ_2, V_2) are equivalent if there is a bounded invertible $A : V_1 \to V_2$ with

$$A \circ \tau_1(g) = \tau_2(g) \circ A,\ g \in G.$$

If the representations are unitary and A is unitary then the representations are called unitarily equivalent.

DEFINITION 6.42. A unitary representation (τ, V) is called irreducible if there is no proper closed subspace of V invariant by all $\tau(g)$, $g \in G$. Clearly this property is independent of representative of the equivalence class.

Notation: Let $\hat{G}$ denote the set of equivalence classes of irreducible unitary representations of G.

For simplicity, henceforth we shall refrain from using the phrase "equivalence class of representation" despite its necessity in the subject.

The collection of representations of G has an algebraic structure.

(i) if (τ_1, V_1), (τ_2, V_2) are representations then so is $(\tau_1 \oplus \tau_2, V_1 \oplus V_2)$;

(ii) if (τ_1, V_1), (τ_2, V_2) are representations then so is $(\tau_1 \otimes \tau_2, V_1 \otimes V_2)$;

(iii) if (τ, V) is a representation then so is (τ^*, V^*) where $\langle \tau^*(g)\lambda, v\rangle := \langle \lambda, \tau(g^{-1})v\rangle$;

(iv) if (τ, V) is a representation then so is $(\tau^k, S^k(V))$ where S^k is the k-th symmetric power;

(v) if (τ, V) is a representation then so is $(\bigwedge^k \tau, \bigwedge^k(V)$ where $\bigwedge^k(V)$ is the k-th exterior power.

For infinite dimensional vector spaces slightly more technical constructions than these given above are necessary.

PROPOSITION 6.43 (Unitary Trick). *Let G be a connected compact Lie group and (τ, V) a representation of G. Then there is an Hermitian inner product on V with respect to which τ is unitary.*

PROOF. Take any Hermitian inner product, $(,)_0$, on V. Define a new one by

$$(v, w) = \int_G (\tau(g)v, \tau(g)\omega)_0 dg.$$

The integral exists by the continuity of τ and the compactness of G. The inner product $(,)$ is G invariant, i.e. τ is unitary, by the left invariance of the Haar integral. □

The analog of harmonic "analysis" versus "synthesis" is the theorem on complete reducibility.

THEOREM 6.44 (Complete Reducibility). *Let G be a connected compact Lie group and (τ, H) a unitary representation on a separable Hilbert space. Then there are closed invariant subspaces $H_i \subseteq H$ with $(\tau_i = \tau|_{H_i}, H_i)$ irreducible so that*

$$(\tau, H) \sim \sum \oplus(\tau_i, H_i).$$

PROOF. From the preceding result, (τ, H) may be assumed to be unitary for G. If (τ, H) is reducible and finite dimensional, take a proper minimal invariant subspace and apply induction to the orthogonal complement. For infinite dimensional H a variant of this argument can be used. □

For compact Lie groups the technicalities arising from infinite dimensional Hilbert space representations disappear because of the previous theorem and the following theorem.

THEOREM 6.45. *Let G be a connected compact Lie group and (τ, H) an irreducible representation. Then* $\dim H < \infty$.

PROOF. We sketch the proof. One can reduce the result to G semisimple and to (τ, H) unitary. For such G the Killing form on $\mathfrak{g}$ is a symmetric definite bilinear form. Then use the Killing form to give G an invariant Riemannian metric. For this metric construct an invariant Laplace operator, Δ, acting on $L^2(G)$. From elliptic theory we get that the eigenspaces of Δ must be finite dimensional, while from the G-invariance of Δ we get that they are preserved by the left action of G. Finally, one constructs an equivalent realization of (τ, H) in an eigenspace of the Laplace operator. □

Given Theorem 6.45, we see that (like $\hat{S^1}$) $\hat{G}$ consists of finite dimensional representations of G. Consequently for the rest of this section a representation will always be on a finite dimensional vector space.

PROPOSITION 6.46. *Let G be a connected compact Lie group. Then G is a matrix group, i.e. G has a faithful finite dimensional representation.*

PROOF. Suppose first that G is semisimple. The natural representation $\mathrm{Ad}: G \to \mathrm{GL}(\mathfrak{g})$ has kernel contained in the center of G. Indeed, recall that Ad is the differential of conjugation, so

$$\exp \mathrm{Ad}(g)(X) = g \exp X g^{-1}.$$

If g is in Ker(Ad) then g commutes with the image of exp X. Since this image contains an open neighborhood of e (6.10) and G is connected, then g is central and Ker(Ad) is closed. But G is also semisimple and so has no non-finite normal subgroups. It follows that Ker(Ad) is finite. Now one argues that one can construct a representation faithful on the finite group. Taking the direct sum of this with the Adjoint representation gives a faithful representation of G. As for possible toral factors, since they are central in G they are easily represented faithfully. □

This Proposition leads us to several observations concerning alternative constructions of representations of compact semisimple groups. We shall take some liberty with technical details in order to give the idea behind them.

We have already seen that to any homomorphism of G there is a homomorphism of its Lie algebra $\mathfrak{g}$. Conversely if G is simply connected, to a homomorphism of the Lie algebra there is a homomorphism of the group. (This is no loss of generality as other groups with Lie algebra $\mathfrak{g}$ are quotients of the simply connected one. Their homomorphisms are those from the simply connected one that factor through the kernel of the covering map.) Thus, for simply connected compact G we have

$$\text{Representations of } G \leftrightarrow \text{ Representations of } \mathfrak{g}.$$

However for (τ, V) a representation of $\mathfrak{g}$, V is a complex vector space. Take the complex Lie algebra $\mathfrak{g}_{\mathbb{C}}$ obtained by base extension, $\mathfrak{g}_{\mathbb{C}} = \mathfrak{g} \otimes_{\mathbb{R}} \mathbb{C}$, and extend τ complex linearly to $\mathfrak{g}_{\mathbb{C}}$ thereby getting a complex linear representation of $\mathfrak{g}_{\mathbb{C}}$. Conversely, a $\mathbb{C}$-linear representation of $\mathfrak{g}_{\mathbb{C}}$ gives by restriction a representation of $\mathfrak{g}$. Thus

$$\text{Representations of } \mathfrak{g} \leftrightarrow \mathbb{C} - \text{linear Representations of } \mathfrak{g}_{\mathbb{C}}.$$

Since G is compact it has a faithful representation on some finite dimensional complex vector space (6.46), V, hence $\mathfrak{g} \hookrightarrow \mathrm{End}(V)$. But then $\mathfrak{g}_{\mathbb{C}} \hookrightarrow \mathrm{End}(V)$ and there is a Lie subgroup $G_{\mathbb{C}}$ of $\mathrm{GL}(V)$ that contains G. For G simply connected one gets in this way a natural complexification, $G_{\mathbb{C}}$, which is a Lie group and also a complex manifold with group operations holomorphic. Thus we obtain yet another equivalence:

$$\text{Representations of } G \leftrightarrow \text{ Holomorphic Representations of } G_{\mathbb{C}}.$$

There is a final piece to all this structure, but I will only mention it and not use it in detail although it is very important. As $\mathfrak{g}$ can be viewed as (left-)invariant vector fields on G, so too can $\mathfrak{g}_{\mathbb{C}}$ be viewed as complex (left-)invariant vector fields on G. Considering vector fields as first order differential operators, it is reasonable to expect there to be a larger object, the collection of all complex (left-)invariant differential operators on G. This is called the universal enveloping algebra, and usually denoted $\mathfrak{U}(\mathfrak{g}_{\mathbb{C}})$. The "enveloping" aspect essentially means $\mathfrak{U}(\mathfrak{g}_{\mathbb{C}})$ is the smallest associative algebra with identity containing $\mathfrak{g}_{\mathbb{C}}$, while the "universal" aspect means in particular that any $\mathbb{C}$-linear representation of $\mathfrak{g}_{\mathbb{C}}$ extends uniquely to one of $\mathfrak{U}(\mathfrak{g}_{\mathbb{C}})$. Conversely, a module for $\mathfrak{U}(\mathfrak{g}_{\mathbb{C}})$ by restriction is a representation for $\mathfrak{g}_{\mathbb{C}}$. Thus we arrive at:

$$\mathbb{C}\text{-linear Representations of } \mathfrak{g}_{\mathbb{C}} \leftrightarrow \text{ Modules for } \mathfrak{U}(\mathfrak{g}_{\mathbb{C}}).$$

Each of these alternatives for studying the representation theory of a compact semisimple group leads to an interesting approach involving different tools:

(i) through G one arrives at classical Weyl theory;

(ii) through $G_{\mathbb{C}}$ one arrives at quantization of Kähler manifolds and representations on their cohomology, epitomized by the Borel-Weil-Bott Theorem;

(iii) through $\mathfrak{g}_{\mathbb{C}}$ one arrives at classical Cartan theory;

(iv) through $\mathfrak{U}(\mathfrak{g}_{\mathbb{C}})$ one arrives at modules introduced by Harish-Chandra and later called Verma modules;

(v) a combination of (ii) and (iii) leads to a Lie algebra cohomology approach identified with Kostant.

Summarizing we obtain a diagram of the various viewpoints toward understanding $\hat{G}$:

$$\begin{array}{ccccc} \mathrm{Rep}^{ns} G & \longleftrightarrow & \mathrm{HoloRep}^{ns} G_{\mathbb{C}} & & \\ \updownarrow & & \updownarrow & & \\ \mathrm{Rep}^{ns} \mathfrak{g} & \longleftrightarrow & \mathbb{C} - \mathrm{Rep}^{ns} \mathfrak{g}_{\mathbb{C}} & \longleftrightarrow & \mathrm{Verma}\ \mathfrak{U}(\mathfrak{g}_{\mathbb{C}}) \end{array}$$

To obtain a parametrization of $\hat{G}$ we will give a brief introduction to (i) and (iii), (i.e. Cartan-Weyl theory). In the next section we conclude with yet another approach involving the Dirac operator.

To put things into perspective it is important in the subject to understand a few good examples that serve as prototypes for the general phenomenon. We take a detailed look at the smallest such G, the group $SU(2)$.

EXAMPLE 6.47. $SU(2) = \left\{ \begin{pmatrix} \alpha & \beta \\ -\overline{\beta} & \overline{\alpha} \end{pmatrix} \middle| |\alpha|^2 + |\beta|^2 = 1 \right\}$

First we follow the Weyl approach which is intimately related to tensor products of standard representations. For each $n \geqslant 0$ set

$$V_n = \{a_n z_1^n + a_{n-1} z_1^{n-1} z_2 + \cdots + a_0 z_2^n\}.$$

If $(\sigma, \mathbb{C}^2)$ is the standard representation of SU(2), then V_n is naturally isomorphic to $(\sigma^n, S^n(\mathbb{C}^2))$. For a polynomial p in V_n and g in $SU(2)$ define

$$\tau_n(g) p \begin{pmatrix} z_1 \\ z_2 \end{pmatrix} = p\left(g^{-1} \begin{pmatrix} z_1 \\ z_2 \end{pmatrix} \right).$$

One can show that (τ_n, V_n) is an irreducible representation of $SU(2)$ of dimension $n + 1$. Moreover any irreducible unitary representation of $SU(2)$ is equivalent to one of the (τ_n, V_n), i.e.

$$\widehat{SU}(2) = \{0, 1, 2, \dots\}.$$

For harmonic analysis, to a smooth function f on $SU(2)$ one defines the operator on V_n by

$$\hat{f}(n) = \int_{SU(2)} f(g) \tau_n(g) dg.$$

Denoting the Hilbert-Schmidt norm of $\hat{f}(n)$ by $\|\hat{f}(n)\|_{HS}$ one has the analog of the Plancherel theorem (in generality due to Peter-Weyl)

$$\|f\|^2_{L^2(SU(2),dg)} = \sum_{\widehat{SU(2)}} (n+1)\|\hat{f}(n)\|^2_{HS},$$

and the analog of the inversion formula

$$f(e) = \sum_{S\widehat{U}(2)} (n+1)\mathrm{Tr}(\widehat{f}(n)).$$

For S^1 the characters $e^{in\theta}$ are also functions of S^1. Here for $SU(2)$, the entries of a matrix realizing $\tau_n(g)$ are functions on $SU(2)$. An especially important expression obtained from these functions is the character of the representation. If g is in $SU(2)$ define a function called the character of τ_n by

$$\chi_n(g) = \mathrm{Tr}\ \tau_n(g).$$

In particular, for a diagonal matrix $t_\theta = \begin{pmatrix} e^{i\theta} & 0 \\ 0 & e^{-i\theta} \end{pmatrix}$, one has two expressions for χ_n:

$$\begin{aligned} \chi_n(t_\theta) &= e^{in\theta} + e^{i(n-2)\theta} + \cdots + e^{-in\theta}, \\ &= \frac{e^{i(n+1)\theta} - e^{-i(n+1)\theta}}{e^{i\theta} - e^{-i\theta}}. \end{aligned}$$

The first is the sum over the eigenspaces of $\tau_n(t_\theta)$; the second is a succinct formula called the Weyl character formula.

Alternatively to working with $SU(2)$, we can examine modules for $\mathrm{su}(2)_{\mathbb{C}}$ i.e. follow Cartan theory. As $SU(2)$ consists of unitary matrices, it is an easy exercise to see that the Lie algebra su(2) consists of skew-Hermitian matrices of trace zero. Since any 2×2 complex matrix of trace zero is uniquely the sum of a skew-Hermitian and a Hermitian, and a Hermitian matrix is i times a skew-Hermitian, one sees that $\mathrm{su}(2)_{\mathbb{C}} = \mathrm{sl}(2, \mathbb{C})$. Take a basis over $\mathbb{C}$ for $\mathrm{sl}(2, \mathbb{C})$ consisting of

$$h = \begin{pmatrix} 1 & 0 \\ 0 & -1 \end{pmatrix}$$

$$e = \begin{pmatrix} 0 & 1 \\ 0 & 0 \end{pmatrix}$$

$$f = \begin{pmatrix} 0 & 0 \\ 1 & 0 \end{pmatrix}$$

These elements satisfy the bracket relations

$$\begin{aligned} [h, e] &= 2e \\ [h, f] &= -2f \\ [e, f] &= h. \end{aligned}$$

Now given (π_n, W_n) an irreducible representation $\mathrm{sl}(2, \mathbb{C})$ of dimension $n + 1$ one shows that there is a basis $\{w_0, \dots w_n\}$ of W_n satisfying

$$\begin{aligned} \pi_n(h)w_i &= (n - 2i)w_i \\ \pi_n(e)w_0 &= 0 \\ \pi_n(f)w_i &= w_{i+1} \\ \pi_n(e)w_i &= i(n - i + 1)w_{i-1}. \end{aligned}$$

This is actually straightforward. One takes any eigenvector for $\pi_n(h)$ and then uses repeatedly the bracket relations and finite dimensionality.

Conversely, given an $n+1$ dimensional vector space, take a basis for it and define operators on this basis using the above identities. In this way one sees that the irreducible finite dimensional representations of $\mathrm{sl}(2,\mathbb{C})$ correspond to $\{n \geqslant 0\}$ with a unique such of dimension $n+1$.

EXERCISE 6.48. Show that the differentiated version of (τ_n, V_n) gives the representation (π_n, W_n).

We conclude the example with the form of the Casimir operator for $\mathrm{sl}(2,\mathbb{C})$.

LEMMA 6.49. *If π is a representation of $\mathrm{sl}(2,\mathbb{C})$ then*

$$\Omega = \frac{1}{2}\pi(h)^2 + \pi(h) + 2\pi(f)\pi(e)$$

commutes with each $\pi(X)$, X in $\mathrm{sl}(2,\mathbb{C})$.

PROOF. One has

$$\begin{aligned}
[\Omega, \pi(X)] &= \Omega\pi(X) - \pi(X)\Omega \\
&= \frac{1}{2}\pi(h)^2\pi(X) - \frac{1}{2}\pi(X)\pi(h)^2 + \pi([h,X]) \\
&\quad + 2\pi(f)\pi(e)\pi(X) - 2\pi(X)\pi(f)\pi(e) \\
&= \frac{1}{2}\pi(h)\pi([h,X]) - \frac{1}{2}\pi([X,h])\pi(h) \\
&\quad + \pi([h,X]) + 2\pi(f)\pi([e,X]) - 2\pi([X,f])\pi(e).
\end{aligned}$$

Now if $X = h$, then $[h,X] = 0$, $[e,X] = -2h$ and $[X,f] = -2f$ and one gets zero. Similarly for $X = e$ and $X = f$. Thus Ω commutes with $\pi(\mathrm{sl}(2,\mathbb{C}))$. □

For any semisimple group there is an analog of Ω called the Casimir operator. It is fundamental to the subject as it commutes with $\pi(\mathfrak{g}_\mathbb{C})$. Also, it appears naturally in Parthasarathy's formula for the square of Dirac which is a refinement of the Schrödinger-Lichnerowicz formula when the manifold is a symmetric space.

6.3.3. Cartan-Weyl Theory. Let $\mathfrak{g}_\mathbb{C}$ be a complex semisimple Lie algebra. The prototype of such is $\mathrm{sl}(2,\mathbb{C})$ and the first goal is to find the analog of the basis $\{h, e, f\}$. To start, take any element X in $\mathfrak{g}_\mathbb{C}$ such that $\mathrm{ad}X$ is diagonalizable. To show such X exists comes down to considering the characteristic polynomial $\det(\mathrm{ad}X - \lambda I)$ and using dimension arguments to show there are X with distinct roots. Given one such then

$$\mathfrak{g}_\mathbb{C} = \sum \oplus \mathfrak{g}(X, \lambda)$$

where $\mathfrak{g}(X,\lambda)$ is the eigenspace of $\mathrm{ad}X$ of eigenvalue λ. $\mathfrak{g}(X,0)$ is the centralizer of X and we refine our choice of X to minimize the dimension of this. Say H is our new choice of diagonalizable element and denote $\mathfrak{g}(H,0)$ by $\mathfrak{h}$. It follows from the Jacobi identity that $\mathfrak{h}$ is a subalgebra. At this point Cartan uses an argument to show that $\mathfrak{h}$ is an abelian subalgebra. Then instead of the eigenspace decomposition with respect to only H we can take a simultaneous eigenspace decomposition relative to $\mathrm{ad}\mathfrak{h}$. The eigenvalues will then be linear functionals on $\mathfrak{h}$, i.e. in $\mathfrak{h}^*$, and are called roots because they come from the characteristic polynomial. Denote this decomposition by

$$\mathfrak{g}_{\mathbb{C}} = \mathfrak{h} \oplus \sum_{\alpha} \oplus \mathfrak{g}_{\alpha}, \quad \alpha \in \mathfrak{h}^*.$$

Then using Cartan's Criterion for semisimple (6.38) one shows

(i) if α is an eigenvalue so is $-\alpha$;

(ii) $\dim \mathfrak{g}_{\alpha} = 1$.

From (i) one can group the roots into two sets Δ_+ and $-\Delta_+$. Using a basis of $\mathfrak{h}$ on which all elements in Δ_+ are real valued, one defines a partial order having a unique highest root α_0. It is not difficult to see from the Jacobi identity and (ii) that if α, β are in Δ_+, $[\mathfrak{g}_{\alpha}, \mathfrak{g}_{\beta}] = 0$ or $\mathfrak{g}_{\alpha+\beta}$. Thus the root spaces for Δ_+ form a subalgebra (as do those for $-\Delta_+$) which is nilpotent. Since each $\mathfrak{g}_{\alpha}$ is an eigenspace for $\mathfrak{h}$ we get a solvable algebra

$$\mathfrak{b} = \mathfrak{h} \oplus \sum_{\Delta_+} \oplus \mathfrak{g}_{\alpha}$$

called a Borel subalgebra. Then the highest root space is characterized by

(iii) $\mathfrak{g}_{\alpha_0}$, the highest root space, is an eigenspace for the Borel algebra $\mathfrak{b}$ of multiplicity one.

The final piece of structure is a natural group of permutations of $\Delta_+ \cup -\Delta_+$ which is called the Weyl group. For sl$(2, \mathbb{C})$ it amounts to sending e to f and f to e, and in general it is generated by similar permutations for each root. So we have

(iv) the Weyl group W permutes the roots.

What has just been described in (i), (ii), (iii), (iv) is the structure of the representation $(\mathrm{ad}, \mathfrak{g}_{\mathbb{C}})$ for the Lie algebra $\mathfrak{g}_{\mathbb{C}}$. The beautiful accomplishment of Cartan is that for any irreducible representation (τ, V) of a complex semisimple Lie algebra there is a very similar structure. The basic features are:

(i)' V has a basis of simultaneous eigenvectors of $\mathfrak{h}$, called weights;

(iii)' there is a highest weight space $\mathfrak{g}_{\lambda}$ which is an eigenspace of multiplicity one for the Borel algebra $\mathfrak{b}$;

(iv)' the Weyl group W acts on the weights of V.

We still need to obtain a parametrization of the irreducible representations. This will be done using the highest weight λ. By considering the collection of sl$(2, \mathbb{C})$'s contained in $\mathfrak{g}_{\mathbb{C}}$ associated to each of the roots $\alpha \in \Delta_+$, and from the representation theory of sl$(2, \mathbb{C})$ sketched before, it is easy to see that λ must have integral properties and positivity properties relative to all the roots in Δ_+. The collection of such is called the set of dominant integral weights. They are a discrete *cone* in $\mathfrak{h}^*$. Conversely, as we saw for sl$(2, \mathbb{C})$, given a dominant integral weight one shows that there is an irreducible representation for $\mathfrak{g}_{\mathbb{C}}$ with this as highest weight. Summarizing the preceding discussion we have the theorem of the highest weight:

THEOREM 6.50. *Let $\mathfrak{g}_{\mathbb{C}}$ be a complex semisimple Lie algebra. Let $\mathfrak{h}$ be a Cartan subalgebra and Δ_+ a set of positive roots. Let $\mathfrak{b}$ be the corresponding Borel subalgebra. Let (τ, V) be an irreducible finite dimensional representation of $\mathfrak{g}_{\mathbb{C}}$. Then*

(i) V is a direct sum of weight spaces for $\mathfrak{h}$;

(ii) there is a unique highest weight λ which is dominant integral with respect to Δ_+, and so that the weight space $\mathfrak{g}_\lambda$ of V has dimension 1;

(iii) the Weyl group acts on the weights of V.

Conversely, to each dominant integral form λ there is a unique irreducible finite dimensional representation of $\mathfrak{g}_{\mathbb{C}}$ with highest weight λ.

6.3.4. Weyl Character Formula. Let G be compact connected semisimple Lie group. Let (τ, V) be a finite dimensional representation of G. The function on G given by

$$\chi_\tau(g) = \mathrm{Tr}\ \tau(g)$$

is called the character of τ. Since $\mathrm{Tr}(ABA^{-1}) = \mathrm{Tr}(B)$, a character depends only on the conjugacy class of g and also not on any matrix realization of τ. Recall that the complexification $\mathfrak{g}_{\mathbb{C}}$ of the Lie algebra of $\mathfrak{g}$ contains a maximal abelian subalgebra $\mathfrak{h}$. Then $\mathfrak{h} \cap \mathfrak{g} = \mathfrak{t}$ is also maximal abelian. Let T be the corresponding group of G. T is called a maximal torus.

EXAMPLE 6.51. $\mathrm{G} = \mathrm{SU}(n)$

Set

$$T = \left\{ \begin{pmatrix} e^{i\theta_1} & & \\ & \ddots & \\ & & e^{i\theta_n} \end{pmatrix} \middle| \sum_{i=1}^{n} \theta_i = 0 \right\}.$$

From the spectral theorem we know that if $g \in \mathrm{SU}(n)$ then there exists $t \in T$ with $g = xtx^{-1}$. So a character of $\mathrm{SU}(n)$ is specified by its values on T. Indeed for an arbitrary G, any element of G is conjugate to an element in a maximal torus.

Let $(\tau_\lambda, V_\lambda)$ be an irreducible representation for G. Then for $H \in \mathfrak{t}$

$$\begin{aligned} \chi_{\tau_\lambda}(\exp H) &= \Sigma n_\nu e^{\nu(H)} \\ &= \frac{\sum_W \epsilon(\omega) e^{\omega(\lambda+\rho)(H)}}{\sum \epsilon(\omega) e^{\omega\rho(H)}} \end{aligned}$$

6.3.5. Tensor Products. For irreducible representations $(\tau_\lambda, V_\lambda)$ and (τ_μ, V_μ) their tensor product $V_\lambda \otimes V_\mu$ usually is not irreducible. So from 6.44, it decomposes into

$$V_\lambda \otimes V_\mu \cong \sum \oplus n_\nu V_\nu, \ n_\nu \geq 0.$$

The exact summands with their multiplicity are not known in general although there is an algorithm due to Littleman (see [**BE**]) and a combinatorial formula for them due to Kostant. However, there are some constitutents for which we have precise information.

(1) *Cartan Composition.* Let $(\tau_\lambda, V_\lambda)$ and (τ_μ, V_μ) be irreducible representations of $\mathfrak{g}_\mathbb{C}$ with λ, μ dominant integral. Then

$$V_\lambda \otimes V_\mu \cong \sum \oplus n_\nu V_{\lambda+\nu},\ n_\nu \geq 0.$$

Here ν ranges over the weights of V_μ, while one has for the highest weight μ that $n_\mu = 1$.

PROOF. The possible eigenspaces are sums of those from V_λ and those from V_μ. Since the eigenspace for λ and μ have dimension 1 the eigenvalue $\lambda + \mu$ has dim $= 1$. For other ν one uses SL(2) theory. □

Notation: The irreducible component $V_{\lambda+\mu}$ is called the Cartan Composition of V_λ and V_μ. It is used to construct differential operators called Stein-Weiss operators. See [**Bra97a**] for a classification of which of these give elliptic operators.

(2) *P-R-V Composition* (see [**PRV67**]). Given $(\tau_\lambda, V_\lambda)$ and (τ_μ, V_μ) irreducible representations of $\mathfrak{g}_\mathbb{C}$. Let $w \in W$ and suppose $\lambda + w\mu$ is dominant integral. Generalizing the result in (1) where w is the identity, we have $V_{\lambda+w\mu}$ occurs in $V_\lambda \otimes V_\mu$ with multiplicity one.

6.4. Compact Reductive Homogeneous Spaces: Differential Operators

Classical Example: $\mathbb{P}^1(\mathbb{C})$.

$\mathbb{P}^1(\mathbb{C}) = \{\text{lines in } \mathbb{C}^2 \text{ thru } (0,0)\}$, and $\mathrm{SU}(2) = \left\{\begin{pmatrix} \alpha & \beta \\ -\overline{\beta} & \overline{\alpha}\end{pmatrix} \middle| |\alpha|^2 + |\beta|^2 = 1\right\}$ acts transitively on the lines in $\mathbb{C}^2$ thru $(0,0)$. Fix one, say $\{z_2 = 0\}$. The stabilizer in SU(2) of $\{z_2 = 0\}$ is

$$T = \left\{\begin{pmatrix} e^{i\theta} & 0 \\ 0 & e^{-i\theta}\end{pmatrix}\right\}.$$

So

$$\mathbb{P}^1(\mathbb{C}) \cong \mathrm{SU}(2)/T.$$

$\mathrm{SL}(2,\mathbb{C})$ similarly acts transitively on $\mathbb{P}^1(\mathbb{C})$, and the stabilizer of $\{z_2 = 0\}$ is $B = \left\{\begin{pmatrix} a & b \\ 0 & a^{-1}\end{pmatrix}\right\}$. Thus

$$\mathrm{SL}(2,\mathbb{C})/B \cong \mathbb{P}^1(\mathbb{C}) \cong \mathrm{SU}(2)/T.$$

From these two incarnations of $\mathbb{P}^1(\mathbb{C})$ we see distinct properties. As a quotient of SU(2) it is obviously compact and can be given an invariant Riemannian metric by means of the Killing form. As a quotient of $\mathrm{SL}(2,\mathbb{C})$ by a complex subgroup it is obviously a complex manifold with a $\overline{\partial}$ operator coming from $\mathrm{SL}(2,\mathbb{C})$. As to which homogeneous spaces of a compact simple group have such structure there is the following result.

THEOREM 6.52 (H. C. Wang). *Let G be a compact, connected, simple Lie group with a maximal torus T. Let $T_1 < T$ be a closed subgroup. Set*

$$C(T_1) = Z_G(T_1) = \{g \in G | gt_1 = t_1 g\}.$$

Then G/H is compact complex $\Leftrightarrow H = C(T_1)$, some T_1.

To give the realization of $G/C(T_1)$ as a quotient of complex groups we recall, a consequence of Proposition 6.46, that G has a complexification $G_{\mathbb{C}}$ whose Lie algebra we denote by $\mathfrak{g}_{\mathbb{C}}$. In $\mathfrak{g}_{\mathbb{C}}$ we have a Cartan subalgebra $\mathfrak{h}$ with positive roots Δ_+. Then $\mathfrak{h}\cap\mathfrak{g} = \mathfrak{t}$ is also maximal abelian and the corresponding group of G, T, is a maximal torus. To a subgroup $T_1 < T$ consider the subset of the roots Δ_+ that are trivial on T_1. The same is true for the negative of these roots. Supplementing the Borel subalgebra with the eigenspaces spaces corresponding to these negative roots one obtains a subalgebra of $\mathfrak{g}_{\mathbb{C}}$ which we denote $\mathfrak{p}$. By construction $\mathfrak{p}$ contains the Borel subalgebra $\mathfrak{b}$. Such a subalgebra is called a parabolic subalgebra, with the corresponding subgroup P of $G_{\mathbb{C}}$ called a parabolic subgroup. With this in mind we have

THEOREM 6.53. *Let G be a compact, connected, simple Lie group with complexification $G_{\mathbb{C}}$. Let $T < G$ be a maximal torus and let $H_{\mathbb{C}}$ be its complexification. Fix a Borel subgroup $B \supseteq H_{\mathbb{C}}$. Then $G/T \overset{diffeo}{\cong} G_{\mathbb{C}}/B$ and is a complex manifold. More generally, for each $T_1 < T$ there is a parabolic subgroup $P \supseteq B$ with*

$$G/C(T_1) \overset{diffeo}{\cong} G_{\mathbb{C}}/P.$$

We look at two examples more complicated than $\mathbb{P}^1(\mathbb{C})$ to illustrate the above.

EXAMPLE 6.54. $\dim T_1 = 1$, $G(n,k;\mathbb{C})$

Fix n and take $1 \le k \le n$. Let $G(n,k;\mathbb{C})$ denote the Grassmannian of k-planes in $\mathbb{C}^n$. Let $G = \mathrm{SU}(n)$ so that $G_{\mathbb{C}} = \mathrm{SL}(n,\mathbb{C})$, and set

$$T = \left\{ \begin{pmatrix} e^{i\theta_1} & & \\ & \ddots & \\ & & e^{i\theta_n} \end{pmatrix} \,\middle|\, \sum_{i=1}^{n} \theta_i = 0 \right\}.$$

Take the one dimensional subgroup of T given by

$$T_1 = \left\{ \begin{pmatrix} e^{i\theta/k} & & & & & \\ & \ddots & & & & \\ & & e^{i\theta/k} & & & \\ & & & e^{-i\theta/(n-k)} & & \\ & & & & \ddots & \\ & & & & & e^{-i\theta/(n-k)} \end{pmatrix} \right\}.$$

Then

$$C(T_1) = \left\{ \begin{pmatrix} U(k) & 0 \\ 0 & U(n-k) \end{pmatrix} \,\middle|\, \det = 1 \right\}$$

and thus

$$G(n,k;\mathbb{C}) \cong \mathrm{SU}(n)/C(T_1).$$

For the complex subgroups of $\mathrm{SL}(n,\mathbb{C})$ set

$$H_{\mathbb{C}} = \left\{ \begin{pmatrix} z_1 & & 0 \\ & \ddots & \\ 0 & & z_n \end{pmatrix} \middle| \Pi z_i = 1 \right\}$$

$$B = \left\{ \begin{pmatrix} * & ** \\ 0 & * \end{pmatrix} \middle| \det = 1 \right\}$$

$$P = \left\{ \begin{pmatrix} \mathrm{GL}(k,\mathbb{C}) & ** \\ 0 & \mathrm{GL}(n-k,\mathbb{C}) \end{pmatrix} \middle| \det = 1 \right\}$$

Then

$$\mathrm{SL}(n,\mathbb{C})/P \cong G(n,k;\mathbb{C}) \cong \mathrm{SU}(n)/C(T_1).$$

EXAMPLE 6.55. $\dim T_1 = k > 1$

Take $G = \mathrm{SU}(n)$, $G_{\mathbb{C}} = \mathrm{SL}(n,\mathbb{C})$ and the usual maximal torus

$$T = \left\{ \begin{pmatrix} e^{i\theta} & & \\ & \ddots & \\ & & e^{i\theta_n} \end{pmatrix} \middle| \sum_{i=1}^{n} \theta_i = 0 \right\}.$$

Set $T_1 = \left\{ \begin{pmatrix} e^{i\theta_1} & & & & & \\ & \ddots & & & & \\ & & e^{i\theta_k} & & & \\ & & & 1 & & \\ & & & & \ddots & \\ & & & & & 1 \end{pmatrix} \middle| \sum_{i=1}^{k} \theta_i = 0 \right\}$

Then $C(T_1) = \left\{ \begin{pmatrix} e^{i\psi_1} & & & \\ & \ddots & & \\ & & e^{i\psi_k} & \\ & & & U(n-k) \end{pmatrix} \middle| \det = 1 \right\}$

$$H_{\mathbb{C}} = \left\{ \begin{pmatrix} h_1 & & 0 \\ & \ddots & \\ 0 & & h_n \end{pmatrix} \middle| \prod_{i=1}^{n} h_i = 1 \right\}$$

$$P = \left\{ \begin{pmatrix} h_1 & & & *** \\ & \ddots & & \\ & & h_k & \\ & & & \mathrm{GL}(n-k,\mathbb{C}) \end{pmatrix} \middle| \det = 1 \right\}$$

Then $G_{\mathbb{C}}/P \cong G/C(T_1)$.

6.4.1. Associated Bundles.

A good reference for the following is [**Hus94**].

DEFINITION 6.56. Let G be a Lie group. A G-principal fiber bundle is a triple (P,π,M) satisfying:

(i) $\pi : P \to M$ is a smooth submersion

(ii) G act smoothly and freely from the right – i.e.

$$P \times G \to P \text{ with}$$
$$p \cdot g = p \Leftrightarrow g = e.$$

(iii) for each $x \in M$ there is a local trivialization

$$\phi : \pi^{-1}(U) \to U \times G$$
$$p \mapsto (\pi(p), \varphi(p)).$$

In particular, the map $\phi|_{\pi^{-1}(y)} \to \{y\} \times G$ commutes with the action of G.

EXAMPLE 6.57. Let G be a Lie group and H a closed subgroup. Then

$$H \to G \to G/H$$

is an H-principal fiber bundle. Here $M = G/H$ and $P = G$ and H is the "G".

Given a G-principal fiber bundle (P, π, M) and a representation (τ, V) of G, one can form the associated vector bundle $\mathbb{V} \to M$ as follows. Consider $P \times V$ with G action

$$g \cdot (p, v) = (p \cdot g, \tau(g)^{-1} v).$$

Then taking the orbit space of this G-action we obtain the associated vector bundle $\mathbb{V} := P \times V/G$.

In particular for the homogeneous space G/H and a representation (τ, V) of H we have $\mathbb{V} = G \times V/H$, which is usually denoted

$$\mathbb{V} = G \underset{H}{\times} V.$$

Sections of $\mathbb{V}$ are given by taking a class of functions on G, $\mathfrak{F}(G)$; forming $\mathfrak{F}(G) \otimes V$; and taking the H-invariants under the tensor product of the right regular representation, r, of H on $\mathfrak{F}(G)$ and τ, i.e. $r \otimes \tau$,

$$(\mathfrak{F}(G) \otimes V)^H.$$

Since $G \times G$ acts on functions on G by $\ell \otimes r$, i.e.

$$\ell \otimes r(g_1, g_2) f(x) = \ell(g_1) \otimes r(g_2) f(x) = f(g_1^{-1} x g_2),$$

and we have used only the "r" there remains on $[\mathfrak{F}(G) \otimes V]^H$ a representation of G, i.e.

$$\ell(g) \cdot f(x) \otimes v = f(g^{-1} x) \otimes v$$

and this preserves the space of H-invariant vectors.

In topological notation this would be

$$(\mathfrak{F}(G) \otimes V)^H = \Gamma(G/H; \mathbb{V}).$$

6.4.2. Connection. Let G be a Lie group and let H be a closed subgroup. G/H is called a reductive homogeneous space if there is a subspace $\mathfrak{p} \subseteq \mathfrak{g}$ such that

(i) $\mathfrak{g} = \mathfrak{h} \oplus \mathfrak{p}$

(ii) $\mathrm{Ad}|_H : \mathfrak{p} \to \mathfrak{p}$

REMARK 6.58. If H is compact this is always the case for one can apply complete reducibility to $(\mathrm{Ad}|_H, \mathfrak{g})$.

For G/H reductive the tangent bundle is an associated bundle

$$T(G/H) \cong G \underset{H}{\times} \mathfrak{p}.$$

We assume that G/H is reductive and Riemannian. Let (τ, V) be a representation of H and $\mathbb{V}$ the associated bundle. There is a natural connection

$$\nabla : \Gamma(G/H; \mathbb{V}) \to \Gamma(G/H; \mathbb{V} \otimes T^*)$$

obtained from a basis $\{X_i\}$ for $\mathfrak{p}$ and a dual basis $\{X_i^*\}$ by setting

$$\nabla f = \Sigma(r(X_i) \otimes 1) f \otimes X_i^*.$$

Here $f \in (C^\infty(G) \otimes V)^H$ and $r(X_i)f(g) \otimes v = \frac{d}{dt} f(g \exp tX_i)|_{t=0} \otimes v$.

6.4.3. Homogeneous Differential Operators. Let (τ, V), (π, W) be representations of H and $\mathbb{V}$, $\mathbb{W}$ the associated bundles. As we have seen G acts by the left regular representation on sections of $\mathbb{V}$ and sections of $\mathbb{W}$.

DEFINITION 6.59. A differential operator

$$D : \Gamma(G/H; \mathbb{V}) \to \Gamma(G/H; \mathbb{W})$$

is called homogeneous if

$$D \circ \ell_G^{\mathbb{V}} = \ell_G^{\mathbb{W}} \circ D.$$

EXAMPLE 6.60. Dirac operator.

Suppose that G/H is spin, i.e. there is a lift of Ad

$$\begin{array}{ccc} & & \mathrm{Spin}(\mathfrak{p}) \\ & \nearrow & \downarrow \\ H & \xrightarrow{\mathrm{Ad}} & \mathrm{SO}(\mathfrak{p}) \end{array}$$

and $\dim G/H = 2n$. Let (σ, Σ_n) be the spinor module for $Cl(\mathfrak{p})$. Then by composition $\sigma\circ$ lift we get a representation of H, denoted $(\tilde{\sigma}, \Sigma_n)$. The associated bundle $G \underset{H}{\times} \Sigma_n$ is the bundle of spinors on G/H. The Dirac operator is given by the expression

$$Df = \sum_i r(X_i) \otimes c(X_i^*) f$$

EXAMPLE 6.61. Stein-Weiss operators.

Suppose that H is compact and the representation on $\mathfrak{p}$ is irreducible, say with highest weight μ. Let $(\tau_\lambda, V_\lambda)$ be an irreducible representation of H with highest weight λ. Then let

$$P : V_\lambda \otimes \mathfrak{p} \to V_{\lambda+\mu}$$

be the projection on the Cartan Composition.

$$D_\lambda f = P \circ \Sigma r(X_i) f \otimes X_i^*$$

is a homogeneous differential operator from

$$\Gamma(G/H; \mathbb{V}_\lambda) \to \Gamma(G/H; \mathbb{V}_{\lambda+\mu}).$$

EXAMPLE 6.62. $\overline{\partial}$ operator.

Suppose that G/H is a compact complex manifold. Then we have the usual splitting of the complexified tangent space according to the almost complex structure and hence for the associated bundles

$$\mathfrak{p}_{\mathbb{C}} = \mathfrak{p} \underset{\mathbb{R}}{\otimes} \mathbb{C} = T^{1,0} \oplus T^{0,1}.$$

Let $\{Z_i\}$ be an orthonormal basis of $T^{1,0}$ and $\{\overline{Z}_i\}$ one for $T^{0,1}$.

Set

$$\overline{\partial} f = \sum_i (r(\overline{Z}_i) \otimes 1) f \otimes \overline{Z}_i^*.$$

Then $\overline{\partial}$ is a homogeneous differential operator

$$\overline{\partial} : \Gamma_h(G/H; \mathbb{V}) \to \Gamma_h(G/H; \mathbb{V} \otimes T^{0,1*}).$$

6.4.4. Borel-Weil-Bott. Let G be a compact, connected, semisimple Lie group with maximal torus T. Let $G_{\mathbb{C}}$ be a complexification and B a Borel subgroup. Recall that

$$\hat{G} \leftrightarrow \{\text{dominant integral elements of } \mathfrak{h}^*\}.$$

We will assume the "cone" of dominant integral elements is determined by this choice of B. Let λ be dominant integral and let $\mathbb{C}_\lambda$ be the representation of T with character $\xi_\lambda(t)$. Let $\mathbb{L}_\lambda$ be the associated line bundle $G \underset{T}{\times} \mathbb{C}_\lambda$ over G/T.

The $\overline{\partial}$ operator gives a complex

$$\Gamma(\mathbb{L}_\lambda) \xrightarrow{\overline{\partial}} \Gamma(\mathbb{L}_\lambda \otimes T^{0,1*}) \xrightarrow{\overline{\partial}} \dots \xrightarrow{\overline{\partial}} \Gamma(\mathbb{L}_\lambda \otimes \wedge^k T^{0,1*}) \xrightarrow{\overline{\partial}} \dots \Gamma(\mathbb{L}_\lambda \otimes \wedge^n T^{0,1*}) \to 0$$

Here $n = \dim_{\mathbb{C}} G/T$.

The cohomology groups will be denoted $H^{0,k}(G/T; \mathbb{L}_\lambda)$. Notice that since all the bundles are associated bundles G acts on the sections. Also, since $\overline{\partial}$ is a homogeneous differential operator, the action of G on sections descends to one on the cohomology. As G/T is compact, it is known that $H^{0,k}(G/T; \mathbb{L}_\lambda)$ is a finite dimensional vector space.

THEOREM 6.63 (Borel-Weil). *Assume that λ is dominant and integral. Then*

(i) $H^{0,k}(G/T; \mathbb{L}_\lambda) = 0, \ k \geq 1$

(ii) $H^{0,0}(G/T; \mathbb{L}^\lambda)$ is an irreducible representation of G with highest weight λ.

REMARKS 6.64. (1) the proof of (i) usually is obtained through a general result about cohomology and not much Lie theory.

(2) $H^{0,0}(G/T; \mathbb{L}_\lambda)$ can be identified with global holomorphic sections of $\mathbb{L}_\lambda$. Thus we have constructed a geometric model of the general irreducible representation of G. Conversely, from representation theory we can construct the holomorphic sections as follows: take a basis for the irreducible module V_λ, $\{v_\lambda^0, v_\lambda^1, \dots, v_\lambda^n\}$ where v_λ^0 is the eigenvector for the Borel subgroup B. Then form matrix coefficients

$$\varphi_i(g) = \langle \tau_\lambda(g) v_\lambda^0, v_\lambda^i \rangle$$

Notice that $\varphi_i(gt) = \xi_\lambda(t) \varphi_i(g)$ so φ_i are in $\Gamma(G/T; \mathbb{L}_\lambda)$ and holomorphic because v_λ^0 is an eigenvector for the Borel subgroup.

Of course it is circular reasoning to prove the existence of V_λ by starting with V_λ!

Recall that besides $G/T \cong G_{\mathbb{C}}/B$ there are other complex homogeneous spaces for G, namely

$$G/C(T_1) \cong G_{\mathbb{C}}/P.$$

The extension of Borel-Weil to $G/C(T_1)$ was done by Bott. The statement of the theorem is similar to that of the Borel-Weil Theorem and will be omitted.

If the choice of "cone" and Borel subgroup are not compatible then the result becomes

THEOREM 6.65. *Suppose that λ is an integral form and that $w \in W$ is such that $w(\lambda+\rho)-\rho$ is dominant. Let $\#w$ be the minimal number of transpositions needed to express w. Then*

$$H^{0,k}(G/T;\mathbb{L}_\lambda) = \begin{cases} 0 & k \neq \#w \\ V_\lambda & k = \#w \end{cases}$$

6.4.5. Reformulation of Borel-Weil (-Bott). Let $K(G)$ denote formal integral combinations of irreducible representations of G.

EXAMPLE 6.66.

$$\begin{aligned} 2V_\lambda &= V_\lambda \oplus V_\lambda \\ W &= V_{\lambda_1} - V_{\lambda_2} + 2V_{\lambda_3} \\ \Leftrightarrow W \oplus V_{\lambda_2} &= V_{\lambda_1} \oplus V_{\lambda_3} \oplus V_{\lambda_3} \end{aligned}$$

Fix λ integral but not necessarily dominant for the Borel subgroup B. In $K(G)$ let

$$R^*(\mathbb{L}_\lambda) = \sum_k (-1)^k H^{0,k}(G/T;\mathbb{L}_\lambda).$$

Then the theorem can be summarized in the statement that R^* is a surjective functor from $\hat{T}$ to $\hat{G}$, called the holomorphic induction functor.

6.4.6. Dirac Induction. Having a Riemannian metric on G/T there is a standard procedure to define an adjoint operator, $\overline{\partial}^*$, and show that the cohomology $H^{0,k}(G/T;\mathbb{L}_\lambda)$ is represented by $\mathrm{Ker}(\overline{\partial}+\overline{\partial}^*)$. This $\overline{\partial}+\overline{\partial}^*$ operator is just Dirac coupled with the canonical line bundle. So, in a sense, $\widehat{G}$ can be realized as the kernel of Dirac. This is however simply the beginning of a systematic approach to the parametrization of $\hat{G}$ using a Dirac operator. Unfortunately, the lectures end here. There remains to do the many papers such as [**Agr, AS77, Par72, BW80, Wol73, MS89, Hua01, HP02, Kos99, Kos00, Kos, Lan01, Lan00, Sle85, Sle87a, Sle87b, Sle88**] listed in the Bibliography on Dirac for compact reductive homogeneous spaces, the index theory of Bott for homogeneous differential operators, and the vast material for noncompact reductive homogeneous spaces. The sources in the Bibliography will provide a door to these beautiful topics.

ROBERT J. STANTON, DEPARTMENT OF MATHEMATICS, OHIO STATE UNIVERSITY, 231 W. 18TH AVE. COLUMBUS, OH 43210, USA.
E-mail address: stanton@math.ohio-state.edu

CHAPTER 7

The Atiyah-Singer Index Theorem and Applications

H. Blaine Lawson, Jr.

Abstract. The purpose of this chapter is to present some of the spectacular applications of the Dirac operator in geometry and topology which come from the Atiyah-Singer Index Theorem. We begin with a brief discussion of vector bundles and characteristic classes. We next introduce elliptic operators and present various forms of the Index Theorem. A number of applications are then examined in detail. This exposition is meant to be a cursory sketch. Complete details and full references can be found in [**LM89, Ati67, Gil84, BGV91**].

Contents

7.1. *K*-Theory and Characteristic Classes

We begin with an elementary discussion of vector bundles. Let X be a topological space. By a *family of complex vector spaces* of dimension n over X we mean a topological space E and a continuous map $\pi : E \to X$ such that for each $x \in X$, the fibre $E_x \equiv \pi^{-1}(x)$ has the structure of an n-dimensional complex vector space with the usual topology. A *morphism* between two families $\pi : E \to X$ and $\pi' : E' \to X$ is a continuous map $f : E \to E'$ such that

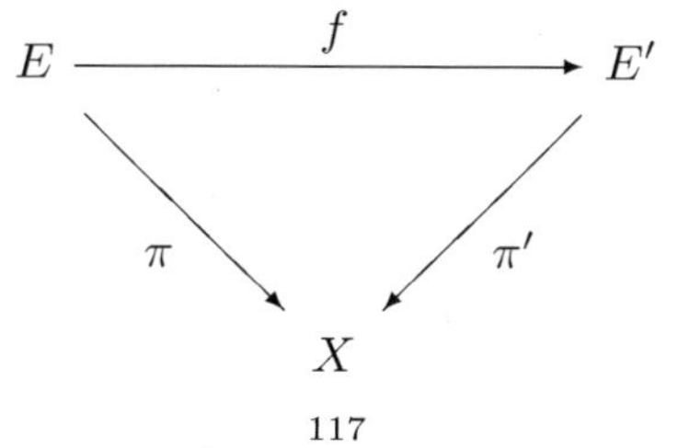

commutes and $f\big|_{E_x} : E_x \to E'_x$ is linear for all $x \in X$. A morphism $f : E \to E'$ is an **isomorphism** if it has a continuous inverse. A family $\pi : E \to X$ is **trivial** if it is isomorphic to the product $\mathrm{pr} : X \times \mathbb{C}^n \to X$.

DEFINITION 7.1. A **complex vector bundle of rank** n on X is a locally trivial family of n-dimensional complex vector spaces $\pi : E \to X$, that is, each point $x \in X$ has a neighborhood U such that $\pi : \pi^{-1}(U) \to U$ is trivial.

EXAMPLE 7.2. **The Tautological Bundle over the Grassmanian.** Fix positive integers $n < N$ and let $G_n(\mathbb{C}^N)$ denote the set of linear subspaces of dimension n in $\mathbb{C}^N$. This space has the structure of a compact complex manifold of dimension $n(N-n)$. We define a vector bundle $\pi : E_n \to G_n(\mathbb{C}^N)$ by setting

$$E_n = \{(P, v) \in G_n(\mathbb{C}^N) \times \mathbb{C}^N : v \in P\}$$

and $\pi(P, v) = P$.

All the standard operations on vector spaces carry over to vector bundles. Given vector bundles E and F over X, one can form: $E \oplus F$, $E \otimes F$, $\bigwedge^k E$, $\mathrm{Hom}(E, F)$, etc. For example, $E \oplus F$ is the bundle whose fibre at $x \in X$ is $E_x \oplus F_x$. There are also notions of subbundles $F \subset E$ and quotient bundles E/F whose definitions should be obvious.

Of fundamental importance is the fact that vector bundles can be pulled back by a continous map of topological spaces.

DEFINITION 7.3. Given a continuous map $\psi : Y \to X$ and a vector bundle $\pi : E \to X$, the **induced bundle** $\psi^*\pi : \psi^*E \to Y$ is defined to be the fibre-product

$$\psi^*E \equiv \{(y, e) \in Y \times E : \psi(y) = \pi(e)\}$$

where $\psi^*\pi$ is given by projection to the first factor of $Y \times E$. Note the commutative diagram

$$\begin{array}{ccc} \psi^*E & \xrightarrow{\ \widetilde{\psi}\ } & E \\ {\scriptstyle \psi^*\pi}\big\downarrow & & \big\downarrow{\scriptstyle \pi} \\ Y & \xrightarrow{\ \psi\ } & X \end{array}$$

where $\widetilde{\psi}$ is given by projection to the second factor of $Y \times E$. Note also that if $\psi \equiv x$ is constant, then $\psi^*E = Y \times E_x$ is trivial. If $\psi : Y \hookrightarrow X$ is an embedding, then ψ^*E is the restriction of E to Y.

Denote by $\mathrm{Vect}_n(X)$ the set of isomorphism classes of complex vector bundles of rank n on X. Denote by $[X, Y]$, the set of homotopy classes of continuous maps from X to Y.

THEOREM 7.4. *Let X be a compact Hausdorff space. There is a natural bijection*

$$Vect_n(X) \cong \lim_{N \to \infty} [X, G_n(\mathbb{C}^N)]. \tag{7.1}$$

NOTE 7.5. This limit is defined as follows. The embeddings given by the "first N coordinates" give inclusions:

$$\begin{array}{rcl} \cdots \subset \mathbb{C}^N & \subset & \mathbb{C}^{N+1} \subset \cdots \\ \cdots \subset G_n(\mathbb{C}^N) & \subset & G_n(\mathbb{C}^{N+1}) \subset \cdots \\ \cdots \subset\ [X, G_n(\mathbb{C}^N)] & \subset & [X, G_n(\mathbb{C}^{N+1})] \subset \cdots \end{array}$$

and we take the usual (co)limit of these spaces. The limit $G_n(\mathbb{C}^\infty) = \lim_{N\to\infty} G_n(\mathbb{C}^N)$ is topologized by declaring that a subset $C \subset G_n(\mathbb{C}^\infty)$ is closed if and only if $C \cap G_n(\mathbb{C}^N)$ is closed for all N. In this topology every compact subset of $G_n(\mathbb{C}^\infty)$ is contained in $G_n(\mathbb{C}^N)$ for some N. It follows that (7.1) can be rewritten as

$$\text{Vect}_n(X) \cong [X, G_n(\mathbb{C}^\infty)]. \tag{7.2}$$

PROOF. Full details of this proof can be found in [**MS74**] or [**Ati67**]. The basic idea is as follows. The tautological bundles introduced in Example 7.2 sit in a sequence of bundle morphisms:

$$\begin{array}{ccccccccc} \cdots & \longrightarrow & E_n & \longrightarrow & E_n & \longrightarrow & E_n & \longrightarrow & \cdots \\ & & \downarrow & & \downarrow & & \downarrow & & \\ \cdots & \longrightarrow & G_n(\mathbb{C}^N) & \longrightarrow & G_n(\mathbb{C}^{N+1}) & \longrightarrow & G_n(\mathbb{C}^{N+2}) & \longrightarrow & \cdots \end{array}$$

which yield a limiting **tautological bundle**

$$\pi : E_n \to G_n(\mathbb{C}^\infty). \tag{7.3}$$

The correspondence (7.2) associates to a map $\psi : X \to G_n(\mathbb{C}^\infty)$ the bundle $\psi^* E_n$. A homotopy $\Psi : [0,1] \times X \to G_n(\mathbb{C}^\infty)$ yields a bundle isomorphism $\psi_0^* E_n \cong \psi_1^* E_n$ (where $\psi_k = \Psi|_{\{k\}\times X}$) via the local triviality of $\Psi^* E_n$ and a direct compactness argument.

The inverse to the above correspondence is constructed as follows. Given a bundle $\pi : E \to X$, choose a finite covering of X by open sets $U_1, \ldots, U_m$ with local trivializations $(\pi, \rho_\alpha) : \pi^{-1}(U_\alpha) \to U_\alpha \times \mathbb{C}^n$ for $\alpha = 1, ..., n$. Let $\{\phi_\alpha\}_{\alpha=1}^m$ be a partition of unity with respect to $\{U_\alpha\}_{\alpha=1}^m$, and define a map

$$\widehat{\psi} : E \longrightarrow \mathbb{C}^n \oplus \cdots \oplus \mathbb{C}^n \qquad (m\text{-times})$$

by

$$\widehat{\psi}(e) = (\phi_1(\pi e)\rho_1(e), ..., \phi_m(\pi e)\rho_m(e)).$$

This map is linear and injective on the fibres of E, and we define $\psi(x) = \widehat{\psi}(E_x) \in G_n(\mathbb{C}^{nm})$ for $x \in X$. Setting $\widetilde{\psi}(e) = (\psi(\pi e), \widehat{\psi}(e)) \in E_n$ gives a commutative diagram

$$\begin{array}{ccc} E & \xrightarrow{\ \widetilde{\psi}\ } & E_n \\ {\scriptstyle\pi}\downarrow & & \downarrow{\scriptstyle\pi} \\ X & \xrightarrow{\ \psi\ } & G_n(\mathbb{C}^{nm}) \end{array}$$

which shows that $E = \psi^* E_n$. □

Theorem 7.4 is basic – it leads directly to the theory of characteristic classes. We adopt the standard notation

$$\mathrm{BU}_n \equiv G_n(\mathbb{C}^\infty)$$

and rewrite (7.2) as

$$\mathrm{Vect}_n(X) \;\cong\; [X, \mathrm{BU}_n] \tag{7.4}$$

DEFINITION 7.6. Let $u \in H^*(\mathrm{BU}_n;\, \Lambda)$ where Λ is a fixed coefficient ring. Then to any complex vector bundle $\pi : E \to X$ we associate the u-**characteristic class**

$$u(E) \;=\; \psi_E^*(u) \in H^*(X;\, \Lambda)$$

where $\psi_E : X \to \mathrm{BU}_n$ is the "classifying map" for E given by (7.4), and

$$\psi_E^* : H^*(\mathrm{BU}_n;\, \Lambda) \;\longrightarrow\; H^*(X;\, \Lambda) \tag{7.5}$$

is the induced map in cohomology (which depends only on the homotopy class of the mapping ψ_E).

LEMMA 7.7. (**Natural Transformation Law**) *Let $f : X' \to X$ be a continuous map of compact Hausdorff spaces and $E \in Vect_n(X)$. Then for each $u \in H^*(BU_n;\, \Lambda)$, one has*

$$u(f^*E) \;=\; f^*u(E).$$

PROOF. If $\psi_E : X \to \mathrm{BU}_n$ classifies E, then $\psi_E \circ f$ classifies f^*E since $(\psi_E \circ f)^*(E_n) = f^*(\psi_E^* E_n) = f^*E$. Hence $u(f^*E) = (\psi_E \circ f)^*u = f^*(\psi_E^* u) = f^*u(E)$. □

The **characteristic map** (7.5) is a **ring homomorphism**, and we have the following.

THEOREM 7.8. *There is an isomorphism of rings*

$$H^*(BU_n;\, \mathbb{Z}) \;\cong\; \mathbb{Z}[c_1, \ldots, c_n]$$

where $\mathbb{Z}[c_1, \ldots, c_n]$ denotes the polynomial ring in classes $c_k \in H^{2k}(BU_n;\, \mathbb{Z})$ for $k = 1, .., n$.

The class c_k is called the **universal kth Chern class**, and $c_k(E)$ is called the **kth Chern class** of the vector bundle E.

EXAMPLE 7.9. Note that $\mathrm{BU}_1 = \mathbb{P}^\infty(\mathbb{C}) = \lim_{N\to\infty} \mathbb{P}^N(\mathbb{C})$, the infinite complex projective space, and $H^*(\mathbb{P}^\infty(\mathbb{C});\, \mathbb{Z}) \cong \mathbb{Z}[c_1]$.

THEOREM 7.10. (**Whitney Duality**) *For $E \in Vect_n(X)$ define the total Chern class of E to be*

$$c(E) \;\equiv\; 1 + c_1(E) + c_2(E) + \cdots + c_n(E).$$

Then for $E \in Vect_n(X)$ and $F \in Vect_m(X)$ one has

$$c(E \oplus F) \;=\; c(E) \cdot c(F).$$

Consider now the "splitting map" $s : G_1(\mathbb{C}^N)\times\cdots\times G_1(\mathbb{C}^N) \longrightarrow G_n(\mathbb{C}^N\oplus\cdots\oplus\mathbb{C}^N)$ given by $s(\ell_1, \ldots, \ell_n) = \ell_1 \oplus \cdots \oplus \ell_n$. Taking the limit as $N \to \infty$ gives a mapping

$$s : \mathrm{BU}_1 \times \cdots \times \mathrm{BU}_1 \;\longrightarrow\; \mathrm{BU}_n$$

THEOREM 7.11. (**The Splitting Principle**) *The induced ring homomorphism*

$$\begin{array}{ccccc} H^*(BU_n;\mathbb{Z}) & \xrightarrow{s^*} & H^*(BU_1;\mathbb{Z})\otimes\cdots\otimes H^*(BU_1;\mathbb{Z}) & & \\ \downarrow & & \downarrow & & \\ \mathbb{Z}[c_1,\ldots,c_n] & \xrightarrow{s^*} & \mathbb{Z}[x_1]\otimes\cdots\otimes\mathbb{Z}[x_n] & \xrightarrow{=} & \mathbb{Z}[x_1,\ldots,x_n] \end{array}$$

is injective, and its image is the subring of symmetric polynomials in $x_1, ..., x_n$. Specifically

$$s^*(c_k) \;=\; \sigma_k(x_1,...,x_n) \;\equiv\; \sum_{i_1<\cdots<i_k} x_{i_1}\ldots x_{i_k}$$

is the kth elementary symmmetric polynomial in $x_1, ..., x_n$. That is, for the total Chern class $c = 1 + c_1 + \cdots + c_n$ one has

$$s^*(c) \;=\; \prod_{k=1}^{n}(1 + x_k) \tag{7.6}$$

See [**Bot69, BH60**] for a proof. Theorem 7.11 clearly implies Theorem 7.8. It also implies Theorem C. To see this consider the map

$$\sigma : G_n(\mathbb{C}^N) \times G_m(\mathbb{C}^M) \;\longrightarrow\; G_{n+m}(\mathbb{C}^{N+M})$$

given by $\sigma(P, Q) = P \oplus Q$. We want to show that

$$\sigma^*(1 + c_1 + \cdots + c_{n+m}) = (1 + c_1' + \cdots + c_n')(1 + c_1'' + \cdots + c_m'').$$

However after pulling back by the splitting maps this becomes

$$\sigma^* \prod_{k=1}^{n+m}(1 + x_k) \;=\; \prod_{k=1}^{n}(1 + x_k') \prod_{k=1}^{m}(1 + x_k'')$$

where $\sigma^* x_k = x_k'$ for $k = 1, ..., n$ and $\sigma^* x_{n+k} = x_k''$ for $k = 1, ..., m$.

An important consequence of Theorem 7.11 is that *formally* we can consider any vector bundle E as a direct sum of line bundles

$$E \;\cong\; \ell_1 \oplus \cdots \oplus \ell_n$$

and write

$$c(E) \;=\; 1 + c_1(E) + \cdots + c_n(E) \;=\; \prod_{k+1}^{n} c(\ell_k) \;=\; \prod_{k+1}^{n}(1 + x_k)$$

where

$$c_k(E) \;=\; \sigma_k(x_1, ..., x_n)$$

We then apply the following basic fact (See [**Lan65**] for example).

PROPOSITION 7.12. *Every symmetric polynomial in $x_1, ..., x_n$ can be written uniquely as a polynomial in the elementary symmetric functions $\sigma_1, ..., \sigma_n$. In other words if one considers the symmetric group Σ_n acting on $\mathbb{Q}[x_1, ..., x_n]$ by permutation of the x_i's, then there is an isomorphism*

$$\mathbb{Q}[x_1, ..., x_n]^{\Sigma_n} \;\cong\; \mathbb{Q}[\sigma_1, ..., \sigma_n]$$

where $\mathbb{Q}[x_1, ..., x_n]^{\Sigma_n}$ denotes the subring of polynomials fixed by Σ_n.

From this we shall construct characteristic classes with nice properties.

EXAMPLE 7.13. **The Chern Character**. Formally splitting the the bundle E as above we define the *Chern character* of E by

$$\begin{aligned} ch(E) &\equiv e^{x_1} + \cdots + e^{x_n} \\ &= n + \sum_{k=1}^{n} x_k + \frac{1}{2}\sum_{k=1}^{n} x_k^2 + \frac{1}{3!}\sum_{k=1}^{n} x_k^3 + \ldots \\ &= n + c_1 + \frac{1}{2}(c_1^2 - 2c_2) + \frac{1}{3!}(c_1^3 - 3c_1c_2 + 3c_3) + \ldots \\ &\equiv n + ch_1(E) + ch_2(E) + ch_3(E) + \ldots \end{aligned}$$

where $ch_k(E) \in H^{2k}(X; \mathbb{Q})$.

PROPOSITION 7.14. *For complex vector bundles E and F on X one has that*

$$\begin{aligned} ch(E \oplus F) &= ch(E) + ch(F) \\ ch(E \otimes F) &= ch(E) \cdot ch(F) \end{aligned}$$

PROOF. The first assertion is obvious from the definition. For the second assertion note that $(\ell_1 \oplus \cdots \oplus \ell_n) \otimes (\ell_1' \oplus \cdots \oplus \ell_m') = \sum_{i,j} \ell_i \otimes \ell_j'$. Since $c_1(\ell_i \otimes \ell_j') = c_1(\ell_i) + c_1(\ell_j') = x_i + x_j'$ we have that

$$(e^{x_1} + \cdots + e^{x_n})(e^{x_1'} + \cdots + e^{x_m'}) = \sum_{i=1}^{n}\sum_{i=j}^{m} e^{x_i + x_j'}$$

as desired. □

Note that

$$\mathcal{V}(X) \equiv \coprod_{n \geq 0} \mathrm{Vect}_n(X) \tag{7.7}$$

(where $\mathrm{Vect}_0(X) = \{0\}$) is a commutative semi-ring under the operations of direct sum and tensor product. Proposition 7.14 asserts that

$$ch : \mathcal{V}(X) \longrightarrow H^{2*}(X; \mathbb{Q}) \tag{7.8}$$

is a (semi-) ring homomorphism.

EXAMPLE 7.15. **Mulitplicative Sequences**. Fix a formal power series

$$m(t) = 1 + a_1 t + a_2 t^2 + a_3 t^3 + \ldots \in \mathbb{Q}[[t]]$$

with constant term 1. Then for any $E \in \mathrm{Vect}_n(X)$ we define

$$\mathbb{M}(E) \equiv m(x_1) \ldots m(x_n) \tag{7.9}$$

$$= 1 + M_1(c_1) + M_2(c_1, c_2) + M_3(c_1, c_2, c_3) + \ldots \tag{7.10}$$

where $M_k(c_1, ..., c_k)$ is a polynomial of pure cohomological degree $2k$ in the Chern classes. The $M_k(c_1, ..., c_k)$ are universal expressions completely determined by $m(t)$ via Proposition 7.12. The series (7.10) is called a *multiplicative sequence of Chern classes.* It has the very useful property that

$$\mathbb{M}(E \oplus F) = \mathbb{M}(E) \cdot \mathbb{M}(F)$$

Note that if $m(t) = 1 + t$, then $\mathbb{M}(E) = c(E)$.

EXAMPLE 7.16. **The Todd Series.** Let

$$m(t) = \frac{t}{1-e^{-t}} = 1 + \frac{1}{2}t + \sum_{k=1}^{\infty}(-1)^{k-1}\frac{B_k}{(2k)!}t^{2k}$$

where B_k is the kth Bernoulli number. The resulting multiplicative sequence

$$\mathbb{M}(E) = \mathbb{T}d(E) = 1 + Td_1(E) + Td_2(E) + \dots$$

is called the **Todd series**, and $Td_k(E)$ is called the kth **Todd polynomial**. This series plays a central role in the generalized Riemann-Roch Theorem.

DEFINITION 7.17. We now fix a compact Hausdorff space X and consider the semi-ring $\mathcal{V}(X)$ given in (7.7). We extend this semi-ring to a ring by taking its **additive group completion**, which is defined as the quotient

$$K(X) \equiv F(\mathcal{V}(X))/R$$

where $F(\mathcal{V}(X))$ is the free abelian group on the elements of $\mathcal{V}(X)$ and R is the subgroup generated by elements of the form $E + F - (E \oplus F)$ for $E, F \in \mathcal{V}(X)$. The tensor product $\otimes$ extends naturally to $K(X)$ making it a ring, called the **K-ring** of X.

The additive group completion is completely characterized as a group $K(X)$ extending $\mathcal{V}(X)$ and having the universal property that any semi-group homomorphism $\varphi : \mathcal{V}(X) \to A$ into an abelian group A extends to a unique group homomorphism $\widetilde{\varphi} : K(X) \to A$.

There is an alternative definition of $K(X)$ given by considering the diagonal embedding $\Delta : \mathcal{V}(X) \to \mathcal{V}(X) \times \mathcal{V}(X)$ and taking $K(X)$ to be the set of "cosets" of $\Delta(\mathcal{V}(X))$ in $\mathcal{V}(X) \times \mathcal{V}(X)$, that is, the set of orbits of the semi-group action of $\Delta(\mathcal{V}(X))$ on $\mathcal{V}(X) \times \mathcal{V}(X)$. This again forms a semi-group under the induced addition, and we have the embedding $\mathcal{V}(X) \hookrightarrow \mathcal{V}(X) \times \mathcal{V}(X)/\Delta\mathcal{V}(X)$ given by $E \mapsto [(E, 0)]$. Note that each element $[(E, F)]$ has an additive inverse given by $[(F, E)]$. Therefore, $K(X) = \mathcal{V}(X) \times \mathcal{V}(X)/\Delta\mathcal{V}(X)$ is a group. The reader can check that these two group completions of $\mathcal{V}(X)$ are uniquely isomorphic.

From this second viewpoint we see that an element of $K(X)$ is an equivalence classe of a pair of bundles (E, F). We think of this class as "$E - F$". Note that by definition $[(E, F)] \cong [(E \oplus G, F \oplus G)]$ for any G. Furthermore, for every $F \in \mathcal{V}(X)$, there exists $G \in \mathcal{V}(X)$ with $F \oplus G \cong \mathbb{C}^N$, the trivial N-plane bundle, for some N. To see this, let $\psi_F : X \to G_n(\mathbb{C}^N)$ be the classifying map and set $G = \psi_F^*(E_n^\perp)$. It follows that every element in $K(X)$ can be written in the form $[(E, \mathbb{C}^N)]$. That is elements of $K(X)$ can be thought of as **stable equivalence classes** of vector bundles, where E and F are stably equivalent if there are trivial bundles $\mathbb{C}^N$ and $\mathbb{C}^M$ such that $E \oplus \mathbb{C}^M \cong F \oplus \mathbb{C}^N$.

Note by Theorem 7.10 that for such a pair the total Chern class satisfies $c(E) = c(F)$, since $c(\mathbb{C}^M) = 1$. Hence, the Chern classes $c_k(\bullet)$ are well defined on elements of K-theory.

Observe now that the induced bundle construction (see 7.3) associates to a continuous map $f : X \to X'$, a ring homomorphism $f^* : K(X') \to K(X)$. Thus $K(\bullet)$ gives a functor from the category of compact Haudorff spaces to the category of rings. The Chern character gives a natural transformation of functors from $K(\bullet)$ to $H^{2*}(\bullet; \mathbb{Q})$.

If X is provided with a distinguished base point $x_0 \in X$, then the **reduced K-ring** of X is defined to be

$$\widetilde{K}(X) \equiv \ker\{K(X) \to K(x_0)\}$$

Then for each integer $n \geq 0$ we define

$$\widetilde{K}^{-n}(X) \equiv \widetilde{K}(\Sigma^n X)$$

where $\Sigma^n X = S^n \wedge X = S^n \times X/(\{t_0\} \times X \cup S^n \times \{x_0\})$ is the n-fold reduced suspension (and t_0 is a fixed base point in S^n). For spaces without base point we set

$$K^{-n}(X) \equiv \widetilde{K}^{-n}(X_+)$$

where $X_+ = X \coprod \{x_0\}$ is the space obtained from X by adding a disjoint base point.

Given a compact pair $Y \subset X$ we define

$$K(X,Y) \equiv \widetilde{K}(X/Y)$$

where $Y/Y \in X/Y$ is the distinguished base point, and we set $K^n(X,Y) = \widetilde{K}^{-n}(X/Y)$ for each n. With these definitions $K(\bullet)$ expands to a full generalized cohomology theory, that is, it satisfies all the Eilenberg-Steenrod axioms of a cohomology theory except the "Dimension Axiom" (cf. [**ES52**]). Of fundamental importance here is the following (See [**Ati67**].)

THEOREM 7.18. (**The Bott Periodicity Theorem**) *For any compact Hausdorff space X and any $n \geq 0$, there is a natural isomormphism*

$$K^{-n}(X) \cong K^{-n-2}(X) \tag{7.11}$$

In particular,

$$\begin{aligned} K^{-2n}(\mathrm{pt}) &= \widetilde{K}(S^{2n}) \cong \mathbb{Z} \qquad \text{and} \\ K^{-2n-1}(\mathrm{pt}) &= \widetilde{K}(S^{2n+1}) \cong 0 \end{aligned}$$

for all $n \geq 0$.

The isomorphism (7.11) allows us to extend the definition of $K^n(X)$ to all integers. The second statement of Theorem 7.18 is related to the classical Bott Periodicity Theorem as follows. Theorem 7.4 leads to the statement that $\widetilde{K}(S^n) \cong \pi_n(\mathrm{BU}_\infty) = \lim_{k\to\infty} \pi_n(\mathrm{BU}_k)$, and the exact sequence of homotopy groups for a fibration (involving Stiefel manifolds, cf. [**Ste51**]) shows that there is an isomorphism $\pi_n(\mathrm{BU}_\infty) \cong \pi_{n-1}(\mathrm{U}_\infty)$.

If X is a locally compact Hausdorff space we can define the K-theory of X with compact supports to be the ring

$$K_{\mathrm{cpt}}(X) \equiv \widetilde{K}(X^*)$$

where $X^* = X \cup \{\infty\}$ is the one point compactification of X and $\{\infty\}$ is the distinguished base point. In this language the Bott Periodicity Theorem takes the form:

THEOREM 7.19. *For any locally compact Hausdorff space X, there is a natural isomorphism*

$$K_{cpt}(X) \cong K_{cpt}(\mathbb{C} \times X) \tag{7.12}$$

In particular,

$$K_{cpt}(\mathbb{C}^n) = \mathbb{Z} \qquad \text{and} \qquad K_{cpt}(\mathbb{C}^n \times \mathbb{R}) = 0 \tag{7.13}$$

for all $n \geq 0$.

REMARK 7.20. **The Atyiah-Bott-Shapiro Construction**. Elements of $K_{\text{cpt}}(X)$ can be represented by pairs of vector bundles together with a fixed isomorphism at infinity. That is the group $K_{\text{cpt}}(X)$ can be constructed as equivalence classes of triples (E, F, σ) where $E, F \in \text{Vect}_n(X)$ for some n and $\sigma : E \to F$ is an isomorphism outside some compact subset of X. (See [**ABS64**] for details.) Let

$$[E, F; \sigma] \in K_{\text{cpt}}(X)$$

denote the class determined by (E, F, σ).

Let $C\ell_n$ denote the Clifford algebra of $\mathbb{R}^n$ with the standard quadratic form $\|x\|^2$. Then $C\ell_{2n}$ has only one irreducible module $\$$. It has a decomposition $\$ = \$^+ \oplus \$^-$ where $\$^+$ and $\$^-$ are invariant under the even subalgebra ${C\ell_{2n}}^0$ and Clifford multiplication $\mu_v : \$ \to \$$, $\mu_v(s) = v \cdot s$ for $v \in \mathbb{R}^n$ interchanges these spaces. That is,

$$\mu_v : \$^+ \to \$^-,$$

and for $v \neq 0$, μ_v is an ismomorphism. It is a fundamental fact that

$$[\$^+, \$^-; \mu] \in K_{\text{cpt}}(\mathbb{R}^{2n}) \cong \mathbb{Z}$$

is a generator of this group. In fact it is shown in [**ABS64**] that this construction leads to an isomorphism between a certain quotient of the ring of $\mathbb{Z}_2$-graded $C\ell_n$-modules, $n = 0, 1, 2, ...$, and $K^*(\text{pt})$.

REMARK 7.21. **The Real Case**. To this point we have considered only complex vector bundles. There is however a completely parallel story for real vector bundles, which are locally trivial families of real vector spaces over a space X. Let $G_n(\mathbb{R}^N)$ denote the Grassmannian of real linear subspaces of dimension n in $\mathbb{R}^N$, and set

$$BO_n \equiv \lim_{N\to\infty} G_n(\mathbb{R}^N).$$

The proof of Theorem 7.4 immediatly adapts to establish the analogue:

$$\text{Vect}_n^{\mathbb{R}}(X) \cong [X, BO_n]$$

where $\text{Vect}_n^{\mathbb{R}}(X)$ denotes the set of isomorphism classes of real n-dimensional vector bundles over the compact Hausdorff space X. Therefore each cohomology class of BO_n determines a characteristic class for real vector bundles. It turns out that there is a ring isomorphism

$$H^*(BO_n; \mathbb{Z}_2) \cong \mathbb{Z}_2[w_1, \ldots, w_n]$$

where $w_k \in H^k(BO_n; \mathbb{Z}_2)$ is called the kth universal Stiefel-Whitney class. The total Stiefel-Whitney class, defined by

$$w = 1 + w_1 + \cdots + w_n$$

satisfies Whitney duality, i.e.,

$$w(E \oplus F) = w(E) \cdot w(F)$$

for real vector bundles E and F over a space X.

There is also a ring isomorphism

$$H^*(BO_n; \mathbb{Q}) \cong \mathbb{Q}[p_1, \ldots, p_{[n/2]}]$$

where $p_k \in H^{4k}(BO_n; \mathbb{Q})$ is called the kth universal Pontrjagin class. The total Pontrjagin class

$$p = 1 + p_1 + \cdots + p_{[n/2]}$$

also satisfies Whitney duality $p(E \oplus F) = p(E) \cdot p(F)$.

The kth Pontrjagin class has the property that for a real vector bundle E,

$$p_k(E) = (-1)^k c_{2k}(E \otimes_{\mathbb{R}} \mathbb{C})$$

Applying the Splitting Principle to $E \otimes_{\mathbb{R}} \mathbb{C} = \ell_1 \oplus \bar{\ell}_1 \oplus \ell_2 \oplus \bar{\ell}_2 \oplus \ell_3 \oplus \bar{\ell}_3 \oplus \ldots$ we can write

$$p_k \;=\; \sigma_k(x_1^2, \ldots, x_{[n/2]}^2)$$

where x_k formally corresponds to $c_1(\ell_k)$. This enables us to define multiplicative sequences of Pontrjagin classes. Beginning with a formal power series $m(t) = 1 + a_1 t + a_2 t^2 + \ldots$ one defines the associated multiplicative sequence of Pontrjagin classes

$$\mathbb{M}(E) \;=\; m(x_1^2) \ldots m(x_{[n/2]}^2).$$

as in the complex case above.

Example 7.22. **The $\widehat{\mathbb{A}}$-series.** The power series

$$\widehat{a}(t) \;=\; \frac{\sqrt{t}/2}{\sinh(\sqrt{t}/2)} \;=\; 1 - \frac{1}{24}t + \frac{7}{2^7 \cdot 3^2 \cdot 5}t^2 + \ldots$$

determines a multiplicative series

$$\widehat{\mathbb{A}} \;=\; 1 + \widehat{A}_1(p_1) + \widehat{A}_2(p_1, p_2) + \widehat{A}_3(p_1, p_2, p_3) + \ldots$$

where $\widehat{A}_k \in H^{4k}(X; \mathbb{Q})$. This sequence plays a fundamental role in the Atiyah-Singer Index Theorem. It is related to the Todd series above:

$$\mathbb{T}d(E \otimes_{\mathbb{R}} \mathbb{C}) \;=\; \widehat{\mathbb{A}}^2(E)$$

as in the complex case above.

Example 7.23. **The $\mathbb{L}$-series.** The power series

$$\ell(t) \;=\; \frac{\sqrt{t}}{\tanh(\sqrt{t})} \;=\; 1 + \frac{1}{3}t - \frac{1}{45}t^2 + \ldots$$

determines a multiplicative series $\mathbb{L} \;=\; 1 + L_1 + L_2 + \ldots$ which is the key to Hirzebruch's formula which expresses the signature of a manifold in terms of its Pontrjagin classes [**Hir66**].

As in the complex case the isomorphism classes of real vector bundles on X form a commutative semi-ring

$$\mathcal{V}^{\mathbb{R}}(X) \;\equiv\; \coprod_{n \geq 0} \mathrm{Vect}_n^{\mathbb{R}}(X)$$

whose additive group completion is denoted $KO(X)$. All the constructions and general statments made about the rings $K(X)$ apply to $KO(X)$ For example, for spaces with a distinguished base point we define the groups $\widetilde{KO}^{-n}(X) = \widetilde{KO}(\Sigma^n X)$, and when X has no such base point we set $KO^{-n}(X) = \widetilde{KO}^{-n}(X_+)$. The deep result in the subject is the following.

Theorem 7.24. (**The Bott Periodicity Theorem for KO-theory**) *For any compact Hausdorff space X and any $n \geq 0$, there is a natural isomorphism*

$$KO^{-n}(X) \;\cong\; KO^{-n-8}(X)$$

Furthermore,

n	*0*	*1*	*2*	*3*	*4*	*5*	*6*	*7*
$KO^{-n}(\mathrm{pt})$	$\mathbb{Z}$	$\mathbb{Z}_2$	$\mathbb{Z}_2$	*0*	$\mathbb{Z}$	*0*	*0*	*0*

This periodicity result can be restated for locally compact spaces X in the form

$$KO_{\text{cpt}}(X) \cong KO_{\text{cpt}}(\mathbb{R}^8 \times X)$$

There is an Atiyah-Bott-Shapiro construction of the generators of the groups $KO^{-n}(\text{pt})$ using $\mathbb{Z}_2$-graded $C\ell_n$-modules [**ABS64**].

The direct sum $KO^{-n}(\text{pt}) = \bigoplus_{n\geq 0} KO^{-n}(\text{pt})$ actually carries the structure of a graded ring (cf. [**Ati67**]). As such it is generated by the elements

$$1 \in KO^0(\text{pt}), \qquad \eta \in KO^{-1}(\text{pt}), \qquad y \in KO^{-4}(\text{pt}), \qquad x \in KO^{-8}(\text{pt})$$

with the relations

$$2\eta = 0, \qquad \eta y = 0, \qquad \eta^3 = 0, \qquad y^2 = 4x$$

7.2. The Atiyah-Singer Index Theorems

In this chapter we present several theorems concerning the index of elliptic differential operators, which were established by Atiyah and Singer in the 1960's [**AS68c**]-[**AS69**]. These theorems have wide and important applications in mathematics and physics.

We begin with the notion of a differential operator on a manifold X. Let $\pi : E \to X$ be a smooth (C^∞) vector bundle. By a **smooth section** of E we mean a C^∞-map $\sigma : X \to E$ such that $\pi \circ \sigma = Id_X$. We denote by $\Gamma(E)$ the vector space of all smooth sections of E.

Note that in a smooth local trivialization $\pi^{-1}U \cong U \times \mathbb{C}^n$, a section can be written as $\sigma(x) = (x, \sigma_0(x))$. Mapping $\sigma \to \sigma_0$ gives an isomorphism $\Gamma(\pi^{-1}U) \cong C^\infty(U, \mathbb{C}^n)$.

DEFINITION 7.25. Let $\pi_E : E \to X$, $\pi_F : F \to X$ be smooth vector bundles over a manifold X. A **differential operator of order** k from E to F is a linear map

$$\mathcal{D} : \Gamma(E) \longrightarrow \Gamma(F)$$

which in all smooth local trivializations

$$\pi_E^{-1}U \cong U \times \mathbb{C}^n \qquad \text{and} \qquad \pi_F^{-1}U \cong U \times \mathbb{C}^m$$

can be written in the form

$$\mathcal{D} = \sum_{|\alpha|\leq k} A^\alpha(x) \frac{\partial^{|\alpha|}}{\partial x^\alpha} \tag{7.14}$$

where each $A^\alpha(x)$ is a smooth $m \times n$ matrix-valued function, and

$$\frac{\partial^{|\alpha|}}{\partial x^\alpha} = \frac{\partial^{\alpha_1}}{\partial x_1^{\alpha_1}} \cdots \frac{\partial^{\alpha_\nu}}{\partial x_\nu^{\alpha_\nu}}$$

where $(x_1, \ldots, x_\nu)$ are local coordinates on U and $|\alpha| = \alpha_1 + \cdots + \alpha_\nu$.

LEMMA 7.26. *The sum*

$$\sigma(\mathcal{D}) \equiv i^k \sum_{|\alpha|=k} A^\alpha(x) \frac{\partial^{|\alpha|}}{\partial x^\alpha}$$

given by the highest order terms in (7.14) represents a well-defined section of the bundle $\text{Sym}^k(TX) \otimes \text{Hom}(E, F)$. *It is called the* **principal symbol** *of* $\mathcal{D}$.

PROOF. Under a change of trivializations $g_E : U \to GL_n(\mathbb{C})$ and $g_F : U \to GL_m(\mathbb{C})$ we have

$$g_F \circ \left(\sum_{|\alpha|\leq k} A^\alpha(x) \frac{\partial^{|\alpha|}}{\partial x^\alpha} \right) \circ g_E^{-1} = \sum_{|\alpha|=k} \left(g_F \circ A^\alpha(x) \circ g_E^{-1}\right) \frac{\partial^{|\alpha|}}{\partial x^\alpha} + \text{lower order terms.}$$

Therefore each A^α with $|\alpha| = k$ transforms as a section of $\mathrm{Hom}(E, F)$.

Now under a change of coordinates $y = y(x)$ on X we have

$$\frac{\partial}{\partial x_j} = \sum_{i=1}^{\nu} \frac{\partial y_i}{\partial x_j} \frac{\partial}{\partial y_i}$$

and so

$$\frac{\partial^{|\alpha|}}{\partial x^\alpha} = \sum_{|\beta|=|\alpha|} \left(\frac{\partial y}{\partial x}\right)_\alpha^\beta \frac{\partial^{|\beta|}}{\partial y^\beta} + \text{ lower order derivatives in } y$$

It follows that

$$\mathcal{D} = \sum_{|\alpha|\leq k} A^\alpha \frac{\partial^{|\alpha|}}{\partial x^\alpha} = \sum_{|\alpha|=|\beta|=k} A^\alpha \left(\frac{\partial y}{\partial x}\right)_\alpha^\beta \frac{\partial^{|\beta|}}{\partial y^\beta} + \text{ lower order...} = \sum_{|\beta|\leq k} \widetilde{A}^\beta \frac{\partial^{|\beta|}}{\partial y^\beta}.$$

Therefore for $|\beta| = k$ we have

$$\widetilde{A}^\beta = \sum_{|\alpha|=k} A^\alpha \left(\frac{\partial y}{\partial x}\right)_\alpha^\beta,$$

i.e., the principal symbol transforms as a section of $\mathrm{Sym}^k(TX)$. □

At each point $x \in X$, $\mathrm{Sym}^k(T_x X)$ is simply the set of homogeneous polynomials of degree k on the cotangent space $T_x^* X$. Given a cotangent vector $\xi = \sum \xi_i dx_i \in T_x^* X$, we have

$$\sigma_{\xi,x}(\mathcal{D}) = i^k \sum_{|\alpha|=k} A^\alpha(x)\, \xi^\alpha.$$

DEFINITION 7.27. A differential operator $\mathcal{D}$ is **elliptic** if $\sigma_{\xi,x}(\mathcal{D}) : E_x \longrightarrow F_x$ is an isomorphism for all $\xi \neq 0$ in $T_x X$ and all $x \in X$.

EXAMPLE 7.28. The euclidean Laplace operator $\Delta = \sum \frac{\partial^2}{\partial x_i^2} : C^\infty(\mathbb{R}^n) \to C^\infty(\mathbb{R}^n)$ has symbol $\sigma_\xi(\Delta) = i^2 \sum \xi_i^2 = -\|\xi\|^2$.

EXAMPLE 7.29. More generally the Laplace-Beltrami operator $\Delta : C^\infty(X) \to C^\infty(X)$ on a riemannian manifold X is expressed in local coordinates by

$$\Delta f = \frac{1}{\sqrt{g}} \sum_{i,j} \frac{\partial}{\partial x_i} \left(\sqrt{g} g^{ij} \frac{\partial f}{\partial x_j}\right) = \sum_{i,j} g^{ij} \frac{\partial^2 f}{\partial x_i \partial x_j} + \text{lower order derivatives}$$

where $ds^2 = \sum_{ij} g_{ij} dx_i dx_j$ is the riemannian metric, $g = det((g_{ij}))$ and $((g^{ij})) = ((g_{ij}))^{-1}$ defines the metric on the cotangent bundle. Thus we see that

$$\sigma_\xi(\Delta) = -\sum_{i,j} g^{ij} \xi_i \xi_j = -\|\xi\|^2$$

EXAMPLE 7.30. For any Dirac operator

$$D = \sum_{i=1}^{\nu} e_i \cdot \nabla_{e_i}.$$

the principal symbol is

$$\sigma_\xi(D) = i\sum_{i=1}^{\nu} \xi_i e_i \cdot = i\xi\cdot = \text{Clifford multiplication by } i\xi$$

Consequently, for the Dirac Laplacian we have

$$\sigma_\xi(D) = i\xi \cdot i\xi\cdot = \|\xi\|^2.$$

EXERCISE 7.31. Show that

$$\begin{aligned} \sigma_\xi(\mathcal{D} \circ \mathcal{D}') &= \sigma_\xi(\mathcal{D}) \circ \sigma_\xi(\mathcal{D}') && \text{and that} \\ \sigma_\xi(t\mathcal{D} + t'\mathcal{D}') &= t\sigma_\xi(\mathcal{D}) + t'\sigma_\xi(\mathcal{D}') && \text{if order}(\mathcal{D}) = \text{order}(\mathcal{D}') \quad \text{and} \quad t, t' \in \mathbb{R} \end{aligned}$$

We now observe that if $\pi : T^*X \to X$ is the cotangent bundle of X and $\mathcal{D} : \Gamma(E) \to \Gamma(F)$ is an elliptic differential operator on X, then

$$\sigma(\mathcal{D}) : \pi^*E \longrightarrow \pi^*F$$

is a bundle morphism over T^*X which is an isomorphism outside the 0-section. Therefore by the Atiyah-Bott-Shapiro construction 7.20 we see that the principal symbol of $\mathcal{D}$ defines a **symbol class**

$$\sigma \equiv [\pi^*E, \pi^*F; \sigma(\mathcal{D})] \in K_{\text{cpt}}(T^*X) \tag{7.15}$$

in the K-theory of T^*X with compact supports.

For $\mathcal{D} : \Gamma(E) \to \Gamma(F)$ as above we define

$$\ker\mathcal{D} = \{\varphi \in \Gamma(E) : \mathcal{D}\varphi = 0\} \qquad \text{and} \qquad \text{coker}\mathcal{D} = \Gamma(F)/\mathcal{D}(\Gamma(E)).$$

THEOREM 7.32. *Let $\mathcal{D} : \Gamma(E) \to \Gamma(F)$ be an elliptic operator on a compact manifold X. Then*

(i) $\dim(\ker\mathcal{D}) < \infty$ *and* $\dim(\text{coker}\,\mathcal{D}) < \infty$

(ii) $\text{index}(\mathcal{D}) \equiv \dim(\ker\mathcal{D})$ - $\dim(\text{coker}\,\mathcal{D})$ *is a "homotopy invariant", i.e., if $\mathcal{D}_t$ is a continuous family of elliptic operators, then* $\text{index}(\mathcal{D}_t) \equiv$ *constant.*

PROOF. For a proof the reader is referred to [**Nir70, Gil84, LM89**] or [**BGV91**]. □

COROLLARY 7.33. *The index of an elliptic operator $\mathcal{D} : \Gamma(E) \to \Gamma(F)$ on a compact manifold depends only on its principal symbol $\sigma(\mathcal{D})$.*

PROOF. Suppose $\mathcal{D}' : \Gamma(E) \to \Gamma(F)$ is another differential operator with $\sigma(\mathcal{D}') = \sigma(\mathcal{D})$. Then $\mathcal{D}_t = t\mathcal{D} + (1-t)\mathcal{D}'$ is a continuous family of elliptic operators whose index must be constant by Theorem 7.32 (ii). □

THEOREM 7.34. (**The Atiyah-Singer Index Theorem**) *let $\mathcal{D} : \Gamma(E) \to \Gamma(F)$ be an elliptic operator on a compact manifold X of dimension n. Let $\sigma = [\pi^*E, \pi^*F; \sigma(\mathcal{D})] \in K_{\text{cpt}}(T^*X)$ be its symbol class. Then*

$$\boxed{\text{index}(\mathcal{D}) = (-1)^n \left\{ch\sigma \cdot \pi^*\widehat{\mathbb{A}}(X)^2\right\}[T^*X]}$$

The symbols in the formula mean the following. The Chern character $ch : K_{\text{cpt}}(T^*X) \longrightarrow H^{2*}_{\text{cpt}}(T^*X;\mathbb{Q})$ is of the form $ch = ch_1 + ch_2 + \dots$ where ch_k is of degree $2k$. The $\widehat{A}$-series is written as $\widehat{\mathbb{A}}(X) = 1 + \widehat{A}_1 + \widehat{A}_2 + \widehat{A}_3 \dots$ with $\widehat{A}_\ell \in H^{4\ell}(X;\mathbb{Q})$. Expanding the expressions above we find

$$\begin{aligned}\left\{ch\sigma \cdot \pi^*\widehat{\mathbb{A}}(X)^2\right\}[T^*X] &= \{(ch_1\sigma + ch_2\sigma + \dots)\cdot(1+\pi^*\widehat{A}_1(X) \\ &\quad +\pi^*\widehat{A}_2(X)+\dots)^2\}[T^*X] \\ &= \sum_{k+2\ell+2\ell'=n}\left\{ch_k\sigma\cdot\widehat{A}_\ell(X)\cdot\widehat{A}_{\ell'}(X)\right\}[T^*X]\end{aligned}$$

The pairing $\Phi[T^*X]$ means evaluation of the cohomology class Φ on the fundamental class homology class of the canonically oriented $2n$-dimensional manifold T^*X. For example, if we represent $ch\sigma$ and $\widehat{\mathbb{A}}(X)$ by differential forms, where each $ch_k\sigma$ has compact support on T^*X, the last term above could be written as

$$\sum_{k+2\ell+2\ell'=n}\int_{T^*X} ch_k\sigma\wedge\widehat{A}_\ell(X)\wedge\widehat{A}_{\ell'}(X)$$

The proof of this formula appears in [**AS68c, AS68a, AS68b**]. Other references include [**Pal65, Gil84, LM89, BGV91**].

The Atiyah-Singer formula above can be transferred down to the following formula on X, whenever X is oriented.

$$\boxed{\text{index}(\mathfrak{D}) = (-1)^{\frac{n(n+1)}{2}}\left\{\pi_!(ch\sigma)\cdot\widehat{\mathbb{A}}(X)^2\right\}[X]} \tag{7.16}$$

where $\pi_!$ denotes "integration over the fibre" of the projection $\pi : T^*X \to X$. In cohomological terms $\pi_!$ is the composition

$$H^{2k}(T^*X) \xrightarrow{\cong} H_{2(n-k)}(T^*X) \xrightarrow{\pi_*} H_{2(n-k)}(X) \xrightarrow{\cong} H^{2k-n}(T^*X)$$

where the first and third arrows are the Poincaré duality maps.

For the Dirac operator this formula simplifies beautifully.

PROPOSITION 7.35. *Suppose X is a compact spin manifold of dimension $2m$, and let $\not{D}^+ : \Gamma(\not{S}^+) \longrightarrow \Gamma(\not{S}^-)$ be the Atiyah-Singer Dirac operator on spinors. Then*

$$\pi_!\left\{ch\,\sigma(\not{D}^+)\right\} = (-1)^m\widehat{\mathbb{A}}(X)^{-1}$$

PROOF. (Outline) A general argument involving the Thom isomorphism shows that

$$\pi_!\,ch\,\sigma = \frac{ch(\not{S}^+ - \not{S}^-)}{\chi(TX)} = \frac{ch\not{S}^+ - ch\not{S}^-}{\chi(TX)}$$

where $\chi(TX)$ denotes the Euler class of X.

Case 1: $\dim(X) = 2$. When X is an oriented surface, TX has the structure of a complex line bundle, which we denote by ℓ. Since $c_1(\ell)$ is even, there exist square roots

$$\not{S}^+ = \ell^{-\frac{1}{2}} \quad \text{and} \quad \not{S}^- = \ell^{\frac{1}{2}}$$

which are the spinor bundles of X. If $x = \chi(\ell) = c_1(\ell)$,

$$\frac{ch\not{S}^+ - ch\not{S}^-}{\chi(TX)} = \frac{ch\ell^{-\frac{1}{2}} - ch\ell^{\frac{1}{2}}}{\chi(\ell)} = \frac{e^{-\frac{x}{2}} - e^{\frac{x}{2}}}{x} = -\frac{\sinh(\frac{x}{2})}{\frac{x}{2}} = -\widehat{\mathbb{A}}(X)^{-1}$$

Case 2: $\dim(X) = 2m$. Here we apply the splitting principle and write TX formally as a direct sum of complex line bundles

$$TX = \ell_1 \oplus \cdots \oplus \ell_m.$$

This gives a formal decomposition of the spinor bundle

$$\$(TX) = \$^+ \oplus \$^- = \$(\ell_1) \otimes \cdots \otimes \$(\ell_m) = (\ell_1^{-\frac{1}{2}} + \ell_1^{\frac{1}{2}}) \otimes \cdots \otimes (\ell_m^{-\frac{1}{2}} + \ell_m^{\frac{1}{2}}).$$

Using the oriented volume form one sees that $\$^+ = \bigoplus \ell_1^{\pm\frac{1}{2}} \otimes \cdots \otimes \ell_m^{\pm\frac{1}{2}}$ where the sum is taken over all products with an even number of positive exponents. The bundle $\$^-$ is the (remaining) sum over products with an odd number of positive exponents. Consequently we have that

$$\$^+ - \$^- = \prod_{k=1}^{m} \left(\ell_k^{-\frac{1}{2}} - \ell_k^{\frac{1}{2}}\right).$$

Setting $x_k = c_1(\ell_k)$, we find that

$$\frac{ch\$^+ - ch\$^-}{\chi(TX)} = \frac{\prod_{k=1}^m \left(e^{-\frac{x_k}{2}} - e^{\frac{x_k}{2}}\right)}{x_1 \ldots x_m} = (-1)^m \prod_{k=1}^{m} \frac{\sinh(\frac{x_k}{2})}{\frac{x_k}{2}} = (-1)^m \widehat{\mathbb{A}}(X)^{-1}.$$

□

Suppose now that

$$\not{D}_E^+ : \Gamma(\$^+ \otimes E) \longrightarrow \Gamma(\$^- \otimes E)$$

is the operator of Proposition 7.35 twisted by a coefficient bundle with connection E. Then

$$\sigma\left(\not{D}_E^+\right) = \pi^* E \cdot \sigma\left(\not{D}^+\right)$$

and therefore

$$ch\left(\sigma\left(\not{D}_E^+\right)\right) = \pi^* chE \cdot ch\left(\sigma\left(\not{D}^+\right)\right)$$

and

$$\pi_! \left\{ch\left(\sigma\left(\not{D}_E^+\right)\right)\right\} = chE \cdot \pi_! \left\{ch\left(\sigma\left(\not{D}^+\right)\right)\right\} = (-1)^m chE \cdot \widehat{\mathbb{A}}(X)^{-1}.$$

Combined with (7.16) this gives the following.

THEOREM 7.36. *Let X be a compact spin manifold of even dimension, and $\not{D}_E^+$ the twisted Atiyah-Singer Dirac operator as above. Then*

$$\boxed{\operatorname{index}(\not{D}_E^+) = \left\{chE \cdot \widehat{\mathbb{A}}(X)\right\}[X]}$$

The Index Theorem stated above has an important generalization with some striking applications [**AS71a, AS71b, AS69, Hit74**]. For this we consider the following objects. A **bundle of finite-dimensional graded $C\ell_k$-modules** is a real vector bundle $\pi : E \to X$ with a decomposition $E = E^0 \oplus E^1$ and an action $C\ell_k \subset \operatorname{Hom}(E, E)$ making each fibre into a $\mathbb{Z}_2$-graded $C\ell_k$-module. By a **graded $C\ell_k$-linear differential operator** we mean a differential operator $\mathcal{D} : \Gamma(E) \longrightarrow \Gamma(E)$ such that $\mathcal{D}(\varphi \cdot \sigma) = \varphi \cdot \mathcal{D}(\sigma)$ for all $\varphi \in C\ell_k$ and all $\sigma \in \Gamma(E)$. This operator is called **graded** if with respect to the decomposition $E = E^0 \oplus E^1$ it has the form

$$\mathcal{D} = \begin{pmatrix} 0 & \mathcal{D}^1 \\ \mathcal{D}^0 & 0 \end{pmatrix}.$$

We assume that $\mathfrak{D}$ is elliptic and self adjoint. For graded operators this means that $\mathfrak{D}^0 : \Gamma(E^0) \to \Gamma(E^1)$ is elliptic and $(\mathfrak{D}^0)^* = \mathfrak{D}^1$. In this case $\ker\mathfrak{D}^0$ is a finite dimensional module over $C\ell_k^0 \cong C\ell_{k-1}$.

DEFINITION 7.37. Let $\mathfrak{D}$ be a graded $C\ell_k$-linear differential operator which is elliptic and self-adjoint. The **Clifford index** of $\mathfrak{D}$ is the residue class

$$\text{index}_k(\mathfrak{D}) \equiv [\text{Ker}\mathfrak{D}^0] \in \widetilde{\mathfrak{M}}_k \tag{7.17}$$

where

$$\widetilde{\mathfrak{M}}_k = \mathfrak{M}_{k-1}/i^*\mathfrak{M}_k \cong KO^{-k}(\text{pt}) \tag{7.18}$$

is defined as follows: $\mathfrak{M}_k$ denotes the Grothendieck group of equivalence classes of finite-dimensional, real $C\ell_k$-modules ($=$ either $\mathbb{Z}$ or $\mathbb{Z}\oplus\mathbb{Z}$ in all cases), and $i^* : \mathfrak{M}_k \to \mathfrak{M}_{k-1}$ is the homomorphism given by restriction of scalars under the injection $i_* : C\ell_{k-1} \to C\ell_k$ induced by the isometric inclusion $i : \mathbb{R}^{k-1} \hookrightarrow \mathbb{R}^k$. The equivalence with the KO^*-groups of a point given in (7.18) is from [**ABS64**] and was alluded to in Section 7.1. We recall that

k	1	2	3	4	5	6	7	8
$\widetilde{\mathfrak{M}}_k$	$\mathbb{Z}_2$	$\mathbb{Z}_2$	0	$\mathbb{Z}$	0	0	0	$\mathbb{Z}$

and that

$$\widetilde{\mathfrak{M}}_{k+8} \cong \widetilde{\mathfrak{M}}_k$$

for all k. One can show that this index is constant for any continuous family $\mathfrak{D}_t$, $0 \le t \le 1$ of such operators.

Basic Example Let X be a compact spin manifold of dimension n, and let P_{Spin} denote its principal Spin_n bundle. Consider the associated bundle

$$\mathcal{S} = P_{\text{Spin}} \times_\lambda C\ell_n$$

where $\lambda : \text{Spin}_n \longrightarrow \text{GL}(C\ell_n)$ is given by left multiplication in the algebra (with $\text{Spin}_n \subset C\ell_n^0$). The grading $C\ell_n = C\ell_n^0 \oplus C\ell_n^1$ determines a grading $\mathcal{S} = \mathcal{S}^0 \oplus \mathcal{S}^1$. Since right and left multiplication commute as actions of $C\ell_n$ on itself, right multiplication by elements of $C\ell_n$ make $\mathcal{S}$ into a bundle of graded $C\ell_n$-modules. This multiplication is parallel for the Levi-Civita connection. Therefore the Dirac operator $\not{D}$ on $\mathcal{S}$ is $C\ell_n$-linear since for $\sigma \in \Gamma(\mathcal{S})$ and $\varphi \in C\ell_n$ we have

$$\not{D}(\sigma\cdot\varphi) = \sum_k e_k\cdot\nabla_{e_k}(\sigma\cdot\varphi) = \sum_k e_k\cdot\{(\nabla_{e_k}\sigma)\cdot\varphi\} = \left\{\sum_k e_k\cdot(\nabla_{e_k}\sigma)\right\}\cdot\varphi = (\not{D}\sigma)\cdot\varphi.$$

This attaches a fundamental invariant

$$\widehat{A}(X) \equiv \text{index}_n(\not{D}) \in KO^{-n}(\text{pt}) \tag{7.19}$$

to every compact Spin-manifold X.

Atiyah and Singer showed that $\widehat{A}(X)$ coincides with the following topological invariant introduced by Milnor. Choose an embedding into the sphere $X^n \subset S^{n+8k}$ for some k, and let N denote the normal bundle of this embedding. Then there is a fibre square of mappings

$$\begin{array}{ccc} N & \xrightarrow{\Psi_N} & \xi_{8k} \\ \downarrow & & \downarrow \pi \\ X & \xrightarrow{\psi_N} & \mathrm{BSpin}_{8k} \end{array}$$

where $\xi_{8k} \to \mathrm{BSpin}_{8k}$ is the universal $8k$-plane bundle with Spin-structure over the classifying space for principle Spin_{8k}-bundles, and ψ_N is the classifying map for the principle Spin_{8k}-bundle of N. Embed N into S^{n+8k} as a tubular neighborhood of X. Then we obtain a mapping

$$\Psi : S^{n+8k} \longrightarrow \mathrm{MSpin}_{8k} \equiv \xi_{8k} \cup \{\infty\} = \text{the one-point compactification of } \xi_{8k}$$

which extends Ψ_N by sending $(S^{n+8k} - N)$ to ∞.

Now let $\mathcal{S}_{\mathbb{R}}^{\pm} \to \mathrm{BSpin}_{8k}$ be the bundles associated to the two distinct irreducible real Spinor representations of Spin_{8k}. The Atiyah-Bott-Shapiro construction gives an element

$$s \equiv [\pi^*\mathcal{S}_{\mathbb{R}}^{+}, \pi^*\mathcal{S}_{\mathbb{R}}^{-}; \mu] \in KO_{\mathrm{cpt}}(\xi_{8k}) = KO(\mathrm{MSpin}_{8k}).$$

We define

(7.20) $$\widehat{\alpha}(X) \equiv \psi_N^*(s) \in KO(S^{n+8k}) = KO^{-n-8k}(\mathrm{pt}) = KO^{-n}(\mathrm{pt})$$

THEOREM 7.38. *For any compact spin manifold X of dimension n one has that*

$$\widehat{A}(X) = \widehat{\alpha}(X)$$

PROOF. See [**AS69, Hit74, LM89**]. □

The invariant $\widehat{\alpha}(X)$ depends only on the Spin-cobordism class of the manifold X and yields a ring homomorphism

(7.21) $$\widehat{\alpha} : \Omega_{\mathrm{Spin}}^{*} \longrightarrow KO^{-*}(\mathrm{pt})$$

from the Spin cobordism ring to the ring $KO^{-*}(\mathrm{pt})$ (See [**Hit74**] or [**LM89**] for example). We shall see in the next chapter that this characteristic homomorphism plays a crucial role in the study of manifolds of positive scalar curvature.

An interesting fact concerning the invariant $\widehat{\alpha}$ is that it is non-zero on certain exotic spheres. Let Θ_n denote the diffeomorphism classes of smooth n-dimensional manifolds which are homeomorphic to the n-sphere. This set forms an abelian group under the geometric construction of connected sum. The following is a deep result of Adams and Milnor. (See [**Hit74**] or [**LM89**] for a discussion.)

THEOREM 7.39. *The $\widehat{\alpha}$-homomorphism $\widehat{\alpha} : \Theta_n \to KO^{-n}(\mathrm{pt}) = \mathbb{Z}_2$ is surjective for $n \equiv 1$ or $2 \pmod{8}$.*

7.3. Applications

The Atiyah-Singer Index Theorem for Dirac Operators has important applications in many areas of geometry and topology. We present here a series of results which exemplify the nature and diversity of these applications. We want to give a clear notion of how the thinking goes, so we will keep to simpler examples to illustrate the ideas. Full results are easily found in the literature.

Topics to be discussed include:

(1) Integrality theorems for rational characteristic classes.
(2) The vector field problem on manifolds.
(3) The immersion problem for manifolds.
(4) Group actions on manifolds.
(5) The existence of metrics of positive scalar curvature.

Other areas such as representation theory, number theory, special holomony metrics, mathematical physics, etc., are covered in other lectures.

Let us step back for a moment and review how the Atiyah-Singer Dirac construction retrieves some classical results in geometry.

The Euler characteristic operator Suppose that (X, g) is a compact riemannian manifold, and let $C\ell(X) = C\ell(TX, g)$ denote the Clifford bundle (associated to the tangent bundle) of X. This is a bundle of algebras with a $\mathbb{Z}_2$-grading

$$C\ell(X) \;=\; C\ell^0(X) \oplus C\ell^1(X)$$

into even and odd parts, and the associated Dirac operator gives a mapping

$$D^0 : \Gamma\left(C\ell^0(X)\right) \;\longrightarrow\; \Gamma\left(C\ell^1(X)\right).$$

Now there is a canonical isomorphism of vector bundles

$$C\ell(X) \;\cong\; \Lambda^* T^* X \tag{7.22}$$

(which is *not* an isomorphism of algebra bundles) such that $C\ell^0(X) \cong \Lambda^{\mathrm{ev}} T^* X$ and $C\ell(X) \cong \Lambda^{\mathrm{odd}} T^* X$. Under this indentification we have that

$$D = d + d^* \tag{7.23}$$

where d^* denotes the formal adjoint of the exterior derivative operator. Hence the Dirac Laplacian coincides with the Hodge Laplacian

$$D^2 \;=\; dd^* + d^*d \;=\; \Delta \tag{7.24}$$

and the kernel of D is exactly the subspace $\mathbb{H}^*(X)$ of harmonic forms on X. It follows that

$$\mathrm{index}(D^0) \;=\; \dim \mathbb{H}^{\mathrm{ev}}(X) - \dim \mathbb{H}^{\mathrm{odd}}(X)$$

and the Index Theorem equates this with the Euler characteristic of X.

The signature operator Suppose now that X is a compact oriented riemannian manifold of dimension $4n$. Then the Clifford bundle of X has a parallel decomposition

$$C\ell(X) \;=\; C\ell^+(X) \oplus C\ell^-(C) \tag{7.25}$$

where $C\ell^{\pm}(X)$ are the ± 1-eigenbundles under left Clifford multiplication by the (twisted) volume element

$$\widetilde{\omega} \;=\; (-1)^n e_1 \ldots e_{4n}$$

and where at each point x, $e_1, \ldots, e_{4n}$ represents any oriented orthonormal basis of $T_x X$. One sees easily that $\widetilde{\omega}^2 = 1$ and $C\ell^{\pm}(X) = \frac{1}{2}(1 \pm \widetilde{\omega}) \cdot C\ell(X)$. Since $e_i \widetilde{\omega} = -\widetilde{\omega} e_i$ and $\nabla \omega = 0$, the Dirac operator $D = \sum_i e_i \cdot \nabla_{e_i}$ gives mappings

$$D^{\pm} : \Gamma\left(C\ell^{\pm}(X)\right) \;\longrightarrow\; \Gamma\left(C\ell^{\mp}(X)\right),$$

and with respect to the decomposition (7.25) it has the form

$$D = \begin{pmatrix} 0 & D^- \\ D^+ & 0 \end{pmatrix}. \tag{7.26}$$

Since D is self-adjoint, we see that D^- is the adjoint of D^+ and so

$$\operatorname{index}(D^+) = \dim \operatorname{Ker}(D^+) - \dim \operatorname{Ker}(D^-)$$

Now under the canonical identifications (7.22) – (7.24) above, $\ker(D) = \mathbb{H}^*(X)$, the space of harmonic forms on X, and we see that

$$\operatorname{index}(D^+) = \dim \mathbb{H}_+(X) - \dim \mathbb{H}_-(X)$$

where $\mathbb{H}_\pm(X) = \frac{1}{2}(1 \pm \widetilde{\omega}) \cdot \mathbb{H}^*(X)$. Now an easy calculation shows that up to sign Clifford multiplication corresponds to the Hodge star operator $* : \mathbb{H}^k(X) \xrightarrow{\sim} \mathbb{H}^{4n-k}(X)$. Therefore, all contributions to the index coming from $\mathbb{H}^k(X)$ for $k \neq 2n$ cancel, and we find that

$$\operatorname{index}(D^+) = \dim \mathbb{H}^{2n}_+(X) - \dim \mathbb{H}^{2n}_-(X)$$

where $\mathbb{H}^{2n}_\pm(X)$ are exactly the ± 1-eigenspaces of the Hodge $*$-operator acting on $\mathbb{H}^{2n}(X)$. From this we conclude that

$$\operatorname{index}(D^+) = \operatorname{sig}(X)$$

where $\operatorname{sig}(X)$ denotes the signature of the quadratic form Q on $H^{2n}(X;\mathbb{R}) \cong \mathbb{H}^{2n}(X)$ given by

$$Q(\varphi, \psi) = \int_X \varphi \wedge \psi \cong ([\varphi] \cup [\psi])[X]. \tag{7.27}$$

To see this note that

$$\int_X \varphi \wedge \varphi = \int_X \varphi \wedge * * \varphi = \pm \int_X \varphi \wedge * \varphi = \pm \|\varphi\|^2 \qquad \text{for } \varphi \in \mathbb{H}^{2n}_\pm(X).$$

Applying the Index Theorem to D^+ retrieves the following fundamental result.

THEOREM 7.40. (Hirzebruch) *For any compact oriented manifold of dimension* $4n$,

$$\operatorname{sig}(X) = \mathbb{L}(X)[X]$$

where $\mathbb{L}(X)$ *is the multipliative series in the Pontrjagin classes of* X *given in 1.15 above, and* $\Phi[X]$ *denotes evaluation on the fundamental homology class of* X*. For example,*

$$\operatorname{sig}(X) = \frac{1}{3} p_1(X)[X] \qquad \text{when } \dim(X) = 4$$

$$\operatorname{sig}(X) = \frac{1}{45}\left\{7p_2(X) - p_1(X)^2\right\}[X] \qquad \text{when } \dim(X) = 8.$$

Integrality theorems for rational characteristic classes. Suppose again that X is a compact oriented manifold of dimension $4n$. Associated to X is topological invariant called the **$\widehat{A}$-genus** given by

$$\widehat{A}(X) \equiv \widehat{\mathbb{A}}(X)[X]$$

where $\widehat{\mathbb{A}}(X)$ is the multiplicative series in the Pontrjagin classes of X given in Example 7.22. This is a rational number which is often not an integer. For example, on complex projective spaces one has

$$\widehat{A}(\mathbb{P}^{2n}(\mathbb{C})) = (-1)^n 2^{-4n} \binom{2n}{n}.$$

In particular $\widehat{A}(\mathbb{P}^2(\mathbb{C})) = -\frac{1}{8}$ and $\widehat{A}(\mathbb{P}^4(\mathbb{C})) = \frac{3}{2^7}$. More generally let $\mathbb{H}^{2n}(d)$ be the non-singular complex hypersurface of degree d in $\mathbb{P}^{2n+1}(\mathbb{C})$. For fixed d, all such hypersurfaces are diffeomorphic, and one could choose $\mathbb{H}^{2n}(d)$ to be the variety determined by the homogeneous polynomial equation $\sum_k Z_k^d = 0$. Then one has that

$$\widehat{A}(\mathbb{H}^{2n}(d)) = \frac{2^{-2n}d}{(2n+1)!}\prod_{k=1}^{n}(d^2-(2k)^2). \tag{7.28}$$

Furthermore, the first Chern class is given by

$$c_1(\mathbb{H}^{2n}(d)) = (2n+2-d)g \quad \text{where } g \text{ is a generator of } H^2(\mathbb{H}^{2n}(d);\mathbb{Z}) \cong \mathbb{Z}.$$

It follows that

$$\mathbb{H}^{2n}(d) \text{ is spin} \iff d \text{ is even.}$$

THEOREM 7.41. *Let X be a compact spin manifold of dimension $4n$. Then $\widehat{A}(X) \in \mathbb{Z}$. Furthermore, if n is odd, then $\widehat{A}(X) \in 2\mathbb{Z}$.*

PROOF. Let $\not{D}^+ : \Gamma(\$^+) \longrightarrow \Gamma(\$^-)$ be the Atiyah-Singer Dirac operator on X. Then by Theorem 7.36 we have that $\widehat{A}(X) = \operatorname{index}(\not{D}) = \dim_{\mathbb{C}} \operatorname{Ker}(\not{D}^+) - \dim_{\mathbb{C}} \operatorname{Ker}(\not{D}^-)$, and so $\widehat{A}(X)$ is an integer. When $\dim(X) \equiv 4 \pmod 8$, the spinor bundles are quaternionic bundles and $\not{D}$ is quaternion linear. Hence, $\operatorname{Ker}(\not{D}^+)$ and $\operatorname{Ker}(\not{D}^-)$ are quaternionic vector spaces, and therefore their complex dimension is even. $\square$

The expression in (7.28) is not manifestly an integer for any d, and indeed it is not an integer for d odd. However, for $d = 2m \geq 2n$

$$\begin{aligned}\widehat{A}(\mathbb{H}^{2n}(d)) &= \frac{2m}{(2n+1)!}\prod_{k=1}^{n}(m^2-k^2) = \frac{2}{(2n+1)!}\prod_{k=0}^{2n}(m-n+k)\\ &= \begin{cases} 2\binom{m+n}{2n+1}(m-n)(m-n+1)\dots(2n+1) & \text{if } m < 3n+2\\[2ex] 2\binom{m+n}{m-n-1}(2n+2)(2n+3)\dots(m-n-1) & \text{if } m \geq 3n+2\end{cases}\end{aligned}$$

is indeed an integer as asserted.

COROLLARY 7.42 (Rochlin). *If X is a compact smooth spin 4-manifold, then*

$$\operatorname{sig}(X) \equiv 0 \pmod{16}. \tag{7.29}$$

PROOF. From Example 7.22 and (7.22) we see that $\widehat{A}(X) = \frac{1}{24}p_1(X) = \frac{1}{8}\operatorname{sig}(X)$. $\square$

This seemingly harmless statement is quite deep. It is a Theorem of Wu that the Stiefel-Whitney classes of a manifold are topological invariants and can be defined in fact on any topological manifold [**MS74**]. Furthermore, on any spin manifold of dimension 4, even a topological one, the intersection form (7.27) is *even*, that is, $(u \cup u)[X] \equiv 0 \pmod 2$ for all $u \in H^2(X;\mathbb{Z})$. (See [**LM89**], p. 86 for example.) Then from the theory of quadratic forms over $\mathbb{Z}$ (cf. [**MH73**]), we know that

$$\operatorname{sig}(X) \equiv 0 \pmod{8}.$$

Now for a compact oriented topological 4-manifold X we can define a **Rochlin invariant**

$$\rho(X) = \frac{1}{8}\mathrm{sig}(X) \pmod 2 \in \mathbb{Z}/2\mathbb{Z}$$

and note that by Corollary 3.3, $\rho(X) = 0$ if X is smooth. In his celebrated work of 1982 [**Fre82**], Mike Freedman showed that there exist compact topological 4-manifolds with non-zero Rochlin invariant. In fact there exists such a manifold X with $\mathrm{sig}(X) = 8$. These manifolds are *not smoothable.*

Note that generally if X is a compact spin manifold of dimension $2n$ then by Theorem 7.36

$$\left\{chE \cdot \widehat{\mathbb{A}}(X)\right\}[X] \in \mathbb{Z}$$

for every complex vector bundle $E \to X$. This embodies a large family of integrality results. Here is a simple but important case. When X is the sphere S^{2n},

$$\left\{chE \cdot \widehat{\mathbb{A}}(S^{2n})\right\}[S^{2n}] = \{chE \cdot 1\}[S^{2n}] = \frac{1}{(n-1)!}c_n(E)[S^{2n}].$$

This proves the following [**Bot59**].

THEOREM 7.43 (Bott). *For every complex vector bundle E on S^{2n}, $c_n(E)$ is divisible by $(n-1)!$ in $H^{2n}(S^{2n}; \mathbb{Z}) \cong \mathbb{Z}$.*

Vector fields on manifolds. Let X be a compact smooth n-manifold. We pose the

Question: What is the largest integer $s = s(X) \equiv \mathrm{span}(X)$ such that there exist s pointwise linearly independent vector fields on X, i.e. vector fields $v_1, \ldots, v_s$ such that $\dim \mathrm{span}_{\mathbb{R}}\{v_1(x), \ldots, v_s(x)\} = s$ for all $x \in X$.

Note that if the Euler characteristic $\chi(X) \neq 0$, then $\mathrm{span}(X) = 0$.

Following ideas of Atiyah [**Ati70**] the Dirac operator can be used to study this question. Let $v_1, \ldots, v_s$ be pointwise linearly independent vector fields on X. Then there is a riemannian metric with respect to which $v_1, \ldots, v_s$ are pointwise orthonormal and there is a Dirac operator D on $C\ell(X)$ associated to that metric. When $n = 4m$, this operator decomposes according to (7.25) and (7.26) to define the signature operator $D^+ : \Gamma(C\ell^+(X)) \to \Gamma(C\ell^+(X))$ whose index is the signature of X.

We now consider the finite subgroup $G \subset \Gamma(C\ell(X))$ generated by $v_1, \ldots, v_s$,

$$G = \{\pm 1\} \cup \{\pm v_{i_1} \ldots v_{i_p} : 1 \leq i_1 < \cdots < i_p \leq s\}.$$

whose group algebra is $\mathbb{R} \cdot G = C\ell_s$, and observe that:

(i) Right multiplication by elements of G commutes with left multiplication by the volume form $\widetilde{\omega}$ defined above. Therefore,

$$C\ell^{\pm}(X) \cdot G = C\ell^{\pm}(X).$$

(ii) The symmetrized operator

$$\widetilde{D} \equiv \frac{1}{|G|}\sum_{g \in G} R_g \circ D \circ R_{g^{-1}}, \tag{7.30}$$

where R_g denotes right Clifford multiplication by g, decomposes exactly as in (7.26) with respect to the decomposition (7.25), and therefore defines an elliptic operator

$$\widetilde{D}^+ : \Gamma(C\ell^+(X)) \to \Gamma(C\ell^+(X)).$$

with **the same principal symbol as** D. Therefore

$$\text{index}(\widetilde{D}^+) \ = \ \text{sig}(X).$$

(iii) The operator $\widetilde{D}^+$ commutes with R_g for all $\in G$. Hence, $\ker \widetilde{D}^+$ and $\text{coker}\widetilde{D}^+ = \ker \widetilde{D}^-$ are finite-dimensional $C\ell_s$-modules.

This immediately proves the following (See [**Fra72, May85, Ati70**])

THEOREM 7.44. *Let X be a compact oriented manifold of dimension $4m$. If there exist s pointwise linearly independent vector fields on X, then*

$$\text{sig}(X) \ \equiv \ 0 \,(\text{mod} \ \ a_s)$$

where a_s is the dimension of an irreducible real $C\ell_s$-module.

Using $\mathbb{Z}_2$-grading refinements (see [**LM80**]) one can show that

$\text{sig}(X) \ \equiv \ 0 \,(\text{mod} \ \ 2a_{s-1})$ and $s \equiv 0 \,(\text{mod} \ \ 4) \quad \Rightarrow \quad \text{sig}(X) \ \equiv \ 0 \,(\text{mod} \ \ 4a_{s-1})$.

Stonger results obtained by considering $C\ell(X) \otimes E$ for vector bundles E over X.

Immersions of manifolds. Let X^n be a compact, smooth n-dimensional manifold. By the classical theorem of Whitney (see [**Hir76**] for example), there always exists an immersion of X^n into $\mathbb{R}^{2n-1}$.

Question: What is the smallest integer q such that there exists an immersion of X^n into $\mathbb{R}^{n+q}$?

For the sphere S^n, $q = 1$ of course. More generally, any manifold X^n which is *stably parallelizable*, i.e., which has the obviously necessary property that $T(X^n) \oplus \mathbb{R} \cong \mathbb{R}^{n+1}$, also has $q = 1$. This follows from Smale-Hirsch Theory (see [**Gro86b**] for example).

For the general case we would like some effective numerical criteria for giving lower bounds to q. Classical results of this kind follow from the theory of characteristic classes as follows. Let N be the normal bundle to an immersion of X^n into $\mathbb{R}^{n+q}$. From the bundle equivalence $T \oplus N \cong \mathbb{R}^{n+q}$ where $T = T(X^n)$, we have that

$$w(T) \cdot w(N) \ = \ 1$$

where $w = 1 + w_1 + w_2 + \ldots$ is the total Stiefel-Whitney class. Thus if $w(N) = w(T)^{-1} = 1 + u_1 + \cdots + u_k$ where $u_k \in H^k(X; \mathbb{Z}_2)$ is non-zero, then $q \geq k$.

Sometimes one gets better results from real characteristic classes and the Dirac operator. The idea is as follows. Suppose that X^n immerses into $\mathbb{R}^{n+q}$, and write as above

$$T \oplus N \cong \mathbb{R}^{n+q} \equiv \tau. \tag{7.31}$$

Step 1. Consider $C\ell(\tau)$ as a bundle of left-modules over $C\ell(T)$. Using the splitting (7.31), introduce the connection $\nabla^T \oplus \nabla^N$ on τ by orthogonal projection of the standard flat orthogonal connection on the trivialized bundle τ. This extends canonically to a tensor product connection $\nabla \equiv \nabla^T \otimes 1 + 1 \otimes \nabla^N$ on $C\ell(\tau) \cong C\ell(T) \otimes C\ell(N)$.

Associated to this operator is a Dirac operator $D = \sum e_i \nabla_{e_i}$ on $C\ell(\tau)$. When X^n is oriented and $n = 4k$, we use the parallel volume element $\omega = e_1 \dots e_n$ to split $C\ell(\tau) = C\ell^+(\tau) \oplus C\ell^-(\tau)$ where $C\ell^{\pm}(\tau) = (1 \pm \omega)C\ell^+(\tau)$. With respect to this splitting, D decomposes as in (7.26). From the Atiyah-Singer Index Theorem one calculates that

$$\text{index}(D^+) \;=\; 2^{q+n}\widehat{A}(X^n).$$

Step 2. Consider $C\ell(\tau)$ as a bundle of right $C\ell_{n+q}$-modules. We do this as follows. Let $\epsilon_1, \dots, \epsilon_{n+q}$ be the orthonormal sections of τ given by restricting the standard parallel vector fields in $\mathbb{R}^{n+q}$. Let $G \subset \Gamma(C\ell(\tau))$ be the finite group generated by these (as in the vector field discussion above). Then G acts on $C\ell(\tau)$ by *right* Clifford multiplication. However, this action does not commute with D since the sections ϵ_i are not parallel in the connection ∇ we have introduced on $C\ell(\tau)$. However, we can produce an operator $\widetilde{\nabla}$ which does commute with right multiplication R_g for $g \in G$ by averaging as in (7.30). Since left and right Clifford multiplication commute, $\widetilde{D}$ also decomposes according to (7.26) and the resulting operator $\widetilde{D}^+ : \Gamma C\ell^+(\tau) \to \Gamma C\ell^-(\tau)$ has the same principal symbol as D^+. Hence,

$$\text{index}(D^+) \;=\; \text{index}(\widetilde{D}^+) \;=\; 2^{q+n}\widehat{A}(X^n).$$

Now by the $C\ell_{n+q}$-invariance of $\widetilde{D}$ we see that $\ker(\widetilde{D}^+)$ and $\text{coker}(\widetilde{D}^+)$ are finite-dimensional $C\ell_{n+q}$-modules. In fact they are $\mathbb{Z}_2$-graded $C\ell_{n+q}$-modules. The dimension of such a module is $2a_{n+q-1}$. This proves the following (see [**LM80**]).

THEOREM 7.45. *Let X^n be a compact oriented manifold of dimension n. If there exists a smooth immersion $X^n \longrightarrow \mathbb{R}^{n+q}$, then*

$$a_{n+q-1} \,\big|\, 2^{q+n-1}\widehat{A}(X^n).$$

where a_{n+q} is the dimension of an irreducible real $C\ell_{n+q}$-module.

Recall that $\widehat{A}(X^n)$ is not in general an integer (although $2^n\widehat{A}(X^n)$ is). As a special case of this result suppose that $n = 4$ and recall that $\text{sig}(X^4) = 8\widehat{A}(X^4)$ and $a_5 = 8$. This gives

COROLLARY 7.46. *Let X^4 be a compact oriented 4-manifold. If there exists a smooth immersion $X^4 \longrightarrow \mathbb{R}^6$, then sig$(X)$ is even.*

Using coefficient bundles makes these results stronger ([**LM80**]). As an application of the general results, let $\mathbb{H}^n(d)$ be the complex hypersurface of odd degree d in $\mathbb{P}^{n+1}(\mathbb{C})$ where n is even, and suppose X^{2n} is oriented cobordant to $\mathbb{H}^n(d)$. Then X^{2n} cannot be immersed into $\mathbb{R}^{4n-2\alpha(n)-1}$ where $\alpha(n)$ denotes the number of 1's in the dyadic expansion of n. (See [**LM80**].)

Group actions. The Dirac operator has direct applications to the existence of symmetries in a riemannian manifold. A major result in this direction is the following.

THEOREM 7.47 (Atiyah and Hirzebruch [**AH70**]). *Let X be a compact spin manifold which admits a non-trivial, smooth action of the group S^1. Then $\widehat{A}(X) \;=\; 0$.*

PROOF. (Sketch) By averaging a riemannian metric over the group, one may assume that S^1 acts by isometries on X. The Atiyah-Singer-Dirac operator then becomes an S^1-invariant operator, and thereby has an index in the representation ring

R_{S^1} which is isomorphic to Laurent polynomials $\mathbb{Z}[z, z^{-1}]$ with integer coefficients. Using localization formulas for the this index given by Atiyah and Segal [**AS68a**], a direct computation shows that the S^1-index (and therefore also the ordinary index) is zero. □

THEOREM 7.48 (Lawson and Yau [**LY74**]). *Let X be a compact spin manifold which admits a non-trivial, smooth action of the group S^3. Then $\widehat{\alpha}(X) = 0$.*

PROOF. It is shown that the existence of a (non-trivial) S^3-action implies the existence of a riemannian metric of positive scalar curvature on X. One then applies Theorem 7.52. □

Every compact connected non-abelian Lie group contains a subgroup isomorphic to SU_2 or SO_3. Hence we conclude the following.

COROLLARY 7.49. *Let X be a compact spin manifold with $\widehat{\alpha}(X) \neq 0$. Then every compact connected Lie group $G \subset \mathrm{Diff}(X)$ is abelian.*

Among manifolds X with $\widehat{\alpha}(X) \neq 0$ are half of the exotic spheres in dimensions 1 and 2 mod 8 (See Theorem 7.39).

Metrics of positive scalar curvature. Some of the most spectacular applications of the Dirac operator involve its link to the scalar curvature κ of a riemannian manifold. It has been shown in previous lectures that on any riemannian spin manifold the Atyiah-Singer-Dirac operator

$$\not{D} : \Gamma(\not{S}) \longrightarrow \Gamma(\not{S})$$

satisfies the identity

$$\boxed{\not{D}^2 = \nabla^*\nabla + \frac{1}{4}\kappa} \tag{7.32}$$

due to Lichnerwicz [**Lic63**]. For any $\phi \in \Gamma(\not{S})$ over a compact riemannian spin manifold X, equation (7.32) implies that

$$\int_X \|\not{D}\phi\|^2 = \int_X \left\{ \|\nabla\phi\|^2 + \frac{1}{4}\kappa\|\phi\|^2 \right\}. \tag{7.33}$$

THEOREM 7.50. *Suppose X is a compact spin manifold of dimension $4k$. If X admits a riemannian metric of positive scalar curvature, then*

$$\widehat{A}(X) = 0$$

PROOF. Recall that in dimension $4k$, the spinor bundle decomposes as $\not{S} = \not{S}^+ \oplus \not{S}^-$ and with respect to this decomposition $\not{D}$ can be written as in (7.26) where $\not{D}^+ : \Gamma(\not{S}^+) \to \Gamma(\not{S}^-)$. By Theorem 7.36 we have

$$\widehat{A}(X) = \mathrm{index}(\not{D}^+) = \dim \mathrm{Ker}(\not{D}^+) - \dim \mathrm{Ker}(\not{D}^-)$$

since $\not{D}^-$ is the formal adoint of $\not{D}^+$.

Now if $\kappa > 0$ and $\phi \in \mathrm{Ker}\not{D}$, then $\phi = 0$ by (7.33). Since $\mathrm{Ker}\not{D} = \mathrm{Ker}(\not{D}^+) \oplus \mathrm{Ker}(\not{D}^-)$, $\widehat{A}(X) = 0$ as claimed. □

The above argument actually proves more.

THEOREM 7.51. *Let X be a compact connected spin manifold of dimension $4k$ which admits a riemannian metric of non-negative scalar curvature κ. If $\kappa(x) > 0$ at any point $x \in X$, then $\widehat{A}(X) = 0$. If on the other hand $\widehat{A}(X) \neq 0$, then there parallel spinors on X.*

This second possiblity has important consequences since the existence of a parallel spinor field implies a drastic reduction of the holonomy group of the riemannian manifold X. For manifolds with exceptional holonomy groups, which play an important role in M-theory in modern physics, such structure-reducing parallel spinor fields are always present.

The $C\ell_k$-index Theorem 7.38 yields a profound generalization of the above results.

THEOREM 7.52. *Let X be a compact spin manifold of dimension n which admits a riemannian metric of positive scalar curvature. Then*

$$\widehat{\alpha}(X) = 0 \qquad \text{in} \quad \Omega^n_{\text{spin}}(\text{pt}).$$

PROOF. The arguments above show immediately that the kernel, and therefore also the Clifford index, of the operator introduced in the Basic Example above must be zero. One then applies Theorem 7.38. □

Alternatively, as was shown by Atiyah and Singer, the analytic Clifford index $\widehat{\mathcal{A}}(X)$ can be computed as follows. Let $\mathbf{H} = \text{Ker}(\not{D})$ denote the kernel of the Atiyah-Singer Dirac operator on X. Then

$$\widehat{\alpha}(X) = \widehat{\mathcal{A}}(X) = \begin{cases} \dim_{\mathbb{C}} \mathbf{H} \pmod 2 & \text{if } n \equiv 1 \pmod 8 \\ \dim_{\mathbb{H}} \mathbf{H} \pmod 2 & \text{if} n \equiv 2 \pmod 8 \\ \frac{1}{2}\widehat{A}(X) & \text{if} n \equiv 4 \pmod 8 \\ \widehat{A}(X) & \text{if} n \equiv 0 \pmod 8). \end{cases}$$

One then observes as before that when $\kappa > 0$, we have $\mathbf{H} = 0$ and $\widehat{A}(X) = 0$.

Combining Theorem 7.52 with Theorem 7.39 gives the following.

COROLLARY 7.53. *Half of the exotic spheres in each dimension $n > 8$ with $n \equiv$ 1or2 $\pmod 8$ do not carry a riemannian metric with positive scalar curvature.*

This brings the $\widehat{\alpha}$-invariant defined in (7.21) to centerstage in the discussion of positive scalar curvature. Recall that

$$\widehat{\alpha} : \Omega^*_{\text{Spin}} \longrightarrow KO^{-*}(\text{pt})$$

is a ring homomorphism. The following was proved in [**GL80b**] using surgery techniques.

THEOREM 7.54. *Let X be a compact simply-connected manifold of dimension $n \geq 5$.*

1. If X is not spin, then there exists a metric of positive scalar curvature on X.

2. If X is spin and is spin cobordant to a manifold which admits a metric of positive scalar curvature, then X itself admits a metric of positive scalar curvature.

This led to the conjecture that for simply-connected manifolds of dimension ≥ 5, the homomorphism $\widehat{\alpha}$ gave a complete set of invariants for the existence of positive scalar curvature. In [**GL80b**] it was shown that up to rational cobordism this was true. The full conjecture, which was much harder, was established by S. Stolz in [**Sto92**].

THEOREM 7.55. (Stolz) *Let X be a compact simply-connected spin manifold of dimension $n \geq 5$. Then X admits a metric of positive scalar curvature if and only if $\widehat{\alpha}(X) = 0$.*

The Dirac operator can also be used to relate positive scalar curvature metrics to the fundmental group. The main tool is the following analogue of equation (7.32) for the Atiyah-Singer-Dirac operator $\not{D}_E : \Gamma(\not{S} \otimes E) \to \Gamma(\not{S} \otimes E)$ twisted by a "coefficient bundle" E:

$$\boxed{\not{D}^2 = \nabla^*\nabla + \frac{1}{4}\kappa + \mathfrak{R}_E} \tag{7.34}$$

where $\mathfrak{R}_E : \not{S} \otimes E \to \not{S} \otimes E$ is a symmetric bundle map given by

$$\mathfrak{R}_E = \sum_{i<j} e_i \cdot e_j \cdot R_{e_i,e_j}. \tag{7.35}$$

The key idea for applying this tool to the study of the fundamental group is the following.

DEFINITION 7.56. A compact spin n-manifold X is said to be **enlargeable** if for all $\epsilon > 0$ there exists a finite covering $\widetilde{X} \to X$ and a map $f : \widetilde{X} \to S^n$ of non-zero degree which is ϵ-contracting, i.e., for which $\|f_*v\| \leq \epsilon\|v\|$ for all tangent vectors v to $\widetilde{X}$.

EXERCISES 7.57.

(i) Show that this concept is independent of the riemannian metric on X.
(ii) Show that the n-torus in enlargeable.
(iii) Show that if X is enlargeable, so is $X \times S^1$.
(iv) Show that any hyperbolic n-manifold is enlargeable. (Use the fact that the fundamental group is residually finite, i.e., the intersection of all subgroups of finite index is the identity.)
(v) Show that a compact spin n-manifold Y, which admits a mapping $F : Y \to X$ of non-zero degree onto an enlargeable n-manifold, is enlargeable.

THEOREM 7.58. ([**GL80a, GL83**]) *There exists no metric of positive scalar curvature on an enlargeable spin manifold.*

PROOF. Suppose X is a compact enlargeable spin manifold which admits a metric of scalar curvature $\kappa > 0$, and let $\kappa_0 = \inf_X \kappa > 0$. We assume that $\dim(X) = 2m$, since otherwise by Exercise (iii) we may replace X with $X \times S^1$. Fix a complex vector bundle $E \to S^{2m}$ with $c_m(E) \neq 0$. For any finite covering $\widetilde{X} \to X$ and mapping $f : \widetilde{X} \to S^{2m}$ we have the Dirac operator $\not{D}_{f^*E} : \Gamma(\not{S} \otimes f^*E) \to \Gamma(\not{S} \otimes f^*E)$. It satisfies (7.34) with

$$\sup_X \|\mathfrak{R}_{f^*E}\| \leq c \sup_{\|v\|=1} \|f_*v\|^2$$

where c is a constant which is independent of the covering. We now choose $\epsilon > 0$ such that

$$c\epsilon^2 < \frac{1}{4}\kappa_0.$$

Then as in the proof of Theorem 7.50 we have that $\mathrm{Ker}(\not{D}_{f^*E}) = 0$, and so $\mathrm{index}(\not{D}^+_{f^*E}) = 0$. However, by the Atiyah-Singer Index Theorem

$$\begin{aligned}
\mathrm{index}(\not{D}^+_{f^*E}) &= \left\{\mathrm{ch}(f^*E)\cdot\widehat{\mathbb{A}}(\widetilde{X})\right\}[\widetilde{X}] = \left\{f^*\,\mathrm{ch}(E)\cdot\widehat{\mathbb{A}}(\widetilde{X})\right\}[\widetilde{X}] \\
&= \left\{\left(1+\frac{1}{(m-1)!}f^*c_m(E)\right)\cdot\left(1+\widehat{A_1}(\widetilde{X})+\widehat{A_2}(\widetilde{X})+\dots\right)\right\}[\widetilde{X}] \\
&= \frac{1}{(m-1)!}\left(f^*c_m(E)\right)[\widetilde{X}] + \widehat{A}(\widetilde{X}) \\
&= \frac{1}{(m-1)!}\left(f^*c_m(E)\right)[\widetilde{X}] \\
&= \frac{1}{(m-1)!}\deg(f)\,c_m(E)[S^{2m}] \\
&\neq 0.
\end{aligned}$$

□

There are many elaborations of this argument which prove much more general theorems. See for example [**GL80a, GL83, Gro86a**]. Many results mentioned here have also been proved by Schoen and Yau using the techniques of stable minimal hypersurfaces. (See [**SY79a, SY79c, SY89**].) They were also applied by Schoen and Yau to solve the Positive Mass Conjecture in general relativity [**SY79b**].

For each finitely generated group G there is ring homomorphism

$$\widehat{\alpha} : \Omega^*_{\mathrm{Spin}}(BG) \longrightarrow KO^{-*}(BG)$$

which must vanish on any compact Spin manifold X of positive scalar curvature with $\pi_1(X) \cong G$ (cf. [**Ros83**]). This was conjectured to be a complete set of obstructions for the existence of positive scalar curvature on manifolds with $\pi_1 = G$. For many groups this conjecture holds; for some it fails. See [**RS01**] for a recent account.

Applications to Kähler Geometry The Clifford formalism is often particularly effective for doing long calculations in riemannian geometry. It is even more effective in Kähler geometry. Details appear in [**Mic80**], and we sketch the ideas here.

Suppose that X is a Kähler manifold with almost complex structure $J : TX \to TX$. Recall that J is orthogonal for the given metric and satisfies $\nabla J \equiv 0$ with respect to the Levi-Civita connection. Take a local orthonormal frame field on X of the form $e_1, Je_1, \dots, e_n, Je_n$ and define

$$\epsilon_k = \frac{1}{2}(e_k - iJe_k) \qquad \text{and} \qquad \overline{\epsilon}_k = \frac{1}{2}(e_k + iJe_k).$$

We then have well-defined operators

$$\mathfrak{D} = \sum_{k+1}^{n} \epsilon_k \cdot \nabla_{\overline{\epsilon}_k} \qquad \text{and} \qquad \overline{\mathfrak{D}} = \sum_{k+1}^{n} \overline{\epsilon}_k \cdot \nabla_{\epsilon_k}$$

on any bundle S of complex $C\ell(X)$-modules with a connection ∇. Suppose S is holomorphic and that ∇ is a hermitian connection compatible with Clifford multiplication. (This occurs standardly in practice.) Then these operators satisfy

$$\mathfrak{D}^2 = (\overline{\mathfrak{D}})^2 = 0 \qquad \text{and} \qquad \overline{\mathfrak{D}} \text{ is the formal adjoint of } \mathfrak{D}$$

PROPOSITION 7.59 ([**Mic80**]). *When S is the bundle of complex spinors for X, one has that*

$$\mathfrak{D}\overline{\mathfrak{D}} + \overline{\mathfrak{D}}\mathfrak{D} = \nabla^*\nabla + \frac{1}{4}\sum_{k=1}^{n} \lambda_k \omega_k = \overline{\nabla}^*\overline{\nabla} + \frac{1}{4}\sum_{k=1}^{n} \lambda_k \overline{\omega}_k$$

where $e_1, ..., e_n$ is a complex hermitian orthonormal basis diagonalizing the Ricci form with $Ric(e_k, e_k) = \lambda_k$ and $\omega_k = -\epsilon_k \overline{\epsilon}_k$ for each k.

This result simplifies many calculations. It leads to the following.

THEOREM 7.60 ([**Mic80**]). *Let X be a compact Kähler manifold with spin structure and Λ a hermitian line bundle over X. If the scalar curvature κ of X satisfies*

$$\frac{1}{4}\kappa > \sum_{k+1}^{n} |\lambda_k|$$

where $\lambda_1, ..., \lambda_n$ are the eigenvalues of the Ricci form of Λ, then

$$\{\mathrm{ch}\,\Lambda \cdot \widehat{\mathbb{A}}(X)\}[X] = 0$$

THEOREM 7.61 ([**Mic80**]). *Let X be a compact Káhler manifold with $c_1(X) = 0$. Then there exists a finite covering space which is biholomorphically equivalent to a product of compact Káhler manifolds $\widetilde{X} = T \times X_1 \times \dots X_m \times Y$ where T is a complex torus, $X_1, \dots, X_m$ and Y are simply-connected, and Y has a metric with Sp_m-holonomy. Furthermore, the Todd genus $Td(X_i) = 0$ if $dim(X_i)$ is odd and $Td(X_i) = 2$ if $dim(X_i)$ is even.*

The proof of this last result uses results of Cheeger-Gromoll on manifolds of positive Ricci curvature, the results of Yau on the Calabi Conjecture, and the methods of spin geometry.

H. BLAINE LAWSON, JR., MATHEMATICS DEPARTMENT, STATE UNIVERSITY OF NEW YORK, STONY BROOK, NY 11794, USA.
E-mail address: blaine@math.sunysb.edu

CHAPTER 8

The Spectrum of the Dirac Operator

CHRISTIAN BÄR

ABSTRACT. We expose the analytic basics concerning the spectral properties of the Dirac operator on compact and on complete noncompact Riemannian manifolds. We list the known explicitly computable examples and look at some simple cases in detail. For compact manifolds we discuss eigenvalue estimates. For noncompact manifolds we focus on qualitative questions such as the one for discreteness of the spectrum.

Contents

8.1. The Spectrum of the Dirac Operator on Compact Manifolds

8.1.1. Analytic Basics. Let M denote a compact n-dimensional Riemannian spin manifold without boundary. The Dirac operator maps smooth spinor fields to smooth spinor fields,

$$D : C^\infty(M, \Sigma M) \to C^\infty(M, \Sigma M).$$

A number $\lambda \in \mathbb{C}$ is called an *eigenvalue* of D if there exists a nontrivial spinor field $\varphi \in C^\infty(M, \Sigma M)$ (called *eigenspinor*) such that

$$D\varphi = \lambda\varphi. \tag{8.1}$$

Why are eigenvalues of the Dirac operator interesting? The physical interest stems from the observation that if (8.1) holds, then the time-dependent spinor field $\Phi(t, x) := e^{it\lambda} \cdot \varphi(x)$ satisfies the physical Dirac equation

$$\frac{\partial \Phi}{\partial t} = iD\Phi$$

on the space-time $\mathbb{R} \times M$. Hence λ can be interpreted as the frequency or, equivalently, as the energy of the particle whose wave function is Φ.

The standard theory of self-adjoint elliptic differential operators (see e. g. [**LM89**, Chapter III]) now tells us the following:

- All eigenvalues of D are real, $\lambda \in \mathbb{R}$.
- The eigenvalues form a discrete subset of $\mathbb{R}$, unbounded from above and from below,
$$-\infty \leftarrow \cdots < \lambda_{-2} < \lambda_{-1} < \lambda_0 < \lambda_1 < \lambda_2 < \cdots \rightarrow \infty.$$
- The multiplicities of the eigenvalues are finite,
$$\dim(E_\lambda) < \infty$$
where $E_\lambda = \{\varphi \in C^\infty(M, \Sigma M) \mid D\varphi = \lambda\varphi\}$ is the corresponding eigenspace.
- The space of finite linear combinations of eigenspinors $\bigoplus_{j\in\mathbb{Z}} E_{\lambda_j}$ is dense in $C^\infty(M, \Sigma M)$ with respect to the L^2-norm.
- The eigenspaces for different eigenvalues λ and μ are perpendicular, $E_\lambda \perp E_\mu$, with respect to the L^2-scalar product.

Here the L^2-scalar product on the space of all spinor fields is defined by
$$(\varphi, \psi) = \int_M \langle \varphi, \psi \rangle \, dV$$
and the L^2-norm is the corresponding norm, $\|\varphi\| = \sqrt{(\varphi, \varphi)}$. The totality of all eigenvalues of D together with their multiplicities is called the *spectrum* of D on M or the *Dirac spectrum* of M.

8.1.2. Examples for Explicit Computation. Now given a compact Riemannian spin manifold M what is its Dirac spectrum? In general, it is totally hopeless to try to explicitly compute even one single eigenvalue. Only manifolds of a high degree of symmetry are accessible for explicit computation.

8.1.2.1. *The circle.* Already the simplest manifold in question, the circle $M = S^1 = \mathbb{R}/2\pi\mathbb{Z}$, is interesting to consider. Give M the standard (counter-clockwise) orientation. At each point $t \in M$ there is exactly one positively oriented orthonormal tangent basis consisting of the unique positively oriented unit tangent vector. Hence the total space of the oriented frame bundle is again equal to a circle, $P_{SO}(M) = S^1$, the bundle mapping onto M given by the identity.

Now we observe that M has two different spin structures. Recall that a spin structure is a two-fold covering of $P_{SO}(M)$. We can either choose the trivial two-fold covering $P_1 := S^1 \sqcup S^1$ with $P_1 \to P_{SO} = S^1$ given by two copies of the identity or we may choose $P_2 := S^1$ and $P_2 \to P_{SO} = S^1$ given by the mapping $z \mapsto z^2$ in complex notation.

Let us first analyze the situation for the trivial spin structure P_1. In one dimension spinor space is simply $\Sigma_1 = \mathbb{C}$. Clifford multiplication by the positively oriented unit vector is multiplication by i. Associating this to the trivial spin structure yields the trivial complex vector bundle of rank one over S^1. In other words, spinor fields are the same as complex-valued functions. With respect to the standard coordinate t of S^1 the Christoffel symbol of the Levi-Civita connection vanishes and thus the Dirac operator is
$$D = i\frac{d}{dt}.$$
Now the Fourier decomposition of functions on S^1 is exactly the eigenspace decomposition for D, the eigenvalues are $\lambda_k = k$ and the corresponding eigenfunctions are $\varphi_k(t) = e^{-ikt}$. Hence the spectrum is $\mathbb{Z}$, each eigenvalue having multiplicity 1.

In order to discuss the nontrivial spin structure P_2 let us rephrase the observations for the trivial spin structure as follows. Spinor fields were nothing but complex-valued functions on S^1. Equivalently, we can say that spinor fields on S^1 with respect to P_1 are given by periodic complex-valued functions on $\mathbb{R}$ with period 2π. When we replace P_1 by the nontrivial spin structure P_2 the following will change: When we move around $M = S^1$ one time, then we can lift this path continuously to P_1 and we will return to the same point in P_1. But when we do this in P_2 we will return to the opposite point. Thus spinor fields will now correspond to *anti-periodic* complex-valued functions on $\mathbb{R}$:

$$\varphi(t + 2\pi) = -\varphi(t).$$

Nonetheless a Fourier decomposition is also available in this situation, the eigenfunctions are now $\varphi_k(t) = e^{-i(k+1/2)t}$ and the eigenvalues are $\lambda_k = k + 1/2$. Thus the spectrum is $\mathbb{Z} + \frac{1}{2}$, each eigenvalue again having multiplicity 1.

It is remarkable that this simple example already shows that the spectrum of the Dirac operator will in general depend on the choice of spin structure.

Using Fourier analysis one can treat n-dimensional flat tori in a similar fashion, see [**Fri84**].

8.1.2.2. *Homogeneous Spaces.* Now we turn to spaces which have so many symmetries that they look the same at all points. Such space are called *homogeneous.* More precisely, let G be a compact Lie group acting transitively by orientation preserving isometries on our compact n-dimensional Riemannian spin manifold M. Choose $x_0 \in M$. Then we can write M in the form

$$M = G/H$$

where $H \subset G$ is the subgroup of all elements keeping x_0 fixed. The differentials of the isometries in G act on $P_{SO}(M)$. We assume here that this action lifts to the spin structure of M. Then there is a generalization of the Fourier decomposition of the previous section which we now describe.

The idea is to regard the space of spinor fields $C^\infty(M, \Sigma M)$ (or rather its Hilbert space completion with respect to the L^2-scalar product) as a representation space for G and to decompose it into irreducible subrepresentations. Denote the set of all equivalence classes of irreducible unitary representations of G by $\widehat{G}$. For every $\gamma \in \widehat{G}$ denote the corresponding representation space by V_γ. It is a fact that $\dim(V_\gamma) < \infty$. By the so-called *Frobenius reciprocity* there is a finite dimensional space W_γ such that

$$\bigoplus_{\gamma \in \widehat{G}} V_\gamma \otimes W_\gamma$$

embeds densely into $C^\infty(M, \Sigma M)$ as a subrepresentation. The group G acts on a summand $V_\gamma \otimes W_\gamma$ by $\gamma \otimes id$. The Dirac operator commutes with this G-action and leaves the decomposition invariant. It acts on a summand $V_\gamma \otimes W_\gamma$ by $id \otimes D_\gamma$ where the endomorphism D_γ of W_γ can be computed explicitly [**Bär92b**, Prop. 1]. Hence determining the spectrum of D is now the same as computing the eigenvalues of D_γ on W_γ for all $\gamma \in \widehat{G}$. We have thus reduced the problem to finite dimensional linear algebra.

In practice this linear algebra can still be quite hard. Moreover, one has to deal with an infinite number of finite dimensional eigenvalue problems, one for each $\gamma \in \widehat{G}$. This can be carried out only if one finds some uniform pattern in all the endomorphisms D_γ. Formulas get a lot simpler if G/H is a symmetric space. The Dirac spectrum for many of them is known by now. In the following table we collect the compact Riemannian spin manifolds for which the Dirac spectrum has been computed. Not all of them are symmetric but they are all locally homogeneous. Some of them (spheres and complex projective spaces) have been studied with various different methods.

$\mathbb{R}^n/\Gamma$	flat tori	**[Fri84]**
$\mathbb{R}^3/\Gamma$	3-dim. Bieberbach manifolds	**[Pfä00]**
S^n	spheres of constant curvature	**[Sul79]**, **[Bär96b]**, **[Tra93]**, **[CH96]**
S^n/Γ	spherical space forms	**[Bär96b]**
S^{2m+1}	spheres with Berger metrics	**[Hit74]** for $m=1$ **[Bär96a]** for general m
$S^3/\mathbb{Z}_k$	3-dim. lens spaces with Berger metric	**[Bär92b]**
G	simply connected compact Lie groups	**[Feg87]**
$\mathbb{C}P^{2m-1}$	complex projective spaces	**[CFG89, CFG94]**, **[SS93]**,**[AB98]**
$\mathbb{H}P^m$	quaternionic projective spaces	**[Bun91b]** for $m=2$ **[Mil92]** for general m
$Gr_2(\mathbb{R}^{2m})$	certain real Grassmannians	**[Str80b]** for $m=3$ **[Str80a]** for general m
$Gr_{2p}(\mathbb{R}^{2m})$	certain real Grassmannians	**[See97]**
$Gr_2(\mathbb{C}^{m+2})$	certain complex Grassmannians	**[Mil98]**
$G_2/SO(4)$		**[See97, See99]**
H^3/Γ	3-dim. Heisenberg manifolds	**[AB98]**

TABLE 1

8.1.3. Eigenvalue Estimates. Even if it is not possible to explicitly compute the Dirac spectrum of a manifold M one may still hope to get some control on the eigenvalues in terms of geometric data. For large classes of manifolds one can at least give geometric estimates. Estimating the modulus of an eigenvalue from above or from below turns out to be quite different stories.

8.1.3.1. *Lower Bounds.* We will start by estimating the Dirac eigenvalues from below on manifolds with positive scalar curvature. A naïve bound can be obtained as follows: Recall the Schrödinger-Lichnerowicz formula for the square of the Dirac operator

$$D^2 = \nabla^*\nabla + \frac{\mathrm{Scal}}{4}. \tag{8.2}$$

Now assume $\mathrm{Scal} \geq S$ for some positive constant S and let $D\varphi = \lambda\varphi$. We compute

$$\begin{aligned} \lambda^2(\varphi,\varphi) &= (D^2\varphi,\varphi) = \left(\left(\nabla^*\nabla + \frac{\mathrm{Scal}}{4}\right)\varphi,\varphi\right) \\ &\geq (\nabla\varphi,\nabla\varphi) + \frac{S}{4}(\varphi,\varphi) \geq \frac{S}{4}(\varphi,\varphi). \end{aligned}$$

Therefore each eigenvalue λ of the Dirac operator must satisfy

$$\lambda^2 \geq \frac{S}{4}.$$

What is naïve about this estimate? It is not optimal in sense that there are no manifolds where equality in the estimate is attained. The reason is that we wasted too much when we estimated $(\nabla\varphi, \nabla\varphi)$ by 0. The right approach was found by Friedrich in [**Fri80**]. One defines a new connection for spinors by

$$\widehat{\nabla}_X\varphi := \nabla_X\varphi + \frac{\lambda}{n}X\cdot\varphi.$$

Now one computes

$$\widehat{\nabla}^*\widehat{\nabla} = \nabla^*\nabla - 2\frac{\lambda}{n}D + \frac{\lambda^2}{n} = D^2 - \frac{\mathrm{Scal}}{4} - 2\frac{\lambda}{n}D + \frac{\lambda^2}{n}.$$

If φ is an eigenspinor for the eigenvalue λ, then

$$\begin{aligned} 0 &\le (\widehat{\nabla}\varphi, \widehat{\nabla}\varphi) = (\widehat{\nabla}^*\widehat{\nabla}\varphi, \varphi) \\ &= \lambda^2(\varphi,\varphi) - \frac{1}{4}(\mathrm{Scal}\,\varphi,\varphi) - 2\frac{\lambda^2}{n}(\varphi,\varphi) + \frac{\lambda^2}{n}(\varphi,\varphi) \\ &\le \left(\frac{n-1}{n}\lambda^2 - \frac{S}{4}\right)(\varphi,\varphi) \end{aligned}$$

and therefore

$$\lambda^2 \ge \frac{n}{4(n-1)}S.$$

This estimate is sharp because equality is attained e. g. for spheres. In case M is a Kähler manifold or quaternionic-Kähler it is possible to further improve the estimate. The basic idea is still the same; one tries to find refined Schrödinger-Lichnerowicz formulas in order to loose less in the estimate. Technically things become a lot more complicated. We summarize the results:

THEOREM 8.1. *Let M be a compact Riemannian spin manifold of (real) dimension n. Suppose the scalar curvature satisfies $Scal \ge S > 0$. Then all Dirac eigenvalues λ satisfy*

$$\lambda^2 \ge c_n\frac{S}{4}$$

where

a) (Friedrich [**Fri80**]) *In general,*

$$c_n = \frac{n}{n-1}.$$

b) (Kirchberg [**Kir86, Kir90**]) *If M is Kähler, then*

$$c_n = \begin{cases} \frac{n+2}{n}, & \text{if } \frac{n}{2} \text{ is odd} \\ \frac{n}{n-2}, & \text{if } \frac{n}{2} \text{ is even.} \end{cases}$$

c) (Kramer, Weingart, Semmelmann [**KSW99**]) *If M is quaternionic-Kähler, then*

$$c_n = \frac{n+12}{n+8}.$$

The estimate is sharp in all cases. Recalling the proof of the estimate in the general case we see that if a Dirac eigenvalue satisfies $\lambda^2 = \frac{n}{4(n-1)}S$, then the corresponding eigenspinor must satisfy $\widehat{\nabla}\varphi = 0$, or equivalently,

$$\nabla_X\varphi = -\frac{\lambda}{n}X\cdot\varphi$$

for all tangent vectors X. Such spinor fields are called *Killing spinors*. The Killing spinor equation is overdetermined, thus a generic manifold will not have nontrivial Killing spinors. The simplest example of a manifold with Killing spinors is the sphere S^n with the standard metric of constant curvature. In even dimensions $n \neq 6$ this is the only example. In the other dimensions there are different geometric types of manifolds admitting nontrivial Killing spinors. A complete classification of those types has been achieved by Bär in [**Bär93**] after previous work on construction methods for examples and partial classification results in [**FG85, CGLS86, Hij86b, Fra87, FK88, FK89, FK90, Gru90, BFGK91**].

Complex-projective space $\mathbb{CP}^{\frac{n}{2}}$ is a Kähler manifold and it is spin if its complex dimension $\frac{n}{2}$ is odd. In this case it indeed provides an example where equality is attained in case b). If $\frac{n}{2} \equiv 1 \bmod 4$, then this is the only example but if $\frac{n}{2} \equiv 3 \bmod 4$, then equality is attained if and only if M is the twistor space of a quaternionic-Kähler manifold. For example, $\mathbb{CP}^{\frac{n}{2}}$ is the twistor space of quaternionic-projective space $\mathbb{HP}^{\frac{n}{4}-\frac{1}{2}}$.

If $\frac{n}{2}$ is even, then complex-projective space is not spin and this is the reason why the case distinction is necessary for Kähler manifolds. The manifolds for which equality holds in the estimate are then the twisted products of T^2 and twistor spaces. These results were achieved by Moroianu [**Mor95, Mor99**] after preliminary work by Kirchberg, Lichnerowicz, and Friedrich [**Kir88, Lic90a, Fri93**].

For quaternionic-Kähler manifolds Kramer, Semmelmann, and Weingart have shown [**KSW98a**] that equality is attained only for quaternionic-projective space $\mathbb{HP}^{\frac{n}{4}}$.

The estimates obtained so far yield nontrivial results only if the manifold has strictly positive scalar curvature. One may suspect that if the scalar curvature is negative on a small part of the manifold and sufficiently large on the rest one should still be able to obtain a positive lower bound. In a way this is true. To formulate the result suppose $n \geq 3$ and denote the *Yamabe operator* acting on smooth functions on M by

$$Y = 4\frac{n-1}{n-2}\Delta + \mathrm{Scal}.$$

Here Δ is the usual Laplace-Beltrami operator. Let $\mu_1(Y)$ be the smallest eigenvalue of Y.

THEOREM 8.2 (Hijazi [**Hij86a**]). *Let M be a compact Riemannian spin manifold of dimension $n \geq 3$. Then all Dirac eigenvalues λ of M satisfy*

$$\lambda^2 \geq \frac{n}{n-1}\frac{\mu_1(Y)}{4}.$$

Note that $\mu_1(Y) \geq \min_M \mathrm{Scal}$ so that Theorem 8.2 implies Friedrich's estimate. But $\mu_1(Y)$ is still positive if there is a little bit of negative scalar curvature. The equality case turns out to be the same as in Friedrich's case; equality holds in Hijazi's estimate if and only if the corresponding eigenspinor is a Killing spinor.

Hijazi's original proof combined Friedrich's modification of the spinor connection with a clever conformal change of the metric. There is now a completely different proof available based on a so-called refined Kato inequality [**CGH00b**, Prop. 3.4]. This is a general principle allowing to estimate spectral data for operators acting on sections in a vector bundle (here D^2) against spectral data of a comparison operator acting on functions (here Y).

There is one manifold where one can give a nontrivial estimate for the Dirac eigenvalues without any geometric assumption. This is the 2-dimensional sphere.

THEOREM 8.3 (Bär [**Bär91, Bär92a**]). *Let $M = S^2$ be equipped with any Riemannian metric. Then all Dirac eigenvalues satisfy*

$$\lambda^2 \geq \frac{4\pi}{\text{area}(M)}.$$

Equality holds for the smallest eigenvalue if and only if M has constant curvature.

This estimate was conjectured by Lott in [**Lot86**] where he had shown that for some positive constant C the estimate $\lambda^2 \geq C/\text{area}(M)$ must hold. It turned out that Theorem 8.3 can also be deduced from Hijazi's results, see [**Hij91**]. Interestingly, there is a similar estimate for the Laplace operator on the 2-sphere going in the opposite direction. Hersch [**Her70**] proved that the first positive Laplace eigenvalue $\mu_1(\Delta)$ of S^2 equipped with any Riemannian metric satisfies

$$\mu_1(\Delta) \leq \frac{8\pi}{\text{area}(M)}.$$

It is believed that an estimate like in Theorem 8.3 is impossible for any manifold of dimension $n \geq 3$. More precisely, it is conjectured that given any compact spin manifold M of dimension $n \geq 3$ and given a spin structure on M, then one can find a Riemannian metric such that 0 is a Dirac eigenvalue. Eigenspinors for the eigenvalue 0 are called *harmonic spinors.* This conjecture is known to be true in dimensions $n \equiv 0, 1, 3, 7 \bmod 8$ [**Hit74, Bär96a**]. Moreover, one knows that S^n carries metrics with harmonic spinors also in dimensions $n \equiv 0 \bmod 4$ [**See01**, Thm. 3.27].

The situation for surfaces of higher genus is more subtle. The spin structure is then no longer unique. It turns out that there are two types of spin structures on surfaces which can be distinguished by the so-called *Arf-invariant.* For one type (Arf $= -1$) one has metrics with harmonic spinors. Hence a result like Theorem 8.3 is out of question. For the other type (Arf $= 1$) there is a lower estimate similar to the one in Theorem 8.3 but it is not sharp [**AB02a**, Thms. 5.1 and 6.1].

Let us now sketch the proof of Theorem 8.3.

PROOF. We modify the spinor connection once more. Let $f : M \to \mathbb{R}$ be a smooth function. We set $\widetilde{\nabla}_X\varphi := \nabla_X\varphi + \frac{\lambda}{2}X\cdot\varphi - X\cdot\nabla_{\text{grad}f}\varphi - 2\langle\text{grad}f, X\rangle\varphi$. Let K be the Gauss curvature of M. An elementary but tedious calculation yields

$$\begin{aligned}\widetilde{\nabla}^*(e^{-2f}\widetilde{\nabla}\varphi) &= e^{-2f}\Big\{D^2 - \frac{K}{2} - \lambda D + \frac{\lambda^2}{2} + \Delta f - 2\,\text{grad}f\cdot D + \lambda\,\text{grad}f\cdot \\ &\quad -2\left[\nabla_{\text{grad}f} + (\nabla_{\text{grad}f})^*\right]\Big\}\varphi. \qquad (8.3)\end{aligned}$$

Now let λ be a Dirac eigenvalue and φ a corresponding eigenspinor, $D\varphi = \lambda\varphi$. By a partial integration and by (8.3) we obtain

$$\begin{aligned}
0 \leq \|e^{-f}\widetilde{\nabla}\varphi\|^2 &= (\widetilde{\nabla}^* e^{-2f}\widetilde{\nabla}\varphi, \varphi) \\
&= \int_M e^{-2f}\Big\{\lambda^2|\varphi|^2 - \frac{K}{2}|\varphi|^2 - \lambda^2|\varphi|^2 + \frac{\lambda^2}{2}|\varphi|^2 + (\Delta f)|\varphi|^2 - 2\lambda\langle \mathrm{grad} f\cdot\varphi,\varphi\rangle \\
&\quad +\lambda\langle \mathrm{grad} f\cdot\varphi,\varphi\rangle - 2\langle[\nabla_{\mathrm{grad} f} + (\nabla_{\mathrm{grad} f})^*]\varphi,\varphi\rangle\Big\}dV \\
&= \int_M e^{-2f}\Big\{\frac{\lambda^2}{2}|\varphi|^2 - \frac{K}{2}|\varphi|^2 + (\Delta f)|\varphi|^2 - \lambda\langle \mathrm{grad} f\cdot\varphi,\varphi\rangle \\
&\quad -2\langle[\nabla_{\mathrm{grad} f} + (\nabla_{\mathrm{grad} f})^*]\varphi,\varphi\rangle\Big\}dV. \qquad (8.4)
\end{aligned}$$

For the last term we get

$$\begin{aligned}
\int_M e^{-2f}\langle[\nabla_{\mathrm{grad} f} + (\nabla_{\mathrm{grad} f})^*]\varphi,\varphi\rangle dV &= (\nabla_{\mathrm{grad} f}\varphi, e^{-2f}\varphi) + (\nabla_{\mathrm{grad} f})^*\varphi, e^{-2f}\varphi) \\
&= (\nabla_{\mathrm{grad} f}\varphi, e^{-2f}\varphi) + (\varphi, \nabla_{\mathrm{grad} f}(e^{-2f}\varphi)) \\
&= \int_M \langle \mathrm{grad} f, \mathrm{grad}\langle\varphi, e^{-2f}\varphi\rangle\rangle dV \\
&= \int_M (\Delta f)e^{-2f}|\varphi|^2 dV. \qquad (8.5)
\end{aligned}$$

Since the term $\langle \mathrm{grad} f\cdot\varphi,\varphi\rangle = -\langle\varphi, \mathrm{grad} f\cdot\varphi\rangle = \overline{\langle \mathrm{grad} f\cdot\varphi,\varphi\rangle}$ is purely imaginary and all other terms are real we conclude from (8.4) and (8.5)

$$0 \leq \int_M e^{-2f}\Big\{\frac{\lambda^2}{2}|\varphi|^2 - \frac{K}{2}|\varphi|^2 - (\Delta f)|\varphi|^2\Big\}dV. \qquad (8.6)$$

Now we will make an optimal choice for the function f. Define the function

$$h(x) := \frac{K(x)}{2} - \frac{1}{2\,\mathrm{area}(M)}\int_M K(y)\,dV(y)$$

and observe

$$\int_M h(x)dV(x) = 0.$$

This means that h is perpendicular to the constant functions with respect to the L^2-scalar product. The constant functions form the kernel of the Laplace operator, a self-adjoint differential operator. Thus h (and hence $-h$) lies in the image of the Laplace operator, i. e. there exists a smooth function f such that

$$\Delta f = -h.$$

Plugging this into (8.6) yields by the Gauss-Bonnet theorem

$$\begin{aligned}
0 &\leq \int_M e^{-2f}\Big\{\frac{\lambda^2}{2} - \frac{1}{2\,\mathrm{area}(M)}\int_M K(y)\,dV(y)\Big\}\,|\varphi|^2\,dV \\
&= \frac{1}{2}\int_M e^{-2f}\Big\{\lambda^2 - \frac{2\pi\chi(M)}{\mathrm{area}(M)}\Big\}\,|\varphi|^2\,dV \\
&= \frac{1}{2}\int_M e^{-2f}\Big\{\lambda^2 - \frac{4\pi}{\mathrm{area}(M)}\Big\}\,|\varphi|^2\,dV
\end{aligned}$$

We have proved the estimate

$$\lambda^2 \geq \frac{4\pi}{\mathrm{area}(M)}.$$

It remains to discuss the equality case. If the curvature K is constant, then one easily checks that equality holds. Conversely, if there is an eigenvalue satisfying

$$\lambda^2 = \frac{4\pi}{\text{area}(M)},$$

then the previous proof shows that the corresponding eigenspinor φ must satisfy $\widetilde{\nabla}_X \varphi = 0$, i. e.

$$\nabla_X \varphi = -\frac{\lambda}{2} X \cdot \varphi - \text{grad} f \cdot X \cdot \varphi.$$

Using this one easily computes for the spinorial curvature

$$R^\Sigma(e_1, e_2)\varphi = (\lambda(e_1(f)e_2 - e_2(f)e_1) + (-\lambda^2/2 + \Delta f)e_1 \cdot e_2) \cdot \varphi$$

where e_1, e_2 denotes an orthonormal basis of the tangent plane. On the other hand, using the usual formula relating spinorial curvature to the standard Riemann curvature tensor we obtain

$$R^\Sigma(e_1, e_2)\varphi = -\frac{1}{2} K e_1 \cdot e_2 \cdot \varphi.$$

Combining these two equations we get

$$[\lambda(e_1(f)e_2 - e_2(f)e_1) + (K/2 - \lambda^2/2 + \Delta f)e_1 \cdot e_2] \cdot \varphi = 0. \tag{8.7}$$

Taking the (pointwise) scalar product with $e_2 \cdot \varphi$ yields

$$\begin{aligned} 0 &= \langle [\lambda(e_1(f)e_2 - e_2(f)e_1) + (K/2 - \lambda^2/2 + \Delta f)e_1 \cdot e_2] \cdot \varphi, e_2 \cdot \varphi \rangle \\ &= \lambda e_1(f)|\varphi|^2 + \lambda e_2(f)\langle \varphi, e_1 \cdot e_2 \cdot \varphi \rangle + 0, \end{aligned}$$

hence

$$e_1(f)|\varphi|^2 + e_2(f)\langle \varphi, e_1 \cdot e_2 \cdot \varphi \rangle = 0.$$

Similarly, multiplying with $e_1 \cdot \varphi$ we see

$$e_1(f)\langle \varphi, e_1 \cdot e_2 \cdot \varphi \rangle - e_2(f)|\varphi|^2 = 0.$$

In matrix notation

$$\begin{pmatrix} 0 \\ 0 \end{pmatrix} = \begin{pmatrix} |\varphi|^2 & \langle \varphi, e_1 \cdot e_2 \cdot \varphi \rangle \\ \langle \varphi, e_1 \cdot e_2 \cdot \varphi \rangle & -|\varphi|^2 \end{pmatrix} \begin{pmatrix} e_1(f) \\ e_2(f) \end{pmatrix}$$

Since $\det \begin{pmatrix} |\varphi|^2 & \langle \varphi, e_1 \cdot e_2 \cdot \varphi \rangle \\ \langle \varphi, e_1 \cdot e_2 \cdot \varphi \rangle & -|\varphi|^2 \end{pmatrix} = -(|\varphi|^4 + \langle \varphi, e_1 \cdot e_2 \cdot \varphi \rangle^2)$ is nonzero wherever φ does not vanish (hence on an open dense subset of M) we conclude that the gradient of f vanishes everywhere. Thus $\Delta f = 0$ and by multiplying (8.7) with $e_1 \cdot e_2 \cdot \varphi$ we obtain

$$K/2 - \lambda^2/2 = 0.$$

Hence the curvature is constant and the proof is complete. □

Given this estimate on the 2-sphere and the fact $4\pi = 2\pi\chi(S^2) = \frac{1}{2}\int_{S^2} \text{Scal}\, dV$ one might be tempted to conjecture a more general estimate on n-dimensional manifolds of the form

$$\lambda^2 \geq \frac{n}{4(n-1)} \frac{\int_M \text{Scal}\, dV}{\text{vol}(M)}.$$

However such an estimate is not possible as is shown by the following theorem.

THEOREM 8.4 (Ammann-Bär [**AB00**]). *Let M be a closed spin manifold of dimension $n \geq 3$. Then there exist constants $0 < C_1 \leq C_2 \leq C_3 \leq \ldots$ and there exist Riemannian metrics $g_1, g_2, g_3, \ldots$ on M such that*

$$\lambda_k(g_j)^2 \leq C_k$$

for all k and j while

$$\frac{\int_M Scal_{g_j} dV_{g_j}}{\mathrm{vol}_{g_j}(M)} \overset{j\to\infty}{\longrightarrow} \infty.$$

The proof is based on the *variational characterization* of the eigenvalues. In general, we have

$$\lambda_k^2 = \inf_{\mathrm{V}} \mathrm{Sup}_{\varphi \in \mathrm{V}\setminus\{0\}} \frac{(\mathrm{D}\varphi, \mathrm{D}\varphi)}{\|\varphi\|^2}$$

where the infimum is taken over all k-dimensional vector subspaces $V \subset C^\infty(M, \Sigma M)$. This characterization follows easily from the properties of the spectrum as described in the beginning of this section. Let us now prove Theorem 8.4.

PROOF. We start by choosing a Riemannian metric g_0 on M such that (M, g) contains an embedded Euclidean ball B of radius 1.

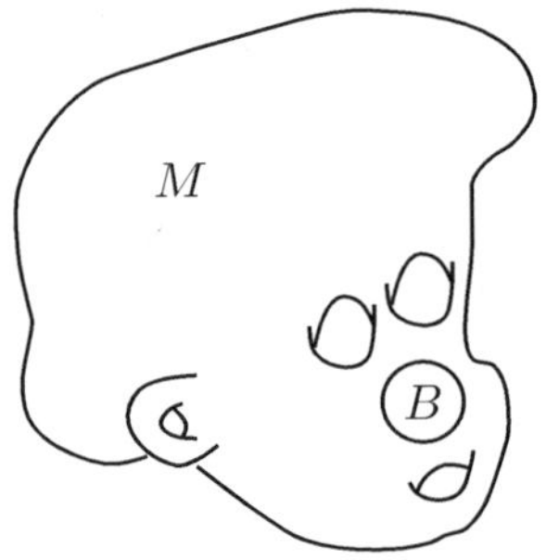

FIG. 1

Write the Euclidean ball B as a union of two annuli and one smaller ball, $B = A_1 \cup A_2 \cup A_3$, where $A_1 = \{x \in \mathbb{R}^n \mid 2/3 \leq |x| \leq 1\}$, $A_2 = \{x \in \mathbb{R}^n \mid 1/3 \leq |x| \leq 2/3\}$ and $A_3 = \{x \in \mathbb{R}^n \mid |x| \leq 1/3\}$. Now fix two parameters $0 < r < 1$ and $L > 0$. Choose a Riemannian metric $g_{r,L}$ on M with the following properties:

- $g_{r,L}$ coincides with g_0 on $M - B$
- $g_{r,L}$ is independent of L on A_1 and on A_3
- $(A_2, g_{r,L})$ is isometric to $S^{n-1}(r) \times [0, L]$ with the product metric where $S^{n-1}(r)$ denotes the round sphere of constant sectional curvature $1/r^2$.

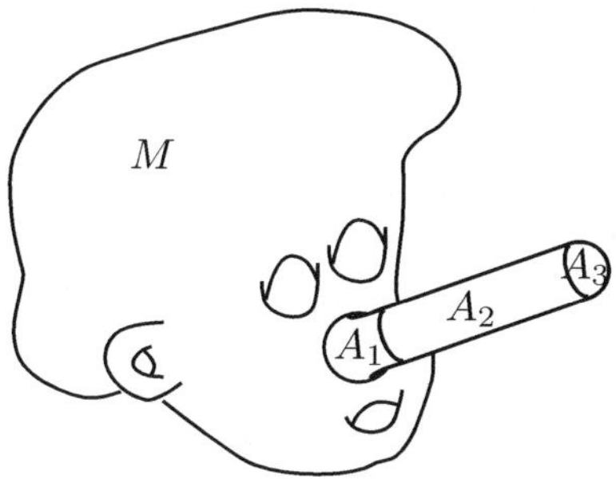

FIG. 2

Heuristically, there is a nose of radius r and length L growing out of the ball. For this reason we call these metrics *Pinocchio metrics.*

Claim 1. *The* k^{th} *eigenvalue* $\lambda_k(r, L)^2$ *of the square* $D^2_{g_{r,L}}$ *of the Dirac operator (w.r.t. the metric* $g_{r,L}$*) is bounded from above by a constant* $C_k > 0$ *independent of* r *and* L.

The proof is very simple. Choose a k-dimensional vector space V_k of spinors ψ on M vanishing on B. Then $\psi \in V_k$ can be considered a spinor for all metrics $g_{r,L}$. We plug it into the Rayleigh quotient for the Dirac operator to get

$$\begin{aligned} \lambda_k(r,L)^2 &\leq \mathrm{Sup}_{\psi \in V_k, \psi \neq 0} \frac{\int_M \langle D^2_{g_{r,L}} \psi, \psi \rangle_{g_{r,L}} \, dV}{\int_M \langle \psi, \psi \rangle_{g_{r,L}} \, dV} \\ &= \mathrm{Sup}_{\psi \in V_k, \psi \neq 0} \frac{\int_M \langle D^2_{g_0} \psi, \psi \rangle_{g_0} \, dV}{\int_M \langle \psi, \psi \rangle_{g_0} \, dV} \\ &=: C_k. \end{aligned}$$

Claim 2. *The normalized total scalar curvature is unbounded from above for* $r \in (0,1)$ *and* $L \in (0, \infty)$.

Let ω_k denote the volume of the k-dimensional unit sphere. We compute

$$\begin{aligned} \frac{\int_M \mathrm{Scal}_{g_{r,L}} \, dV}{vol_{g_{r,L}}(M)} &= \frac{\int_{(M \setminus B) \cup A_1 \cup A_3} \mathrm{Scal}_{g_{r,L}} \, dV + \int_{A_2} \mathrm{Scal}_{g_{r,L}} \, dV}{vol_{g_{r,L}}((M \setminus B) \cup A_1 \cup A_3) + vol_{g_{r,L}}(A_2)} \\ &= \frac{\int_{(M \setminus B) \cup A_1 \cup A_3} \mathrm{Scal}_{g_{r,L}} \, dV + L \cdot \frac{(n-1)(n-2)}{r^2} \cdot r^{n-1} \cdot \omega_{n-1}}{vol_{g_{r,L}}((M \setminus B) \cup A_1 \cup A_3) + L \cdot r^{n-1} \cdot \omega_{n-1}} \\ &\longrightarrow \frac{(n-1)(n-2)}{r^2} \end{aligned}$$

for $L \to \infty$ because on $M \setminus A_2$ the metric $g_{r,L}$ does not depend on L by construction. Hence

$$\mathrm{Sup}_{L>0} \frac{\int_M \mathrm{Scal}_{g_{r,L}} \, dV}{vol_{g_{r,L}}(M)} \geq \frac{(n-1)(n-2)}{r^2}$$

and therefore

$$\mathrm{Sup}_{L>0,\ 0<r<1} \frac{\int_M \mathrm{Scal}_{g_{r,L}} \, dV}{vol_{g_{r,L}}(M)} = \infty.$$

This proves the theorem. □

Another very nice lower eigenvalue estimate can be achieved for manifolds bounding manifolds of nonnegative scalar curvature.

THEOREM 8.5 (Hijazi-Montiel-Zhang [**HMZ01a**, Thm. 6]). *Let N be an $(n+1)$-dimensional compact Riemannian spin manifold with boundary $M = \partial N$. Suppose the scalar curvature of N is nonnegative, $\mathrm{Scal}_N \geq 0$ and the mean curvature of M (w. r. t. the inner normal) is nonnegative, $H \geq 0$. Then all nonnegative eigenvalues of the Dirac operator on M satisfy*

$$\lambda \geq \frac{n}{2} \inf_{\mathrm{M}} \mathrm{H}.$$

This estimate is sharp for the standard sphere bounding a Euclidean ball. Let us sketch the proof of the theorem.

PROOF. From the Schrödinger-Lichnerowicz formula (8.2) applied to the manifold N and the assumption $\mathrm{Scal}_N \geq 0$ we get for each spinor ψ on N

$$\int_N \langle D_N^2 \psi, \psi \rangle \, dV \geq \int_N \langle \nabla^* \nabla \psi, \psi \rangle \, dV.$$

Now one does a partial integration on both sides to remove the second derivatives. The resulting boundary terms can be arranged in such a manner that one obtains

$$\int_M \left(\langle D_M \psi, \psi \rangle - \frac{nH}{2} |\psi|^2 \right) dA \geq -\frac{n}{n+1} \int_N |D_N \psi|^2 \, dV$$

Then one argues that given an eigenspinor φ on M for the eigenvalue λ one can solve the boundary value problem

$$\begin{cases} D_N \psi = 0 & \text{on } N, \\ \pi_+ \psi = \varphi & \text{on } M. \end{cases}$$

Here $\pi_+ : C^\infty(M, \Sigma M) \to C^\infty(M, \Sigma M)$ denotes the projection onto the subspace generated by the D_M-eigenspaces for nonnegative eigenvalues. The theorem then follows immediately. □

A slightly improved version of this estimate can be found in [**HMZ02**].

8.1.3.2. *Upper Bounds.* Theorem 8.4 already gives upper bounds on the Dirac eigenvalues for certain metrics to be constructed. Here is a geometric upper bound for hypersurfaces in Euclidean space.

THEOREM 8.6 (Bär [**Bär98**]). *Let M be a closed oriented immersed hypersurface in $\mathbb{R}^{n+1}$. We give M the induced Riemannian metric and spin structure. Let H be the mean curvature of M w.r.t. the unit normal field ν. Then there are $2^{[n/2]}$ Dirac eigenvalues (counted with multiplicity) satisfying*

$$\lambda^2 \leq \frac{n^2}{4} \frac{\int_M H^2 \, dV}{vol(M)}.$$

The proof is again based on the variational characterization of the eigenvalues. One inserts restrictions of parallel spinors on $\mathbb{R}^{n+1}$ to M into the Rayleigh quotient $\frac{(D\varphi, D\varphi)}{\|\varphi\|^2}$. The general formula

$$D^M \varphi = -\nu \cdot D^{\mathbb{R}^{n+1}} \varphi + \frac{n}{2} H \varphi - \nabla_\nu^{\mathbb{R}^{n+1}} \varphi$$

relating the Dirac operators D^M on M and $D^{\mathbb{R}^{n+1}}$ on $\mathbb{R}^{n+1}$ yields for a parallel φ

$$D^M \varphi = \frac{n}{2} H \varphi$$

and therefore

$$\frac{(D^M\varphi, D^M\varphi)}{\|\varphi\|^2} = \frac{n^2}{4} \frac{\int_M H^2|\varphi|^2 dV}{\int_M |\varphi|^2 dV} = \frac{n^2}{4} \frac{\int_M H^2 \, dV}{vol(M)}$$

since $|\varphi|$ is constant for parallel φ. This proves the theorem up to some technical details concerning the identification of spinors on $\mathbb{R}^{n+1}$ with spinors on M.

The estimate in Theorem 8.6 is sharp in the sense that equality is attained for M a sphere of any radius in $\mathbb{R}^{n+1}$. It is an open question whether or not there are other hypersurfaces M in $\mathbb{R}^{n+1}$ for which equality holds. If one assumes that the mean curvature H is constant, then the lower bound in Theorem 8.5 and the upper bound in Theorem 8.6 agree. One can then deduce the classical Alexandrov Theorem stating that the only embedded compact hypersurfaces in Euclidean space of constant mean curvature are the round spheres, see [**HMZ01a**, Thm. 7].

There are similar results for hypersurfaces of spheres and hyperbolic spaces, see [**Bär98, Gin02**].

8.1.3.3. *Further results.* More involved Dirac eigenvalue bounds can e. g. be found in [**Bär91, Bär92c, Amm00a, Amm03, HZ01a, HZ01b, FK02, Kir02**]. There are also Dirac eigenvalue bounds for compact manifolds with boundary [**HMZ01b, HMR02**].

The spectrum of the Dirac operator does not fully determine the underlying manifold. This can be seen from the existence of *Dirac isospectral manifolds*, i. e. pairs of nonisometric Riemannian spin manifolds having the same Dirac spectrum. Known examples are certain flat tori, certain spherical space forms [**Bär96b**, Thm. 5] and certain nilmanifolds. In [**AB98**, Thm. 5.6] a 1-parameter family of Riemannian metrics is constructed on a 7-dimensional nilmanifold such that the Dirac spectrum is constant (as a function of the parameter) for some spin structures while it changes for the other spin structures.

8.2. The Spectrum of the Dirac Operator on Open Manifolds

From now on let M denote an n-dimensional *complete* noncompact Riemannian spin manifold.

8.2.1. Analytic Basics. The spectral theory is now more involved than in the compact case. Again, a complex number λ is called an *eigenvalue* of the Dirac operator D, if there exists a nonzero $\varphi \in C^\infty(M, \Sigma M) \cap L^2(M, \Sigma M)$ satisfying

$$D\varphi = \lambda\varphi.$$

Note that in the compact case smooth spinors are automatically square integrable while here we additionally impose it. The set of eigenvalues together with their multiplicities is called the *point spectrum* $\mathrm{spec}_p(D)$ of D. Again, eigenvalues are necessarily real and eigenspaces E_λ for different eigenvalues are mutually perpendicular. But the multiplicity $\dim E_\lambda$ of an eigenvalue λ may now be infinite, the point spectrum need no longer be discrete, and $\bigoplus_{\lambda\in \mathrm{spec}_p(D)} E_\lambda$ need no longer be dense in $C^\infty(M, \Sigma M) \cap L^2(M, \Sigma M)$.

We say that a number $\lambda \in \mathbb{C}$ lies in the *essential spectrum* $\mathrm{spec}_e(D)$ if there exists a sequence φ_j of smooth spinors with compact support which are orthonormal with respect to $(\cdot,\cdot)_{L^2}$ such that

$$\|(D-\lambda)\varphi_j\|_{L^2} \to 0$$

for $j \to \infty$. The φ_j can be regarded as approximate eigenspinors for D. Such a sequence is called a *Weyl sequence* for λ. We have $\mathrm{spec}_e(D) \subset \mathbb{R}$.

The point spectrum together with the essential spectrum form the spectrum $\mathrm{spec}(D)$ of D. Note that the point spectrum and the essential spectrum need not be disjoint. For example if λ is an eigenvalue of infinite multiplicity, then it lies in both the point spectrum and in the essential spectrum. Here an orthonormal basis of the infinite dimensional eigenspace can be used as a Weyl sequence. If one removes all essential spectrum (e. g. all eigenvalues of infinite multiplicity) from the point spectrum, then one is left with the *discrete spectrum*,

$$\mathrm{spec}_d(D) := \mathrm{spec}_p(D) \setminus \mathrm{spec}_e(D).$$

Removing all eigenvalues from the essential spectrum yields the *continuous spectrum*,

$$\mathrm{spec}_c(D) := \mathrm{spec}_e(D) \setminus \mathrm{spec}_p(D).$$

Hence we have two disjoint decompositions of the spectrum,

$$\mathrm{spec}(D) = \mathrm{spec}_d(D) \sqcup \mathrm{spec}_e(D) = \mathrm{spec}_p(D) \sqcup \mathrm{spec}_c(D).$$

Given a compact subset $K \subset M$ and $\lambda \in \mathrm{spec}_e(D)$ one can choose a Weyl sequence φ_j for λ such that $K \cap \mathrm{supp}(\varphi_j) = \emptyset$. This shows that $\mathrm{spec}_e(D)$ is unaffected by changes of the manifold in compact sets.

THEOREM 8.7 (Decomposition Principle). *Let M_1 and M_2 be two complete Riemannian spin manifolds, let $K_i \subset M_i$ be compact. Suppose there is a spin structure preserving isometry between $M_1 \setminus K_1$ and $M_2 \setminus K_2$. Then the Dirac operators on M_1 and on M_2 have the same essential spectrum.*

Note that M_1 and M_2 need not even be homeomorphic. This robustness of $\mathrm{spec}_e(D)$ contrasts strongly with the behavior of the discrete spectrum. Eigenvalues are very sensitive to changes of the geometry of the manifold.

8.2.2. Examples for Explicit Computation. The Dirac operator has been studied much less on noncompact manifolds than on compact ones. So we do not have many examples of noncompact manifolds for which we know the Dirac spectrum explicitly.

8.2.2.1. *Euclidean space.* The Dirac operator on Euclidean space is a differential operator with constant coefficients. The classical method to determine the spectrum of such operators is to apply Fourier transformation turning the differential operator into a multiplication operator. The spectrum can then be read off easily.

THEOREM 8.8. *Let $M = \mathbb{R}^n$ with the Euclidean metric. Then the Dirac operator has no eigenvalues, $\mathrm{spec}_p(D) = \emptyset$, and $\mathrm{spec}_c(D) = \mathbb{R}$.*

We give an elementary direct proof in the 1-dimensional case, $n = 1$. Recall from Section 8.1.2.1 that spinors on $M = \mathbb{R}$ can be identified with complex-valued functions and that the Dirac operator is then given by $D = i\frac{d}{dt}$. Suppose $\lambda \in \mathbb{R}$ and

$\varphi \in C^\infty(\mathbb{R}, \mathbb{C})$ such that

$$i\frac{d}{dt}\varphi = \lambda\varphi.$$

This ordinary differential equation has the general solution $\varphi(t) = C \cdot e^{-i\lambda t}$, $C \in \mathbb{C}$. Therefore $|\varphi| \equiv |C|$ and φ cannot be square-integrable unless it is identically zero. This proves $\mathrm{spec}_p(D) = \emptyset$.

Fix $\lambda \in \mathbb{R}$. Choose a nonnegative smooth function $\chi : \mathbb{R} \to \mathbb{R}$ such that $\mathrm{supp}(\chi) \subset [-2, 2]$ and $\chi \equiv 1$ on $[-1, 1]$. For $j, k \in \mathbb{N}$ set $\psi_{j,k}(t) := \chi(\frac{t}{j} - k) \cdot e^{-i\lambda t}$. From $|\psi_{j,k}(t)| = \chi(\frac{t}{j} - k)$ we see

$$\|\psi_{j,k}\|^2_{L^2} = \int_{-\infty}^{\infty} \chi(\tfrac{t}{j} - k)^2 \, dt = j \cdot \int_{-\infty}^{\infty} \chi(u)^2 \, du.$$

Since $\mathrm{supp}(\psi_{j,k}) \subset [j(k-2), j(k+2)]$ we can choose $k = k(j)$ such that $\psi_{j,k(j)}$ have mutually disjoint support. In particular, they are orthogonal with respect to the L^2-scalar product. Thus $\varphi_j := \left(j \cdot \int_{-\infty}^{\infty} \chi(u)^2 \, du\right)^{-1/2} \psi_{j,k(j)}$ are orthonormal.

Moreover, $|(D - \lambda)\psi_{j,k}| = |\frac{i}{j}\chi'(\frac{t}{j} - k)e^{-i\lambda t}| = \frac{1}{j}|\chi'(\frac{t}{j} - k)|$ yields

$$\|(D - \lambda)\psi_{j,k}\|^2_{L^2} = \frac{1}{j^2}\int_{-\infty}^{\infty} \chi'(\tfrac{t}{j} - k)^2 \, dt = \frac{1}{j}\int_{-\infty}^{\infty} \chi'(u)^2 \, du.$$

Hence

$$\|(D - \lambda)\varphi_j\|^2_{L^2} = \frac{\int_{-\infty}^{\infty} \chi'(u)^2 \, du}{j^2 \cdot \int_{-\infty}^{\infty} \chi(u)^2 \, du} \to 0$$

for $j \to \infty$. Thus φ_j is a Weyl sequence for λ. Since $\lambda \in \mathbb{R}$ was arbitrary this shows $\mathrm{spec}_e(D) = \mathbb{R}$ and since the point spectrum is empty also $\mathrm{spec}_c(D) = \mathbb{R}$.

In the physically motivated literatur the Dirac operator with potential, $D + V$, has attracted much attention where the potential V is a Hermitian endomorphism field of $\Sigma\mathbb{R}^n$, see e. g. [**Tha92**, Sec. 4.7] and the references therein.

8.2.2.2. *Hyperbolic space.* In [**Bun91a**] the spectrum of the Dirac operator on real hyperbolic space $\mathbb{R}H^n$ is determined regarding $\mathbb{R}H^n$ as a homogeneous space, $\mathbb{R}H^n = \mathrm{SO}^+(n,1)/\mathrm{SO}(n)$ where $\mathrm{SO}^+(n,1)$ denotes the group of orientation- and timeorientation preserving Lorentz transformations. One can then use representation theoretic methods similar to the ones in Section 8.1.2.2. These methods are technically somewhat involved and in fact there is an incorrect statement about the eigenvalue 0 in [**Bun91a**]. A much simpler and more geometric computation of the Dirac spectrum on $\mathbb{R}H^n$ uses the warped product structure, see e. g. [**Bai97**]. The result is

THEOREM 8.9. *Let $M = \mathbb{R}H^n$ be real hyperbolic space. Then the Dirac operator has no eigenvalues, $\mathrm{spec}_p(D) = \emptyset$, and $\mathrm{spec}_c(D) = \mathbb{R}$.*

So the result is the same as for Euclidean space.

8.2.2.3. *Further examples.* Both Euclidean and real hyperbolic space are examples of a class of manifolds called Riemannian symmetric spaces. These spaces are classified and in principle they are particularly well-suited for representation theoretic methods.

The point spectrum of these spaces is understood. Goette and Semmelmann show in [**GS02**] that the point spectrum of the Dirac operator on a Riemannian symmetric space of noncompact type is either empty or $\{0\}$. They give different characterizations for when the eigenvalue 0 appears.

The continuous spectrum has been determined by Camporesi and Pedon in **[CP02]** for hyperbolic spaces (Riemannian symmetric spaces of noncompact type and rank 1). The result says that if M is a hyperbolic space, then

$$\mathrm{spec}_c(D) = \begin{cases} \mathbb{R}, & \text{if } M \in \{\mathbb{R}H^n, \mathbb{H}H^n, \mathbb{O}H^2\} \\ \mathbb{R}, & \text{if } M = \mathbb{C}H^n \text{ with } n \text{ odd} \\ (-\infty, -\frac{1}{2}] \cup [\frac{1}{2}, \infty), & \text{if } M = \mathbb{C}H^n \text{ with } n \text{ even} \end{cases}$$

The exceptional case $M = \mathbb{C}H^n$ with even complex dimension n is precisely the case when 0 is an eigenvalue.

8.2.3. Qualitative Results.

As for compact manifolds an explicit computation of the eigenvalues of the Dirac operator is possible only in exceptional cases. Due to its robustness the essential spectrum is more accessible. As an illustration let us look at *hyperbolic manifolds* which are, by definition, complete Riemannian manifolds with constant sectional curvature -1. Such manifolds are always of the form $M = \mathbb{R}H^n/\Gamma$ for some suitable discrete group Γ of isometries of real hyperbolic space. There is up to now no hyperbolic manifold of finite volume for which one can compute the point spectrum neither for the Dirac operator nor for any other geometric elliptic operator like the Laplace-Beltrami operator.

Nevertheless, one can say something about the essential spectrum. For the sake of simplicity, let us concentrate on the 3-dimensional case. A 3-dimensional hyperbolic spin manifold of finite volume is known to be decomposable in the form

$$M = M_0 \sqcup E_1 \sqcup \cdots \sqcup E_k$$

where M_0 is compact with boundary and the E_j are the *cusps*. They have the form $E_j = (0, \infty) \times T^2$ with the Riemannian metric $g = dt^2 + e^{-2t} \cdot g_{\text{flat}}$ where t denotes the variable in $[0, \infty)$ and g_{flat} is a flat metric on the 2-torus T^2.

All topological information is contained in M_0 but by the decomposition principle it is irrelevant for the essential spectrum. Hence the essential spectrum can be determined by looking the cusps alone and they are very simple and given completely explicitly. One gets

THEOREM 8.10 (Bär **[Bär00]**). *Every complete oriented hyperbolic manifold M of dimension 3 and finite volume has a spin structure for which*

$$spec(D) = spec_d(D).$$

Some also have a spin structure such that

$$spec(D) = \mathbb{R}.$$

There are no other possibilities.

In other words, either $\mathrm{spec}_e(D) = \emptyset$ or $\mathrm{spec}_e(D) = \mathbb{R}$ and it is the spin structure which is responsible for the presence or the absence of essential spectrum.

One of the main sources for the construction of examples of 3-dimensional hyperbolic manifolds of finite volume is as follows: Take a knot or link L in $\mathbb{R}^3$. Add one point to $\mathbb{R}^3$ to obtain S^3 and regard L as sitting in the 3-sphere, $L \subset S^3$. It can be shown that for "most" links the manifold $M := S^3 \setminus L$ can be given a hyperbolic metric of finite volume. Each component of the link corresponds to one cusp. For such an

M there is a simple criterion for whether or not there exists a spin structure with $\mathrm{spec}(D) = \mathbb{R}$.

THEOREM 8.11 (Bär [**Bär00**, Thm. 4]). *Let $L \subset S^3$ be a link, let $M = S^3 \setminus L$ carry a hyperbolic metric of finite volume.*

If the linking number of all pairs of components (L_i, L_j) of L is even,

$$Lk(L_i, L_j) \equiv 0 \ mod\ 2,$$

$i \neq j$, then the spectrum of the Dirac operator on M is discrete for all spin structures,

$$spec(D) = spec_d(D).$$

If there exist two components L_i and L_j of L, $i \neq j$, with odd linking number, then M has a spin structure such that the spectrum of the Dirac operator satisfies

$$spec(D) = \mathbb{R}.$$

Determining linking numbers modulo 2 is equivalent to counting overcrossings modulo 2 in planar link diagrams, hence extremely simple.

The complements of the following links possess a hyperbolic structure of finite volume. All linking numbers are even. Count e. g. the overcrossings of blue over red. Hence the Dirac spectrum on those hyperbolic manifolds is discrete for all spin structures.

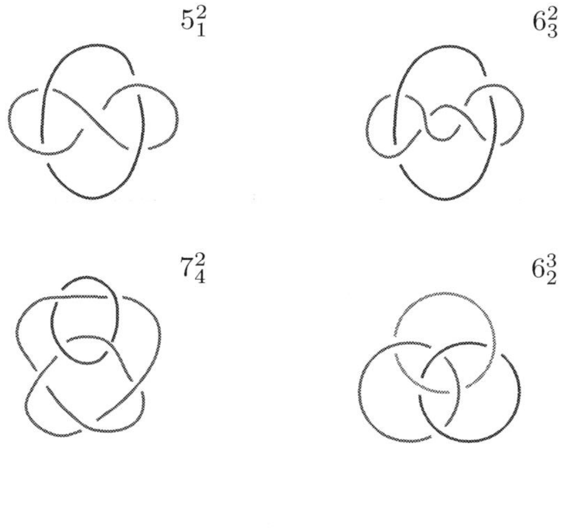

FIG. 3

This example includes the Whitehead link (5_1^2) and the Borromeo rings (6_2^3).

The complements of the following links possess a hyperbolic structure of finite volume. There are odd linking numbers. Hence those hyperbolic manifolds have a spin structure for which the Dirac spectrum is the whole real line.

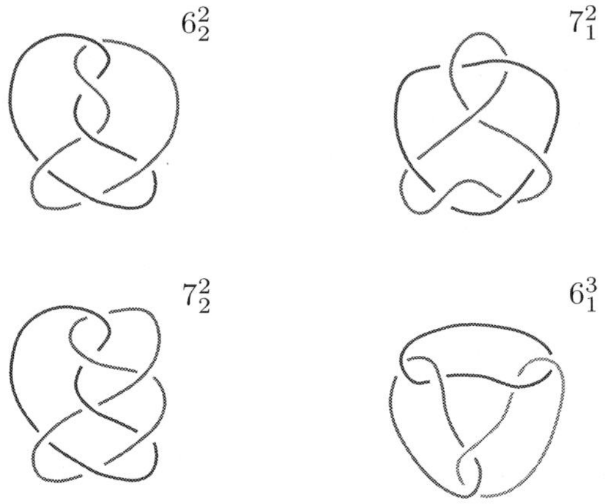

FIG. 4

CHRISTIAN BÄR, UNIVERSITÄT POTSDAM, INSTITUT FÜR MATHEMATIK, PF 60 15 53, D-14415 POTSDAM, GERMANY.
E-mail address: baer@math.uni-potsdam.de

CHAPTER 9

Conformal structure and spin geometry

THOMAS BRANSON

ABSTRACT. The study of the Dirac equation on manifolds is grounded in the need to understand physical fields on curved spacetimes. It is generally believed that the study of Riemannian (positive definite) metrics, or infinitesmial distance functions, will ultimately be relevant to the more directly physical problem of Lorentz signature metrics, via principles of analytic continuation in signature. This adds impetus to the natural mathematical pursuit of studying spin structure, the Dirac operator, and other related operators on Riemannian manifolds, or on manifolds with a positive definite conformal structure. In these notes, we attempt to introduce this subject, with an eye toward fundamental issues that are likely to be important in future work. Among these are Stein-Weiss gradients, Bochner-Weitzenböck formulas, the Hijazi inequality, and the Penrose local twistor idea, leading to the general theory of tractor bundles. All of these ideas are intimately connected with the behavior of differential operators under conformal change of metric, or from another viewpoint, operators that need only a conformal structure (and not a metric) to be well-defined.

Contents

9.1. Spin geometry

The spinor bundle and the Dirac operator are unfamiliar and perhaps intimidating objects to many mathematicians, even some who are quite knowledgeable in the field of differential geometry. Lichnerowicz noted, however, that at an earlier point in history, the same thing could be said of differential forms. What spinors, differential forms, and other species of physical fields have in common is that they take their values in *associated bundles.* A familiar instance of this concept is the fact that the *structure group* of Riemannian geometry is $\mathrm{O}(n)$; this can be detected in the fact that there is a copy of $\mathrm{O}(n)$ acting on each tangent space. These groups and actions fit together into a *principal bundle* [**KN69a**], which then gives rise to many vector bundles (parameterized by representations); in this case, the tensor bundles.

9.1.1. Associated bundles. In *oriented* Riemannian geometry, where we can give ordered local orthonormal frames, consistent up to even permutation, the structure group reduces to $\mathrm{SO}(n)$. It is natural to ask what we get when we now go to the simply connected double cover $\mathrm{Spin}(n)$ of $\mathrm{SO}(n)$, and the answer, in brief, is spin geometry.

In fact, we should really be speaking of covers of $\mathrm{SO}(p,q)$, in order to cover pseudo-Riemannian spin geometry of all signatures. For the most part, we shall ignore this here, however. It turns out that in spin geometry, one has to take a little more care in changing signature than in most of (tensorial) pseudo-Riemannian geometry.

An *associated bundle* is made from two ingredients: a principal bundle $\mathcal{F}$ with structure group H on a manifold M, and a representation (λ, V) of H. The associated bundle is a vector bundle $\mathbf{V} = \mathcal{F} \times_\lambda V$, the sections of which may be naturally identified with smooth functions $\varphi : M \to V$ with the special property that

$$\varphi(x \cdot h) = \lambda(h)^{-1}\varphi(x), \quad x \in M, \ h \in H.$$

Using this parameterization by representations, we can get a kind of inductive understanding of associated vector bundles: the tangent bundle "is" the defining representation of $\mathrm{SO}(n)$; the two-form bundle $\Lambda^2 M$ "is" the alternating two-tensor representation, and so on. Even lacking the details of the associated bundle construction, it is possible to "believe in" these bundles, and to do computations based on their fundamental properties. In brief, the associated bundle construction promotes linear algebraic constructs to the bundle level.

Indeed, much of this can be done on the level of the Lie algebra $\mathfrak{h}$, as opposed to the group H. Part of the genesis of spinors was the realization that the classification of irreducible, finite-dimensional representations of the simple complex Lie algebras, carried out in the cases of $\mathfrak{b}_\ell$ and $\mathfrak{d}_\ell$ ($\mathfrak{so}(n, \mathbf{C})$ for n respectively odd and even), yields much more than simply the irreducible constituents of the tensor algebra over the defining representation. The additional fundamental building block is the *spin representation* Σ, which is irreducible for odd n, and breaks into two irreducible summands $\Sigma_\pm$ when n is even. In fact all irreducible modules may be realized in the tensor algebra over Σ. For example, when n is even, $\Sigma \otimes \Sigma \cong_{\mathfrak{so}(n)} \Lambda$ (the right side being alternating tensors of all orders), so in particular we recover all tensors, since these are generated by Λ^1. The representation Σ may be integrated to a representation of $\mathrm{Spin}(n)$, but not to one of $\mathrm{SO}(n)$.

In going from this Lie algebra fact to the construction of associated bundles for the new modules, however, a new topological obstruction rears its head, in analogy with the orientability obstruction involving $\mathrm{SO}(n)$. This also has to do with the existence of covers by locally framed neighborhoods, in which the frame choices are consistent in a certain combinatorial sense. From now on, we shall assume that this condition is satisfied, so that our manifolds are *spinnable*. An additional caveat at this point is that there are generally many different *spin structures*, or choices of a principal $\mathrm{Spin}(n)$ bundle covering the $\mathrm{SO}(n)$-bundle of orthonormal frames. For the most part, the statements we make will be true for any choice of spin structure. In any case, the old $\mathrm{SO}(n)$ bundles are still with us as we do spin geometry – they are just the ones associated to representations $\lambda : \mathrm{Spin}(n) \to V$ which factor through $\mathrm{SO}(n)$ via the covering homomorphism $\mathrm{Spin}(n) \to \mathrm{SO}(n)$.

Computing some elementary tensor products reveals rich additional structure. First, we find that

$$\Lambda^1 \otimes \Sigma \cong_{\mathrm{Spin}(n)} \Sigma \oplus \mathbf{Tw},$$

where Σ and $\mathbf{Tw}$ have *no joint content* – i.e. no isomorphic Spin(n)-summands. This means there is a nonzero Spin(n)-invariant map – the projection

$$\Lambda^1 \otimes \Sigma \overset{\cong}{\to} \Sigma \oplus \mathbf{Tw} \xrightarrow{\mathrm{Proj}} \Sigma.$$

We may view this, by the usual trick of identifying End(V) with $V \otimes V^*$, as an invariant map

$$\begin{aligned} \Lambda^1 &\to \Sigma \otimes \Sigma^* \cong_{\mathrm{Spin}(n)} \mathrm{End}(\Sigma), \\ \eta &\mapsto \gamma(\eta). \end{aligned}$$

This brings *Clifford multiplication* into view: γ is a canonically defined section of the vector bundle $T \otimes \mathrm{End}(\Sigma)$, called the *Clifford section* or *fundamental tensor-spinor.*

9.1.2. The Clifford relation. The famous *Clifford relation* is forced upon us as follows. Symmetrizing two Clifford multiplications,

$$\frac{1}{2}(\eta \otimes \alpha + \alpha \otimes \eta) \mapsto \frac{1}{2}(\gamma(\eta)\gamma(\alpha) + \gamma(\alpha)\gamma(\eta))$$

gives a well-defined invariant map from the *symmetric 2-tensors* $\odot^2 T^*$ to End(Σ). By the reverse of the trick above, this gives Spin(n)-invariant maps

$$\begin{aligned} \odot^2 T^* &\to \Sigma \otimes \Sigma^*, \\ \odot^2 T^* \otimes \Sigma &\to \Sigma. \end{aligned}$$

But there is only one such thing (up to a constant factor), since

$$\odot^2 T^* \otimes \Sigma \cong_{\mathrm{Spin}(n)} \Sigma \oplus \mathcal{M},$$

where $\mathcal{M}$ has no joint content with Σ. This is a fairly special property of Σ; it is here that attempts to construct Clifford multiplication directly on "higher spin" bundles breaks down, though there are various Clifford multiplications on such bundles by virtue of their residence in the tensor algebra over Σ.

Continuing with the above reasoning, $\odot^2 T^*$ splits as $\mathbf{C}g \oplus \odot 2_0$, where g is the metric (or inner product), and $\odot 2_0$ are the *trace-free* symmetric 2-tensors. Our map above must emanate from just one of these; otherwise we could restrict and contradict uniqueness. But

$$\mathbf{C}g \cong_{\mathrm{Spin}(n)} \Lambda^0$$

is the trivial representation, and of course $\Lambda^0 \otimes \Sigma = \Sigma$. So the map in question must be

$$\frac{1}{2}(\gamma(\eta)\gamma(\alpha) + \gamma(\alpha)\gamma(\eta))\psi = cg^\sharp(\eta, \alpha)\psi$$

for some nonzero constant c, where $g^\sharp$ is the metric on T^*M determined by the metric g on TM. By renormalizing γ, we recover our favorite normalization for the Clifford relation; for example, $c = -2$.

In *abstract index notation* [**PR86**], we have $\gamma(\eta) = \gamma^a \eta_a$. The above argument may be rephrased as

$$\gamma^a \gamma^b + \gamma^b \gamma^a = \underbrace{(\gamma^a \gamma^b + \gamma^b \gamma^a - \frac{2}{n}\gamma^c \gamma_c g^{ab})}_{\odot 2_0 \to \mathrm{End}(\Sigma),\ \text{thus } 0} + \frac{2}{n} \underbrace{\gamma^c \gamma_c}_{-n} g^{ab}.$$

The value $-n$ in the last underbrace is a consequence of our choice of normalization just above.

The Clifford section γ now plays a role analogous to that of the metric tensor g – it is a section which encodes information about the geometry, allowing us to compute with objects of Lie theoretic origin without always referring to things like root systems. That is, it is a vector bundle section that is somehow characteristic of the geometry. More precisely, it is the triple (g, E, γ), where E is a choice of unit volume form, that characterized Riemannian spin geometry – any universal calculation in the theory is a calculation with these objects, subject to a finitely generated set of universal identities. This is underlined by the fact that we may, in hindsight, describe the even-dimensional splitting $\Sigma = \Sigma_+ \oplus \Sigma_-$ using the Clifford section. We contract the volume form $E_{a_1 \ldots a_n}$ with the tensor argument on the gammas to form a *chirality operator*

$$\tilde{\chi} := \frac{1}{n!} E_{a_1 \ldots a_n} \gamma^{a_1} \cdots \gamma^{a_n}.$$

One computes that $\tilde{\chi}2 = (-1)^{\left[\frac{n+1}{2}\right]}$. If n is odd, $\tilde{\chi} = \text{const} \cdot \mathrm{Id}_\Sigma$ by irreducibility (Schur's Lemma). If n is even, there are two eigenmodules for $\tilde{\chi}$, and

$$\chi := \mathrm{i}^{n/2} \tilde{\chi} = \begin{pmatrix} 1 & 0 \\ 0 & -1 \end{pmatrix}$$

in the corresponding block decomposition. This gives exactly the $\Sigma_+ \oplus \Sigma_-$ splitting. We could also have chosen $(-\mathrm{i})^{n/2} \tilde{\chi}$ as our definition of χ, but the choice above turns out to make better contact with weight theoretic conventions for the Lie algebra $\mathfrak{so}(n)$.

9.1.3. Bundles with Spin(n) structure. There are various ways to capture the additional Spin(n) bundles that are not SO(n) bundles. First, one could take bundles associated to other irreducible Spin(n) modules; there are infinitely many of these which are not already SO(n) modules. Second, one could take iterated tensor products of Σ, and pick out the (irreducible) summands; we get infinitely many different isomorphism classes of modules in this way, though we also get infinitely many duplicate copies of each. Finally, we could take the transition functions relating trivializations of the bundle Σ; these will be valued in Spin(n). Composing these transition functions with various representations of Spin(n), we get new sets of transition functions:

$$\mathcal{O}_\alpha \cap \mathcal{O}_\beta \xrightarrow{\text{transition } \tau_{\beta\alpha}} \mathrm{Spin}(n) \xrightarrow{\text{representation } \lambda} \mathrm{Aut}(V). \tag{9.1}$$

These three processes are really equivalent. For an example of the second approach, take

$$\Lambda^1 \otimes \Sigma \cong_{\mathrm{Spin}(n)} \Sigma \oplus \underbrace{\mathbf{Tw}}_{\text{new}}. \tag{9.2}$$

This allows us to discover the *twistor bundle* **Tw**. It also suggests a tensor-spinor *realization* of this bundle. First take spinor-1-forms φ_a, where the index indicates the Λ^1 content. From a general such thing, we may contract to make a spinor $\gamma^a \varphi_a$; this corresponds to the Σ summand in (9.2). A natural way to inject this into the spinor-1-forms is to form $\gamma_a \gamma^b \varphi_b$, or some constant multiple of this. The precise constant multiple is determined by the desire to have the other summand live in an irreducible bundle, and thus be annihilated (by Schur's Lemma) by the contraction with γ:

(9.3)
$$\varphi_a = \varphi_a + \underbrace{\frac{1}{n}\gamma_a\gamma^b\varphi_b}_{\textbf{Tw}\text{ part}} \underbrace{-\frac{1}{n}\gamma_a\gamma^b\varphi_b}_{\Sigma\text{ part}}\,.$$

Thus *twistors* are spinor-1-forms φ_a with $\gamma^a\varphi_a = 0$. Like the spinors, these are irreducible for odd n, and split into two pieces (another chiral decomposition) when n is even.

There are some obvious generalizations of this construction, some of which are under intense scrutiny in current research. For example, one could start with the spinor-k-forms $\Sigma\otimes\Lambda^k$ as in [**Bra84**]. The generalization of the map used just above is the *interior Clifford multiplication*

(9.4)
$$\begin{aligned} \operatorname{int}(\gamma) : \Sigma\otimes\Lambda^k &\to \Sigma\otimes\Lambda^{k-1}, \\ \varphi_{a_1\dots a_k} &\mapsto \gamma^{a_1}\varphi_{a_1\dots a_k}. \end{aligned}$$

Since it is possible for powers of $\operatorname{int}(\gamma)$ to annihilate a spinor-form without $\operatorname{int}(\gamma)$ itself annihilating, $\Sigma\otimes\Lambda^k$ will generally break up into several irreducible pieces [**RDSS92**]. We shall denote the "top piece," i.e. the null bundle for $\operatorname{int}(\gamma)$, by $\mathbf{Y}_k$. Alternatively, we could replace Λ^k above with the symmetric tensors $\odot^k$, and define $\operatorname{int}(\gamma)$ in formally the same way. The null bundle $\mathbf{Z}_k$ for $\operatorname{int}(\gamma)$ will automatically consist of spinor-*trace-free*-symmetric k-tensors, as

(9.5)
$$\gamma^{a_1}\varphi_{a_1\dots a_k} = 0 \;\Rightarrow\; 0 = \gamma^{a_2}\gamma^{a_1}\varphi_{a_1\dots a_k} = -g^{a_1a_2}\varphi_{a_1\dots a_k}\,,$$

by symmetry and the Clifford relation. By symmetry again, if the a_1a_2 trace vanishes, all traces must vanish.

9.1.4. Connections. It is sometimes said that a differential geometer is someone who knows what a connection is. In spin geometry, as in other branches of geometry, an understanding of connections is indeed a kind of gateway to the theory.

The Levi-Civita connection ∇ on the orthonormal frame bundle $\mathcal{F}$ (see, e.g., [**Hel78b**]) lifts to a connection on the bundle $\mathcal{S}$ of spin frames. Transferring to associated bundles, or proliferating the connection to duals and tensor products, we get a connection ∇ on all $\mathrm{Spin}(n)$-bundles, satisfying

$$\nabla g = 0, \quad \nabla E = 0, \quad \nabla\gamma = 0.$$

On the spinor bundle, there is a family of so-called *Friedrich connections* [**Fri80**]

(9.6)
$$\nabla_a + \text{const}\cdot\gamma_a$$

which are arguably just as distinguished as the spin connection, and which are quite useful in problems involving Dirac eigenspinors.

9.1.5. Stein-Weiss gradients. There is a universal construction of first-order differential operators in Riemannian spin geometry due to Stein and Weiss [**SW68**]. First note that if $\mathbf{V}$ is a natural $\mathrm{Spin}(n)$-bundle, then

$$\nabla : \mathbf{V} \to T^*M\otimes\mathbf{V}.$$

One way to think of this is as follows: if a section φ of $\mathbf{V}$ has index structure φ_I, then $\nabla\varphi$ has index structure $\nabla_a\varphi_I$. That is, an extra cotangent index has been added, signalling tensor product with the contangent bundle.

As it happens, the defining representation T of $\mathfrak{so}(n)$ has a multiplicity free *selection rule* governing its tensor products with other $\mathfrak{so}(n)$-modules. This means that if V

is irreducible, then $T \otimes V$ breaks up as $W_1 \oplus \cdots \oplus W_N$, for some N, with the various W_u coming from distinct isomorphism classes. On the bundle level,

$$\begin{aligned} T^*M \otimes \mathbf{V} &\cong_{\mathrm{Spin}(n)} \mathbf{W}_1 \oplus \cdots \oplus \mathbf{W}_{N(\mathbf{V})}, \\ \mathbf{W}_u \cong_{\mathrm{Spin}(n)} \mathbf{W}_v &\Rightarrow u = v. \end{aligned} \tag{9.7}$$

Here we write $N = N(\mathbf{V})$ to highlight the fact that N depends on $\mathbf{V}$ – in fact it is a very useful numerical invariant.

This allows us to speak of projections

$$\mathrm{Proj}_u := \mathrm{Proj}_{\mathbf{W}_u}$$

on the bundle $T^* \otimes \mathbf{V}$, and thus of *gradients*

$$G_u = \mathrm{Proj}_{\mathbf{W}_u} \circ \nabla.$$

These are universal, $\mathrm{Spin}(n)$-equivariant operators, since ∇ and the $\mathrm{Proj}_{\mathbf{W}_u}$ are. These operators are unique up to isomorphic realization of bundles and up to constant factors; because of this, they actually give a *classification* of equivariant first-order operators.

For example, take the spinor bundle Σ, in odd dimensions for now. The decomposition (9.3) exactly realizes the decomposition in (9.7), and the corresponding gradients are

$$\begin{aligned} \mathbf{T} = G_{\mathbf{Tw}} : \psi &\mapsto \nabla_a \psi + \tfrac{1}{n}\gamma_a \gamma^b \nabla_b \psi, \\ G_\Sigma : \psi &\mapsto -\tfrac{1}{n}\gamma_a \gamma^b \nabla_b \psi. \end{aligned} \tag{9.8}$$

In the formula for G_Σ, we are just finding a different realization of the operator

$$\mathcal{D} : \psi \mapsto \gamma^b \nabla_b \psi;$$

i.e. of the *Dirac operator.* (In fact, to get to the standard realization, just left-multiply the formula for G_Σ by γ^a.) $\mathbf{T}$ is called the *twistor operator.* In this sense $\mathcal{D}$ and $\mathbf{T}$ are the two gradients emanating from the spinor bundle, and in particular $N(\Sigma) = 2$.

In even dimensions, we have the chiral decomposition, so we get two Dirac operators and two twistor operators:

$$\mathcal{D} : \Sigma_\pm \to \Sigma_\mp, \quad \mathbf{T} : \Sigma_\pm \to \mathbf{Tw}_\pm.$$

That $\mathcal{D}$ should reverse chirality is clear from the observation that ∇_b (resp. γ^b) should preserve (resp. reverse) chirality.

Other examples of gradients, or (when reducible bundles are involved) operators with a block decomposition whose entries are gradients, are the following. The *exterior derivative* on differential forms is

$$d : \Lambda^k \to \Lambda^{k+1}, \quad (d\varphi)_{a_0 a_1 \ldots a_k} = (k+1)\nabla_{[a_0}\varphi_{a_1 \ldots a_k]}. \tag{9.9}$$

Thus the usual normalization of d is $k+1$ times the gradient from Λ^k to Λ^{k+1}. Note that in abstract index notation, an expression like $\nabla_a \varphi_b$ means $(\nabla\varphi)_{ab}$. The *conformal Killing operator* S carrying Λ^1 to the trace-free symmetric 2-tensors $\odot 2_0$ is

$$(S\eta)_{ab} = \nabla_{(a}\eta_{b)} - \frac{1}{n} g_{ab} \nabla_c \eta^c.$$

As usual, the parentheses around the indices indicate symmetrization. The name of this operator derives from the fact that its null space (viewed as a space of vector fields) consists exactly of the *conformal vector fields* X; i.e. those vector fields for which

$$\mathcal{L}_X g = 2\omega g \tag{9.10}$$

for some smooth function ω, where $\mathcal{L}$ is the Lie derivative. In index notation, (9.10) translates to $\nabla_{(a}X_{b)} = \omega g_{ab}$, and contracting, we find that $\omega = \frac{1}{n}\nabla_c X^c$.

By the basic uniqueness result, the formal adjoint G^* of a gradient G must be a gradient. If G is realized as above, i.e. as going from $\mathbf{V}$ to a subbundle of $T^*M \otimes \mathbf{V}$, then G^* is always a divergence in the additional T^*M tensor argument:

$$G^* = (\mathrm{Proj}_u \nabla)^* = \nabla^* \mathrm{Proj}_u^* = \nabla^* \mathrm{Inj}_u\,,$$

where Inj_u injects the copy of $\mathbf{W}_u$ into $T^*M \otimes \mathbf{V}$. For example, we have the interior derivative $\delta : \Lambda^{k+1} \to \Lambda^k$, defined by

$$(\delta\varphi)_{a_1 \ldots a_k} = -\nabla^{a_0}\varphi_{a_0 \ldots a_k}\,.$$

REMARK 9.1. The alert reader will notice an apparent normalization problem when comparing with (9.9). This is because δ is conventionally computed in the form inner product $(k!)^{-1}\varphi^{a_1\ldots a_k}\psi_{a_1\ldots a_k}$, whereas the above general discussion uses an inner product on $T^*M \otimes \mathbf{V}$ with the property that $|\xi \otimes \varphi|^2 = |\xi|^2|\varphi|^2$. This "default" inner product, when applied to building differential form inner products, leads to an inner product without the $(k!)^{-1}$ formula above. It is easily verified that d and δ are formal adjoints in the form inner product, and that the gradient $(k+1)^{-1}d$ and δ are formal adjoints in the default inner product.

The above implies (as the reader may check explicitly) that

$$(S^*\tau)_b = -\nabla^a \tau_{ab}\,, \quad (\mathbf{T}\varphi)^* = -\nabla^a \varphi_a \tag{9.11}$$

are the formal adjoints of the conformal Killing and twistor operators respectively.

Note that when speaking of formal adjoints, a statement like $\mathcal{D}^* = \mathcal{D}$, while true, hides some subleties. First, recall from above that $\mathcal{D}$ exchanges the two chiralities:

$$\Sigma_+ \underset{\mathcal{D}_-}{\overset{\mathcal{D}_+}{\rightleftarrows}} \Sigma_-\,.$$

In dimension $4k$, each of Σ_+ and Σ_- is self-dual, and $\mathcal{D}_+$ and $\mathcal{D}_-$ are formal adjoints of each other. In dimension $4k+2$, Σ_+ and Σ_- are duals, so each of $\mathcal{D}_+$ and $\mathcal{D}_-$ is formally self-adjoint.

9.2. Conformal structure

The Dirac operator is deeply related to *conformal structure* on manifolds. A conformal structure is an equivalence class of Riemannian metrics, with g equivalent (or *conformal*) to $\hat{g}$ iff $\hat{g} = \Omega^2 g$ for some positive smooth function Ω. In fact, not only the Dirac operator, but *every* gradient depends only on conformal structure, in a suitable sense.

9.2.1. Conformal covariance and weights. Let g and $\hat{g} = \Omega 2g$ be two conformal metrics. Since the volume element and Clifford section have metric-related properties ($|E|^2 = 1$ and the Clifford relation), they need to be given a compatible pointwise scaling:

$$\hat{E} = \Omega^n E, \quad \hat{\gamma} = \Omega^{-1}\gamma.$$

THEOREM 9.2. [**Feg76**] *Each gradient* $G : \mathbf{V} \mapsto \mathbf{W}$ *is* conformally covariant*: there are numbers* a, b *with*

$$\hat{G}\varphi = \Omega^{-b} G(\Omega^{a}\varphi)$$

for all sections φ *of* $\mathbf{V}$.

The *conformal biweight* (a, b) may move, depending on what realizations of the bundles we use. For example, viewing the conformal Killing operator S as acting on tangent, as opposed to cotangent, vector fields results in different conformal biweights. To understand and control this, we need to take a broader view: there's really a structure group larger than $\mathrm{Spin}(n)$ at work in statements like these, namely $\mathbf{R}_{+} \times \mathrm{Spin}(n)$. The $\mathbf{R}_{+}$ part has something to do with the pointwise scaling. To make these statements precise, we need to know something about *weights* of modules for (some) Lie algebras.

Irreducible $\mathrm{Spin}(n)$ modules (thus bundles on spinnable manifolds) are parameterized by the *dominant* $\mathfrak{so}(n)$ *weights*

$$\lambda = (\lambda_1, \dots, \lambda_\ell) \in \mathbf{Z}^{\ell} \cup \left(\frac{1}{2} + \mathbf{Z}\right)^{\ell},$$

where $\ell = [n/2]$. The dominance condition is

$$\begin{array}{ll} \lambda_1 \geq \cdots \geq |\lambda_\ell|, & n \text{ even}, \\ \lambda_1 \geq \cdots \geq \lambda_\ell \geq 0, & n \text{ odd}. \end{array}$$

λ is the *highest weight* of the module, which we may call $V(\lambda)$. We shall call the corresponding bundle $\mathbf{V}(\lambda)$. In general, *weights* parameterize the possible irreducible representations of a (fixed choice of) maximal abelian subalgebra $\mathfrak{t}$. Such irreducible representations are necessarily one-dimensional, and are given by

$$m^1 X_1 + \cdots + m^{\ell} X_{\ell} \mapsto (\mu_1 m 1 + \cdots + \mu_{\ell} m^{\ell})\mathrm{Id},$$

where the X_a form a basis of $\mathfrak{t}$. The ℓ-tuple μ is the *weight* of the representation.

In the case of $\mathfrak{so}(n)$, this plays out as follows [**SW68, Str88**]. In the usual matrix presentation, $\mathfrak{so}(n)$ is the Lie algebra of skew-symmetric $n \times n$ matrices. A basis for this consists of the L_{ab} for $a < b$, where L_{ab} has a 1 in the (a, b) entry, and a -1 in the (b, a) entry. A good choice for $\mathfrak{t}$ is the span of $L_{12}, L_{34}, \dots, L_{2\ell-1,2\ell}$.

Each irreducible $\mathfrak{so}(n)$ module has a lexicographically maximal weight, which is necessarily dominant, and necessarily occurs with multiplicity 1; this is the highest weight. Some Lie algebra theory forces *integrality* conditions on the weights, which, in the present case, work out to the $\mathbf{Z}^{\ell} \cup \left(\frac{1}{2} + \mathbf{Z}\right)^{\ell}$ condition above.

Some familiar modules/bundles and their highest weights are:

$\Lambda^k,\ k < \ell$	$(\overbrace{1, \dots, 1}^{k}, 0, \dots, 0)$
$\Lambda^{\ell}_{\pm}$ (n even)	$(1, \dots, 1, \pm 1)$
$\odot^p_0$	$(p, 0, \dots, 0)$
Σ (n odd)	$(\frac{1}{2}, \dots, \frac{1}{2})$
$\Sigma_{\pm}$ (n even)	$(\frac{1}{2}, \dots, \frac{1}{2}, \pm\frac{1}{2})$
$\mathbf{Tw}$ (n odd)	$(\frac{3}{2}, \frac{1}{2}, \dots, \frac{1}{2})$
$\mathbf{Tw}_{\pm}$ ($4 \leq n$ even)	$(\frac{3}{2}, \frac{1}{2}, \dots, \frac{1}{2}, \pm\frac{1}{2})$
$\mathbf{Tw}_{\pm}$ ($n = 2$)	$(\pm\frac{3}{2})$
algebraic Weyl tensors ($n \geq 5$)	$(2, 2, 0, \dots, 0)$
algebraic $\mathrm{Weyl}_{\pm}$ tensors ($n = 4$)	$(2, \pm 2)$

Recall there is a gradient $\mathbf{V}(\lambda) \to \mathbf{V}(\mu)$ whenever $T^*M \otimes \mathbf{V}(\lambda)$ has a $\mathbf{V}(\mu)$ summand, i.e. whenever

$$(1) \otimes \lambda \cong_{\mathrm{Spin}(n)} \mu \oplus \cdots$$

on the level of representations. This happens iff

$$\begin{gathered} \mu = \lambda \pm e_a, \quad \text{some } a \in \{1, \dots, \ell\}, \\ \underline{\mathbf{or}} \\ n \text{ is odd}, \ \lambda_\ell \neq 0, \ \mu = \lambda. \end{gathered} \tag{9.12}$$

Some important gradients in odd dimensions are the *self-gradients* – operators associated to the exceptional case in (9.12). Among these are the Dirac operator, the *Rarita-Schwinger operator* $\mathcal{S}$ on $\mathbf{Tw}$ (see below), and the operator $\star d$ on Λ^ℓ. In this last example, $\star$ is the *Hodge star operator* taking k-forms to $(n-k)$-forms:

$$(\star\varphi)_{a_{k+1}\dots a_n} := E_{a_1\dots a_n}\varphi^{a_1\dots a_k}.$$

In even dimensions, there will be gradients

$$\mathbf{V}(\sigma, \pm\tfrac{1}{2}) \to \mathbf{V}(\sigma, \mp\tfrac{1}{2})$$

whenever σ is an $(\ell-1)$-tuple with $(\sigma, \frac{1}{2})$ dominant and integral. By the above table of weights, the Dirac operators will be among these. It is then natural to speak of the *Dirac operator* on $\mathbf{V}(\frac{1}{2}, \dots, \frac{1}{2}) \oplus \mathbf{V}(\frac{1}{2}, \dots, \frac{1}{2}, -\frac{1}{2})$.

We can see that there will be an operator on twistors, $\mathcal{R} : \mathbf{Tw} \to \mathbf{Tw}$ in odd dimensions, and for $4 \leq n$ even,

$$\mathbf{Tw}_+ \underset{\mathcal{R}_-}{\overset{\mathcal{R}_+}{\rightleftarrows}} \mathbf{Tw}_-\,.$$

This is the *Rarita-Schwinger* operator. This also makes it clear why there should be no Rarita-Schwinger operator in dimension 2, since this would have to try to take $\mathbf{V}(\frac{3}{2})$ to itself or $\mathbf{V}(-\frac{3}{2})$.

In fact, we may make an elementary construction of the Rarita-Schwinger operator $\mathcal{S}$ by forming the Dirac symbol on φ_a, that is $\gamma^b\nabla_b\varphi_a$, and then adding a *counterterm* designed to attain the condition $\gamma^a(\mathcal{S}\varphi)_a = 0$:

$$(\mathcal{S}\varphi)_a = \gamma^b\nabla_b\varphi_a - \frac{2}{n}\gamma_a\nabla^b\varphi_b.$$

A somewhat better known set of gradients is that emanating from $\mathbf{V}(1) \cong \Lambda^1$. (Here we adopt the convention of omitting terminal strings of zeros in weights.) By the selection rule (9.12), these should go to $\mathbf{V}(2)$, $\mathbf{V}(1,1)$, and $\mathbf{V}(0)$ for $n \geq 5$. These bundles are exactly $\odot 2_0$, $\Lambda 2$, and Λ^0, and the corresponding operators are (up to constants and isomorphic realizations) S, d, and δ. More generally, there is an S-like operator on Λ^k [**Bra96**], in addition to the familiar d and δ.

Returning to our discussion of conformal structure, note that we have less to work with than we did in the Riemannian regime – g is determined only up to a positive function multiple. This may be reflected by taking a *larger* number of trivializing neighborhoods $(\mathcal{O}_\alpha, g_\alpha)$; here g_α is a metric on $\mathcal{O}_\alpha$. On an overlap $\mathcal{O}_\alpha \cap \mathcal{O}_\beta$, we have $g_\beta = \Omega^2_{\beta\alpha}g_\alpha$, and we may build a bundle $\mathbf{V}(w|\lambda)$ with transition functions

$$\Omega^w_{\beta\alpha}\tau_{\beta\alpha}$$

where $\tau_{\beta\alpha}$ are (Riemannian spin) transitions $(\mathcal{O}_\alpha, g_\alpha) \to (\mathcal{O}_\beta, g_\alpha)$. What we get out of this are tensor-spinor *density* bundles – corresponding to each old (Riemannian) bundle $\mathbf{V}(\lambda)$, we now have a whole family $\mathbf{V}[w|\lambda]$ of bundles, parameterized by a complex number w. Since $\mathbf{V}[w|\lambda]$ is naturally isomorphic to $\mathbf{V}[w|0] \otimes \mathbf{V}[0|\lambda]$, it would have been sufficient to define the *density bundles* $\mathcal{E}[w] := \mathbf{V}[w|0]$ (with powers of $\Omega_{\beta\alpha}$ as their transitions), and the tensor-spinors of *conformal weight* $w = 0$.

REMARK 9.3. The tangent bundle is naturally isomorphic to $\mathbf{V}[1|1]$, while the cotangent bundle is $\mathbf{V}[-1|1]$.

In the case of tensors, this means that we can do some of our weight accounting by using index position. For example, if $\mathcal{E}^a$ (resp. $\mathcal{E}_a$) is the tangent (resp. cotangent) bundle, and we put, for example, $\mathcal{E}_a{}^b[w] := \mathcal{E}[w] \otimes \mathcal{E}_a \otimes \mathcal{E}^b$, then

$$\mathcal{E}_a{}^b[w] \cong \mathcal{E}_{ab}[w+2] \cong \mathcal{E}^{ab}[w-2].$$

In fact, this explains why conformal weights and biweights move around as we move among different isomorphic realizations of bundles.

To describe the conformal biweights of the gradients, and of many other operators as well, consider the *rho-shift* of a weight $[w|\lambda]$:

$$\begin{aligned} (\tilde{w}|\tilde{\lambda}) &= (w + n/2|\lambda + \rho_n), \quad \text{where} \\ 2\rho_n &= (n-2, n-4, \ldots, n-2\ell). \end{aligned}$$

The *affine Weyl group* acts on such shifted weights, according to permutations, and

an even number of sign changes,	n even,
any number of sign changes,	n odd.

The equivalence relation of lying in the same affine Weyl orbit will be denoted by $\sim$.

The first necessary condition for having a conformally invariant differential operator $\mathbf{V}[w|\lambda] \to \mathbf{V}[u|\mu]$ is that

$$(\tilde{w}|\tilde{\lambda}) \sim (\tilde{u}|\tilde{\mu}).$$

That is, the bundles must have the same *central character* under the algebra $\mathfrak{so}(n+1,1)$. The relevance of this algebra will be explored in somewhat more detail later, in Section 9.3. (The connection between affine Weyl orbit and central character, which measures the effect of *Casimir elements* of the enveloping algebra, is made via a deep theorem of Harish-Chandra, which we shall not attempt to describe here.) In addition, the drop in conformal weight, $w - u$, must be the order of the differential operator. Note that our "operators" are really functors that assign operators to conformal structures, or to pseudo-Riemannian structures, depending on the context. We adopt the convention that our operators must be nonzero on flat space, so that (for example) $K\Delta$ is not an operator. The condition about the drop in conformal weight is then just a homogeneity condition under uniform scaling of the metric. If our prospective conformally invariant operator is a gradient, we must thus have $u = w - 1$.

Suppose there is a gradient $G : \mathbf{V}(\lambda) \to \mathbf{V}(\mu)$. If we choose any conformal weight w_0 and view G as carrying $\mathbf{V}[w_0|\lambda]$ to $\mathbf{V}[w_0 - 1|\mu]$, some invariant theory shows that the conformal variation of G along the curve of metrics $e^{2\varepsilon\omega} g_0$ must take the form

$$(w_0 + c)[G, m_\omega],$$

where m_ω is multiplication by ω, and c is a constant. Thus we have an invariant operator when $w_0 = -c$. Since we now know that a given gradient is conformally invariant for *some* choice of the initial conformal weight, we may determine *which* choice does the trick by using the above central character considerations. For example, if we have a self-gradient $\mathbf{V}(\star|\lambda) \to \mathbf{V}(\star|\lambda)$ in odd dimensions (where the stars $\star$ represent numbers we do not necessarily know), then we must have

$$\mathbf{V}(\tfrac{1}{2}|\lambda) \to \mathbf{V}(-\tfrac{1}{2}|\lambda).$$

REMARK 9.4. More generally, it is easily verified that if our gradient takes $\mathbf{V}(\star|\cdots\lambda_i\cdots)$ to $\mathbf{V}(\star|\cdots\lambda_i+\varepsilon\cdots)$, where $\varepsilon = \pm 1$ and the corresponding lists $\cdots$ agree, then we must have

$$\mathbf{V}(\varepsilon(\lambda_i+\varepsilon)|\cdots\lambda_i\cdots) \to \mathbf{V}(\varepsilon\lambda_i|\cdots\lambda_{i+1}\cdots).$$

For example, to work out the conformal biweight of the conformal Killing operator, we reason as follows:

$$\begin{aligned}
\mathbf{V}[\star|1] &\to \mathbf{V}[\star|2],\\
\mathbf{V}\left(\star|\tfrac{n}{2},\tfrac{n-4}{2},\dots,\tfrac{n-2\ell}{2}\right) &\to \mathbf{V}\left(\star|\tfrac{n+2}{2},\tfrac{n-4}{2},\dots,\tfrac{n-2\ell}{2}\right),\\
\mathbf{V}\left(\tfrac{n+2}{2}|\tfrac{n}{2},\tfrac{n-4}{2},\dots,\tfrac{n-2\ell}{2}\right) &\to \mathbf{V}\left(\tfrac{n}{2}|\tfrac{n+2}{2},\tfrac{n-4}{2},\dots,\tfrac{n-2\ell}{2}\right),\\
\mathbf{V}[1|1] &\to \mathbf{V}[0|2],\\
\mathcal{E}^a &\to \mathcal{E}_{(ab)_0}[2],
\end{aligned}$$

where $(\ )_0$ is trace-free symmetrization.

When spinors are in play, conventions vary on the treatment of indices. One choice is to suppress spin indices altogether, and another is to write them, assigning weights of $\pm 1/2$ (depending on position) in analogy with the weight ± 1 of a tensor index. We shall have occasion to consider this briefly later, in Subsection 9.3.1.

REMARK 9.5. *Weyl's dimension formula* [**Hum72**] gives a formula for the dimension of the module $V(\lambda)$, and thus for the fiber dimension of $\mathbf{V}(\lambda)$, in terms of the entries of λ. For example in dimension 4,

$$\dim V(\lambda) = \tilde{\lambda}_1^2 - \tilde{\lambda}_2^2. \tag{9.13}$$

For example, one easily sees that the fiber dimensions of the half-spinor bundles $\Sigma_\pm$, as well as the form bundles $\Lambda 1$, $\Lambda 2_\pm$, and $\Lambda 3$ work out as they should. One also gets less obvious but still quite useful information from (9.13); for example the fiber dimensions of the bundles $\mathcal{W}_\pm$ of algebraic Weyl tensors (namely $32 - 22 = 5$), and of $\odot_0^p$, the trace-free symmetric p-tensors (namely $(p+1)2$). In higher dimensions, the best packaging of dimension calculations is often inductive, since the numerator and denominator in Weyl's dimension formula tend to have many common factors.

Though it is not obvious, the above information is almost enough to compute the spectra of all natural operators on bundles over the round sphere S^n; in particular the G^*G when G is a gradient [**Bra97b**]. Implicit in this are the spectra of self-gradients [**Bra99**]. This recovers old results on, for example, the spectrum of the Dirac operator. But it is general, so that one may plug in weight parameters and get, for example, the spectrum of the self-gradient on $\mathbf{V}(p+\tfrac{1}{2},\tfrac{1}{2},\dots,\tfrac{1}{2},\pm\tfrac{1}{2})$; in fact, this is the bundle described in (9.5) above. This operator has been an object of considerable recent study.

This in turn says something about classifying Bochner-Weitzenböck (BW) formulas on *general* Riemannian spin manifolds [**Bra97b**]. There are deep connections between this and estimates that can be made on solutions of first-order elliptic differential systems [**CGH00b, Bra00**], with consequences for decay rates of these solutions on open manifolds. Indeed, this last question actually inspired the original invention of gradients on the part of Stein and Weiss.

9.2.2. Bochner-Weitzenböck formulas and a discrete leading symbol. Here is a partial illustration of all this in a "small" situation. The ideas are closely related to those of the general calculation, but no heavy machinery is required. One may compute (from the conformal variations of the gradients and scalar curvature) that for certain explicit constants k_u,

$$\frac{K}{2(n-1)} - \sum_u k_u G_u^* G_u \tag{9.14}$$

is conformally invariant

$$\mathbf{V}\left(\frac{2-n}{2}\middle|\lambda\right) \to \mathbf{V}\left(\frac{-2-n}{2}\middle|\lambda\right),$$

for each λ [**Bra96**]. In fact, if G_u carries $\mathbf{V}(\lambda)$ to $\mathbf{V}(\sigma_u)$,

$$k_u = 2(1 + \langle \lambda + \sigma_u + 2\rho_n \,, \lambda - \sigma_u \rangle)^{-1}.$$

(In case k_u is undefined, our invariant operator is $G_u^* G_u$.) In the most elementary example, $\lambda = 0$, we get (up to a factor) the *Yamabe operator* or *conformal Laplacian*

$$Y := \Delta + \frac{n-2}{4(n-1)} K. \tag{9.15}$$

The operator (9.14) need not be second-order in general – it could be zero, or at least have order zero, for some bundles $\mathbf{V}(\lambda)$. In fact, it turns out to have order 2 exactly when the numerical invariant $N(\mathbf{V}(\lambda)) =: N(\lambda)$ (recall (9.7)) is *odd.* By some invariant theory, being of order less than 2 and being conformally invariant for $N(\lambda)$ even, expression (9.14) must be an action of the Weyl tensor C. In particular, this discussion applies if our bundle is Σ (n odd) or $\Sigma_\pm$ (n even), since then $N(\lambda) = 2$. Furthermore, the Weyl tensor cannot act, by weight considerations. Thus the expression (9.14) vanishes.

As a result, we have two equations on the two $G_u^* G_u$:

$$\begin{aligned} \frac{K}{2(n-1)} + \frac{2}{n-1} G_1^* G_1 - 2 G_2^* G_2 &= 0, \\ G_1^* G_1 + G_2^* G_2 &= \nabla^* \nabla. \end{aligned} \tag{9.16}$$

The second of these follows directly from the definition of gradients. Here and below, we shall order gradients lexicographically according to their target weights σ_u, the largest weights coming first. In particular, σ_1 is always $\lambda + e_1$. Solving the system (9.16) for the $G_u^* G_u$, we get

$$\begin{aligned} n G_2^* G_2 &= \nabla^* \nabla + \frac{K}{4}, \\ \frac{n}{n-1} G_1^* G_1 &= \nabla^* \nabla - \frac{K}{4(n-1)}. \end{aligned} \tag{9.17}$$

These are, respectively, the *Lichnerowicz formula* and *Friedrich formula.* After taking account of the normalization change resulting from realizing the spinor bundle in different ways (as Σ, and in $T^* \otimes \Sigma$), the Lichnerowicz formula becomes

$$\mathcal{D}^2 = \nabla^* \nabla + \frac{K}{4}. \tag{9.18}$$

The left side of the Friedrich formula is $T^* T$, where T is the twistor operator.

To do this in a more general setting, we use a device called the *discrete leading symbol* [**AB02b**]. To use this, we need a complete understanding of the model case of round S^n. Here the section space of $\mathbf{V}(\lambda)$ is (a completion of) a direct sum of modules,

$$\mathcal{V}(\alpha;\lambda) \cong_{\mathrm{Spin}(n+1)} \overline{\mathbf{V}}(\alpha)$$

for $\mathrm{Spin}(n+1)$, where we use an overline to signal the transition from n to $n+1$. The $\mathrm{Spin}(n+1)$ module labelled by α occurs (with multiplicity 1) iff $\alpha|_{\mathrm{Spin}(n)}$ has a λ summand – this is an instance of *Frobenius reciprocity.* If α occurs, we say that $\alpha \downarrow \lambda$ or $\lambda \uparrow \alpha$. In fact, $\downarrow$ is an example of a *branching rule*, in this case the one describing restriction of $\mathrm{Spin}(n+1)$ representations to a $\mathrm{Spin}(n)$ subgroup (which is imbedded in the standard way). In terms of weight arithmetic, this branching rule is given by the *interlacing* condition

$$\begin{aligned} &\alpha_1 \ge \lambda_1 \ge \alpha_2 \ge \lambda_2 \ge \cdots \alpha_\ell \ge |\lambda_\ell|, && n \text{ even},\\ &\alpha_1 \ge \lambda_1 \ge \alpha_2 \ge \lambda_2 \ge \cdots \ge \lambda_\ell \ge |\alpha_{\ell+1}|, && n \text{ odd}. \end{aligned}$$

We have

$$\Gamma(\mathbf{V}(\lambda)) = \bigoplus_{\alpha\downarrow\lambda} \mathcal{V}(\alpha;\lambda) \cong_{\mathrm{Spin}(n+1)} \bigoplus_{\alpha\downarrow\lambda} \overline{\mathbf{V}}(\alpha).$$

A curious but useful aspect of the interlacing rule reveals itself upon branching twice, from $\mathrm{Spin}(n+1)$ to $\mathrm{Spin}(n)$ to $\mathrm{Spin}(n-1)$. Let $L := [(n+1)/2]$. If

$$\mathring{\alpha} := (\alpha_2, \ldots, \alpha_L),$$

then

$$\alpha \downarrow \lambda \iff \lambda \downarrow \mathring{\alpha} \text{ and } \alpha_1 - \lambda_1 \in \mathbf{N}.$$

Note that $\mathring{\alpha}$ runs through a finite set,

$$\mathcal{A}_\lambda := \{\mathring{\alpha} \mid \alpha \downarrow \lambda\},$$

while α_1 runs through an infinite set.

For a *natural differential operator* on $\mathbf{V}(\lambda)$, there are two descriptions of the discrete leading symbol, one pointwise and one global:

(1) Pick a point (x,ξ) in the cotangent bundle of an *arbitrary* M (with the Riemannian spin structure), with $\xi \neq 0$. This choice of a distinguished direction ξ leaves us with a symmetry group $\mathrm{Spin}(n-1)_\xi$ isomorphic to $\mathrm{Spin}(n-1)$. Take the leading symbol $\sigma_{\mathrm{lead}}(D)(x,\xi) \in \mathrm{End}(\mathbf{V}(\lambda)_x)$, and decompose relative to the $\mathrm{Spin}(n-1)_\xi$-splitting:

$$\mathbf{V}(\lambda)_x = \underline{\mathbf{V}}(\beta_1)_\xi \oplus \cdots \oplus \underline{\mathbf{V}}(\beta_{b(\lambda)})_\xi\,,$$

where

$$\underline{\mathbf{V}}(\beta_i)_\xi \cong_{\mathrm{Spin}(n-1)} \underline{\mathbf{V}}(\beta_i).$$

Here the underline signals the transition from n to $n-1$. We get

$$\sigma_{\mathrm{lead}}(D)(x,\xi) = \sum_i \mu_i(D)\,\mathrm{Id}_{\underline{\mathbf{V}}(\beta_i)_\xi}$$

for some eigenvalues $\mu_i(D)$. A main point is that *the list of* $(\beta_i, \mu_i(D))$ *is independent of* x, ξ, *and* M. In fact the β_i just list the $\beta \uparrow \lambda$. So do the $\mathring{\alpha}$ for $\alpha \downarrow \lambda$, so we can speak of $\mu(D, \mathring{\alpha})$.

(2) Take the special case of the sphere, and take the leading asymptotics of $\mathrm{spec}(D)$, in the sense

$$\mathrm{eig}(D, S^n, \alpha) = \mathbf{k}(D, \mathring{\alpha})\alpha_1^d + o(\alpha_1^d) \quad \text{as } \alpha_1 \to \infty.$$

Of course part of the problem is to show that one *has* such asymptotics.

THEOREM 9.6. **[AB02b]** *These are the same:* $\mu(D, \mathring{\alpha}) = \mathbf{k}(D, \mathring{\alpha})$.

REMARK 9.7. The calculation of the spectrum makes essential use of conformal covariance. The noncompact group $G = \mathrm{Spin}(n+1,1)$ gets involved; see [**BÓO96, Bra97b**].

REMARK 9.8. To oversimplify a bit, the G^*G for gradients G are the basic "building blocks" of natural differential operators. However, these are not usually conformally covariant. The first-order operators G and G^*, on the other hand, are conformally covariant, but don't usually have spectra, since they generally go from one bundle to another. Nevertheless, the calculation works, as G and G^* can be made to live in a diagram of operators which are "simultaneously diagonalizable" (in a suitable loose sense). This is slightly more involved than it looks at first, as some of the operators in these diagrams act on subquotients (under the conformal group) of section spaces, rather than on full section spaces.

REMARK 9.9. By the interlacing rule, the α and $\mathring{\alpha}$ may be identified with lattice points in rectangular boxes. The box $\mathbf{A}$ of α is infinite on one end, and its (finite) base is the box $\mathring{\mathbf{A}}$ of all $\mathring{\alpha}$.

For example, the spinor bundle Σ for n odd gives rise to a two-point set $\mathring{\mathbf{A}}$, and the discrete leading symbol $\mathbf{k}(\mathcal{D})$ of the Dirac operator takes the values ± 1 on these points. The square of the Dirac operator has $\mathbf{k}(\mathcal{D}^2) \equiv 1$, and in fact the discrete leading symbol takes compositions to products:

$$\mathbf{k}(D_1 \circ D_2)(\beta) = \mathbf{k}(D_1)(\beta) \cdot \mathbf{k}(D_2)(\beta).$$

In particular, this shows how to construct an operator D_1 with $D_1 D_2$ of the form $(\nabla^*\nabla)^k + (\text{lower order})$, if such exists.

Generalizing the example of $\mathcal{D}^2$, the discrete leading symbol of any operator of the form

$$(\nabla^*\nabla)^k + (\text{lower order})$$

on any bundle $\mathbf{V}(\lambda)$ is the constant function 1.

Ellipticity is detected by the discrete leading symbol: D is elliptic iff $\mathbf{k}(D)$ is nowhere zero. D is positively elliptic (has positive definite leading symbol) iff $\mathbf{k}(D)$ is a positive function. In fact, the best ellipticity constant c for an operator of order $2p$,

$$\sigma_{2p}(D)(\xi) \geq c|\xi|^{2p}\mathrm{Id}$$

in the sense of endomorphisms, is the minimum value of the discrete leading symbol.

The characteristic function of a point in $\mathring{\mathbf{A}}$ corresponds to a fundamental projection in the algebra of leading symbols. Here the prototypical example is the differential form bundle Λ^k for $k < (n-2)/2$, where the operators δd and $d\delta$ give rise to the fundamental projections. This example is somewhat misleading in its simplicity, however: in general, the operators representing fundamental projections are of higher order, and need not commute.

If $G_1, \ldots, G_N$ are the gradients emanating from $\mathbf{V}(\lambda)$, then

$$\min_{\beta \in \mathcal{B}(\lambda)} \mathbf{k}(G_{a_1}^* G_{a_1} + \cdots + G_{a_p}^* G_{a_p})(\beta)$$

is also the best Kato constant k for solutions of the first-order system $G_{a_1}\varphi = \cdots = G_{a_p}\varphi = 0$, in the sense that

$$\left| d|\varphi| \right|^2 \leq c|\nabla\varphi|^2 \text{ off } \{x \mid \varphi_x = 0\}. \tag{9.19}$$

(See [**Bra00, CGH00b**].)

The presence of a Bochner-Weitzenböck (BW) formula

$$a_1 G_1^* G_1 + \cdots + a_N G_N^* G_N = (0^{\underline{\text{th}}} \text{ order}) \tag{9.20}$$

is signalled by the discrete leading symbol relation

$$a_1 \mathbf{k}(G_1^* G_1) + \cdots + a_N \mathbf{k}(G_N^* G_N) = 0.$$

Given a knowledge (from [**Bra97b**]) of the $\mathbf{k}(G_u^* G_u)$, we can algebraically determine the left sides of BW formulas. Specifically, there are numbers s_u, $\tilde{c}_u$ (built from conformal weights) such that

$$\begin{aligned} &a_1 \mathbf{k}(G_1^* G_1) + \cdots + a_N \mathbf{k}(G_N^* G_N) = 0 \iff \\ &a_1 \tilde{c}_1 s_1^{2j} + \cdots + a_N \tilde{c}_N s_N^{2j} = 0 \quad \text{for } j = 0, 1, \ldots, \left[\frac{N(\lambda) - 1}{2}\right]. \end{aligned}$$

BW formulas are key ingredients in relating local geometry to spectral data, since for a coefficient list (a_u) satisfying (9.20),

$$\underbrace{\sum_{a_u > 0} a_u G_u^* G_u}_{\text{nonneg. op.}} = \underbrace{\sum_{-a_u > 0} (-a_u) G_u^* G_u}_{\text{nonneg. op.}} + \text{Curv}.$$

(Some invariant theory shows that the operator labelled ($0^{\underline{\text{th}}}$ order) in (9.20) is an action of the Riemann curvature tensor; thus the notation "Curv".) The above shows there are $[N(\lambda)/2]$ independent BW formulas on $\mathbf{V}(\lambda)$. If, on a particular manifold, the curvature term is positive (resp. negative) definite, then the operator on the left (resp. the first term on the right) is positive definite. This is only the most obvious of a sophisticated set of techniques for proving vanishing theorems and eigenvalue estimates.

The classification of BW left sides depends on the classification theory of second-order conformally covariant operators, which in turn depends on a complete understanding of second-order contributors to $\nabla^* \nabla$:

$$T^* \otimes \underbrace{(T^* \otimes \mathbf{V}(\lambda))}_{G_u \text{ targets}} \cong \underbrace{(T^* \otimes T^*)}_{\Lambda^0 \oplus \Lambda^2 \oplus \mathrm{TFS}^2} \otimes \mathbf{V}(\lambda).$$

An example of a bundle with $N(\lambda) = 4$ (so that there are 2 independent BW formulas) is the bundle introduced above in (9.2). The gradient target bundles are Σ (or the copy of Σ contained in $T^* \otimes \mathbf{Tw}$), $\mathbf{Tw}$ itself, and the bundles $\mathbf{Y}_2$ and $\mathbf{Z}_2$. (Recall the discussion around (9.4) and (9.5) above.) Denoting the corresponding gradients by G_Σ, $G_{\mathbf{Tw}}$, $G_{\mathbf{Y}}$, and $G_{\mathbf{Z}}$, the discrete leading symbols of the $G_u^* G_u$ on the two-point space $\mathcal{B}(\lambda)$ are:

$$\begin{array}{rcc}
G_\Sigma^* G_\Sigma : & \overset{\frac{1}{n}}{\star} & \overset{0}{\star} \\[1ex]
G_{\mathbf{Tw}}^* G_{\mathbf{Tw}} : & \overset{\frac{n-2}{n(n+2)}}{\star} & \overset{\frac{n}{(n+2)(n-2)}}{\star} \\[1ex]
G_{\mathbf{Y}}^* G_{\mathbf{Y}} : & \overset{0}{\star} & \overset{\frac{n-3}{2(n-2)}}{\star} \\[1ex]
G_{\mathbf{Z}}^* G_{\mathbf{Z}} : & \overset{\frac{n}{n+2}}{\star} & \overset{\frac{n+1}{2(n+2)}}{\star}
\end{array}$$

The corresponding BW formulas are derived in [**BH02**]; a basis is:

$$\begin{aligned}\frac{(n-3)(n-2)}{2n}G_\Sigma^*G_\Sigma & -\frac{(n-3)(n+2)}{2n}G_{\mathrm{Tw}}^*G_{\mathrm{Tw}}+\mathrm{G}_{\mathbf{Y}}^*\mathrm{G}_{\mathbf{Y}}\\ &=-\frac{(n-2)(n-3)}{8n(n-1)}K+\frac{n-3}{2(n-2)}b\cdot+\frac{1}{8}C:,\end{aligned}$$

$$\begin{aligned}-\frac{(n-1)(n+2)}{2n}G_\Sigma^*G_\Sigma & -\frac{(n-2)(n+1)}{2n}G_{\mathrm{Tw}}^*G_{\mathrm{Tw}}+\mathrm{G}_{\mathbf{Z}}^*\mathrm{G}_{\mathbf{Z}}\\ &=-\frac{(n+2)(n+1)}{8n(n-1)}K-\frac{n+1}{2(n-2)}b\cdot+\frac{3}{8}C:.\end{aligned}$$

Here b is the trace-free Ricci tensor, and the curvature actions $b\cdot$ and $C:$ are:

$$\begin{aligned}(b\cdot\varphi)_\mu &= \frac{1}{n}\left((n-2)b_{\mu\lambda}-b_{\alpha\lambda}\gamma^\alpha\gamma_\mu\right)\varphi^\lambda,\\ (C:\varphi)_\mu &= C_{\alpha\beta}{}^\lambda{}_\mu\gamma^\alpha\gamma^\beta\varphi_\lambda\,.\end{aligned}$$

These are the *unique* invariant actions of $\odot_0 2$ (where b lives) and of the algebraic Weyl tensor bundle $\mathcal{W}$ (where C lives) on twistors.

Here is a sample theorem demonstrating what one can do with the resulting estimates:

THEOREM 9.10. [**BH02**] *Suppose $n=8$, and g is an Einstein manifold (i.e. $b=0$). Then any twistor φ in* $\mathrm{Ker}(\mathrm{G}_{\mathbf{\Sigma}})\cap\mathrm{Ker}(\mathrm{G}_{\mathbf{Y}})$ *or in* $\mathrm{Ker}(\mathrm{G}_{\mathbf{T}})\cap\mathrm{Ker}(\mathrm{G}_{\mathbf{Z}})$ *is parallel (i.e. $\nabla\varphi=0$).*

9.2.3. The Hijazi inequality. One of the major touchstones of spin geometry has been the *Hijazi inequality* [**Hij86a**], which relates the bottom eigenvalue of the conformal Laplacian Y (recall (9.15)) with that of the square of the Dirac operator. To explain how this comes about, recall the first equation of (9.16), which integrates to

$$\begin{aligned}0&=\int\left(\frac{K}{2(n-1)}|\psi|^2-2|G_2\psi|^2+\frac{2}{n-1}|G_1\psi|^2\right)dv_g\\ &=\int\left(\frac{K}{2(n-1)}|\psi|^2-\frac{2}{n}|\mathcal{D}\psi|^2\right)dv_g+(\text{nonnegative}).\end{aligned}\tag{9.21}$$

(Recall the normalization change implicit in (9.17,9.18).) The Yamabe operator Y (recall (9.15)) governs the conformal change of the scalar curvature K according to

$$\frac{K_\omega}{2(n-1)}=\frac{2}{n-2}e^{-(n+2)\omega/2}Ye^{(n-2)\omega/2},\tag{9.22}$$

where K_ω is the scalar curvature of the metric $g_\omega=e^{2\omega}g$, and the operator Y is computed in the metric g. A well-known symmetrization argument shows that the bottom eigenfunction of Y does not change sign, so we may assume that it is a positive function, which we may then write as $e^{(n-2)\omega_1/2}$ for some smooth function ω_1. With $\omega=\omega_1$ in (9.22), and μ_1 the bottom eigenvalue of Y, we have

$$\frac{K_{\omega_1}}{2(n-1)}=\frac{2}{n-2}e^{-(n+2)\omega_1/2}\cdot\mu_1Ye^{(n-2)\omega_1/2}=\frac{2}{n-2}\mu_1e^{-2\omega_1}.$$

Substituting into the basic inequality (9.21) at the metric $g_1:=g_{\omega_1}$, we have

$$0=\int\left(\frac{2}{n-2}\mu_1e^{-2\omega_1}|\psi|^2-\frac{2}{n}|\mathcal{D}_{\omega_1}\psi|^2\right)dv_{g_1}+(\text{nonnegative}).\tag{9.23}$$

The obvious thing to do now is to take advantage of the conformal covariance relation for the Dirac operator, in the form

$$\mathcal{D}_{\omega_1}\psi = e^{-(n+1)\omega_1/2}\mathcal{D}(e^{(n-1)\omega_1/2}\psi)$$

If $e^{(n-1)\omega_1/2}\psi$ is an eigenspinor for $\mathcal{D}$, with eigenvalue λ, then

$$0 = \int e^{-2\omega_1}|\psi|^2\left(\frac{2}{n-2}\mu_1 - \frac{2}{n}\lambda^2\right) dv_{g_1} + (\text{nonnegative}),$$

and we have:

THEOREM 9.11. [**Hij86a**] $\lambda^2 \geq n\mu_1/(n-2)$.

It is clear how one would use the derivation above to study the case of equality. There are various improvements in the case of Kähler manifolds, quaternionic Kähler manifolds, and the like; see, for example, [**Kir86**], [**HM97**], and [**KSW98b**].

There are two main obstacles to the attempt to extend this line of reasoning to other bundles. First, we depended on the fact that, in some BW formula, we could have just *one* $G_u^*G_u$ appear with a coefficient of a given sign. In general, the coefficient signs tend to try to split about evenly, as one might expect. Another is that it is difficult to keep the other curvatures (b and C) out of the picture. The spinor and scalar bundles are special in that weight considerations prevent b and C from acting; this simplified state of affairs never occurs again as we go up into the space of bundles. There *is* a BW formula omitting b, whenever $N(\lambda)$ is even – the combination (9.14). Effective estimation using this, however, depends on having some control over C. See [**BH97, BH00**] for some attempts in this direction.

Like central results in other fields, the Hijazi inequality is closely related to other fundamental phenomena. For example, an argument of Bär and Moroianu explained in [**CGH00a**] ties it closely to the sharp Kato estimate for solutions of the Dirac equation, namely

$$\mathcal{D}\psi = \lambda\psi \Rightarrow \left|d|\psi|\right|^2 \leq \frac{n-1}{n}|\tilde{\nabla}\psi|^2 \quad \text{off } \{x \mid \psi_x = 0\},$$

where $\tilde{\nabla}_a = \nabla_a + \lambda\gamma_a/n$. (Recall (9.6) and (9.19) above.) We compute

(9.24)

$$\left(Y - \frac{n-2}{n}\lambda^2\right)|\psi|^{(n-2)/(n-1)} = \frac{n(n-2)}{(n-1)^2}\left(\left|d|\psi|\right|^2 - \frac{n-1}{n}|\tilde{\nabla}\psi|^2\right)|\psi|^{-n/(n-1)}.$$

Modulo some analytic niceties having to do with the possible zero set of ψ, the nonnegativity of the right side of (9.24) establishes the Hijazi inequality.

9.3. Tractor constructions

The constructions of spin-tractors and form-tractors below is joint work of the author and Rod Gover [**BG**].

9.3.1. Enlarging the structure group.

To some extent, the use of the word *twistor* to describe the bundle **Tw** is a misunderstanding that has been solidified by tradition. The Penrose *local twistor bundle* in 4 dimensions is essentially the bundle holding the finitely many independent jets of a solution of the twistor equation $T\psi = 0$; its generalizations are sure to be much-studied in the near future. Recall that we are working in the setting of *Riemannian* spin geometry. The constructions we are about to introduce generalize more or less readily to the pseudo-Riemannian case, including the Lorentzian regime where Penrose's construction finds its most natural home.

A first step is to realize that there is a larger group which is trying to assert itself, namely

$$G = \mathrm{Spin}(n+1,1),$$

and that the structure groups we have dealt with thus far, as well as some others, are natural subgroups of G. When the *Iwasawa* and *Langlands decompositions* of a semisimple Lie group,

$$G = KAN = \overline{N}MAN,$$

are applied to our G, the groups $\overline{N}$ and N are copies of $\mathbf{R}^n$; and the group A is a copy of $\mathbf{R}_+$. (The *conformal weight* lives in the dual of this.) The group M is $\mathrm{Spin}(n)$, and K is $\mathrm{Spin}(n+1)$. The *stereographic injection* $\mathbf{R}^n \to S^n$ is

$$\overline{N} \hookrightarrow \overline{N}MAN \to \overline{N}MAN/MAN.$$

The introduction of conformal weights carried us from M-bundles to MA-bundles. But there is a further step, suggested by Cartan's work: $P = MAN$-bundles – not just on homogeneous spaces for G, but on conformal (spin) manifolds. That is, there is a P-structure to these manifolds that is not usually exploited. It is not generally known that there is a rich supply of MAN-bundles, in which the N factor acts **nontrivially** – in contrast to the situation for the classical tensor-spinor-density bundles. These bundles exist on MA-manifolds; i.e. manifolds endowed with a conformal structure. That is, our manifolds need only have MA structure, and we get bundles with the richer MAN structure free of charge.

The basic concept is one that seems to be rediscovered in different forms at irregular intervals. Besides the work of Penrose [**PR86**], pioneering work was done by T.Y. Thomas [**Tho26**] in the 1920's, but lacking the language of bundles, it was largely forgotten until the explanation of [**BEG94**] in modern terminology. I.E. Segal's *spannors* and *plyors* [**PSV87, OS89**], introduced to explain fundamental physical phenomena, are examples of such objects, although Segal et al. treated them only on homogeneous model spaces.

An important new feature of MAN-bundles, as opposed to M- or MA-bundles, is the lack of *complete reducibility.* That is, MAN-bundles have composition lattices, or (dropping some information) Jordan-Hölder series of the form

$$\mathbf{V}_s \subset\!\!\!\!\!+\, \mathbf{V}_{s-1} \subset\!\!\!\!\!+ \cdots \subset\!\!\!\!\!+\, \mathbf{V}_r\,. \tag{9.25}$$

The $\subset\!\!\!\!\!+$ notation means that each $\mathbf{V}_q \subset\!\!\!\!\!+ \cdots \subset\!\!\!\!\!+\, \mathbf{V}_r$ for $s \geq q \geq r$ is an invariant subbundle, but one which has no invariant direct complement. In other words, any attempt to get just the information from $\mathbf{V}_s, \ldots, \mathbf{V}_{q+1}$ involves taking a quotient. If we reduce the structure group to MA, we may replace the $\subset\!\!\!\!\!+$ in (9.25) with $\oplus$, but the action of N carries us from $\mathbf{V}_q$ to $\mathbf{V}_{q-1}$ for each q.

On the face of it, it would appear appear to be a formidable task to find lists of MA-representations that can be "glued together" using actions of N. But in fact, there are abundant sources of such things, and they are geometrically very natural. First, we can always restrict representations of G to MAN. When we do this for the (half) spin representation in dimension 4, we obtain Penrose local twistors and Segal spannors. When we use the full differential form representation, we get Segal plyors. But there is really not too much that is special about dimension 4 from this point of view – the general setup only really recognizes a difference between even and odd dimensions.

An alternative construction, generalizing the jet construction of Penrose's local twistor bundle, is to form a bundle which "holds" the finitely many independent jets of an overdetermined operator. In fact, one immediately suspects that such bundles will have Jordan-Hölder series, given the way that jets behave. We shall not treat jet constructions in detail here; but see the discussion around (9.46) below for an example of the relation between jet bundle quotients and tractor bundles.

With the above restrictions of G-representations in hand, we can always take MAN-*subquotients*. This means we iterate and alternate the processes of taking subrepresentations and quotients. Following the standard (after [**BEG94**]) modern parlance, we shall call these *tractor* (or tractor-density) bundles. The *full* tractor bundles play a central role in the theory – these are obtained by restricting a (usually irreducible) G-representation (and *not* passing to subquotients).

One benefit of having all these bundles available is a cure for the *composition problem* for conformally invariant operators. Such operators on conventional tensor-spinor-density bundles almost never compose, since forming the composition AB would mean that the weights a and d in

$$\hat{A} = \Omega^{-b} A \Omega^{a}, \quad \hat{B} = \Omega^{-d} B \Omega^{c}$$

must match up – in practice, this "almost never" happens. With all these MAN-bundles, though, there are many compositions, and in fact this is a good way of constructing new operators. In particular, one may start in a conventional tensor-spinor-density bundle (where N acts trivially), travel through some tractor bundles, and emerge at a conventional bundle. In doing so, we are performing a curved version of *Jantzen-Zuckerman translation*, or a *curved translation* for short [**ER87**].

Let us concentrate on local twistors and spannors, or what we might call *spin-tractors* from the current viewpoint. Suppose n is even, realize the positive spin representation of $\mathrm{Spin}(n+2, \mathbf{C})$ in the real form $\mathrm{Spin}(n+1, 1)$, and restrict to $M = \mathrm{Spin}(n)$. We get an object "containing" both positive and negative spinors:

$$\Sigma_{+}^{M}[\tfrac{1}{2}] \mathbin{\oplus\!\!\!\!\!\supset} \Sigma_{-}^{M}[-\tfrac{1}{2}].$$

Here the first summand is an MAN-quotient, and the second is an MAN-submodule; recall that this is the meaning of the semidirect sum sign $\mathbin{\oplus\!\!\!\!\!\supset}$. The numbers in brackets encode the conformal weights. A more Penrosian notation would be

$$\mathcal{E}^{P} \mathbin{\oplus\!\!\!\!\!\supset} \mathcal{E}^{P'}[-1].$$

Here the M-modules involved are indicated by the indices adorning the generic bundle symbol $\mathcal{E}$, the P indicating positive spinors, and the P' negative. In this notation, the index positions carry conformal weights. Recall that this is also true in the tensor case, where, for example, the cotangent bundle $\mathcal{E}_a$ is isomorphic to $\mathcal{E}^a[2]$ (the vector field 2-densities) as an MA-bundle. The difference is that reversal of a spinor index position incurs only half the "penalty" (or "reward") of a tensor index move. One point that must be observed is that in dimension $4k$, the positive and negative spinors are self-contragredient (their own dual modules), while in dimension $4k+2$, they are dual to each other. The upshot is that, for example,

$$\mathcal{E}_P = \mathcal{E}^P[-1], \quad n = 4k,$$
$$\mathcal{E}_P = \mathcal{E}^{P'}[-1], \quad n = 4k+2.$$

Let us use $\mathbf{A}, \mathbf{B}, \cdots$ as spin-tractor indices. As a consequence of the discussion immediately above,

$$\mathcal{E}^{\mathbf{B}} \cong_{MAN} \mathcal{E}_{\mathbf{B}} \cong \begin{cases} \mathcal{E}^{P} \mathbin{\subset\!\!\!\!+} \mathcal{E}_{P'}, & n = 4k, \\ \mathcal{E}^{P} \mathbin{\subset\!\!\!\!+} \mathcal{E}_{P}, & n = 4k+2. \end{cases} \tag{9.26}$$

What do form-tractors (plyors) look like? To see, one could use the relation to spinors in even dimensions: as M-bundles,

$$\Lambda \cong \Sigma \otimes \Sigma.$$

On the tractor level, we can take the tensor square of the spin-tractors $\mathcal{E}^{\mathbf{B}}$. In dimension $4k$, this gives

$$\mathcal{E}^{PQ} \mathbin{\subset\!\!\!\!+} (\mathcal{E}^{P}{}_{Q'} \oplus \mathcal{E}^{P'}{}_{Q}) \mathbin{\subset\!\!\!\!+} \mathcal{E}_{P'Q'}.$$

This reflects the usual notational convention on tensor products; for example, $\mathcal{E}^{PQ} = \mathcal{E}^{P} \otimes \mathcal{E}^{Q}$. Thus we are getting forms of (reduced) weights 1, 0, and -1. This is different from what we get in the de Rham complex with its natural conformal weights:

$$\mathcal{E}_{[a_1 \cdots a_k]} = \Lambda^k[-k].$$

In dimension 4, the subquotients of the above are:

$$\begin{aligned} \mathcal{E}^{P}{}_{Q'} &\cong \mathcal{E}^{P'}{}_{Q} \cong \mathcal{E}_a[1], \\ \mathcal{E}^{PQ} &\cong \mathcal{E}^{\mathrm{SD}}_{[ab]}[3] \oplus \mathcal{E}[1], \\ \mathcal{E}_{P'Q'} &\cong \mathcal{E}^{\mathrm{ASD}}_{[ab]}[1] \oplus \mathcal{E}[-1], \end{aligned}$$

where SD (resp. ASD) means self-dual (resp. anti-self-dual). The first line of this is one of the things that makes 2-component spinor calculus in dimension 4 so appealing: a pair of spinor indices, one primed and one unprimed, is "equivalent" to a tensor index.

This doesn't quite tell the whole story of the MAN-structure of the form-tractors, even in dimension 4. In fact, what one is getting is a direct sum of 1-form tractors and self-dual 3-form tractors:

$$\mathcal{E}^{\mathbf{BC}} \cong \mathcal{E}_A \oplus \mathcal{E}^{\mathrm{SD}}_{[ABC]}.$$

The duality referred to here is under the structure group $G = \mathrm{Spin}(5,1)$. This basically comes down to the fact that the full tractor bundle based on a tensor product of G-modules is the tensor product of the relevant full tractor bundles – tractorization and tensor product commute.

A form-tractor bundle $\mathcal{E}_{[A_1 \cdots A_k]}$ (for k small enough to safely avoid considerations of duality in the middle order of form) looks like

$$\begin{gathered} \Lambda^{k-1}[1] \mathbin{\subset\!\!\!\!+} (\Lambda^k[0] \oplus \Lambda^{k-2}[0]) \mathbin{\subset\!\!\!\!+} \Lambda^{k-1}[-1] \cong \\ \mathcal{E}_{[a_1 \cdots a_{k-1}]}[k] \mathbin{\subset\!\!\!\!+} (\mathcal{E}_{[a_1 \cdots a_k]}[k] \oplus \mathcal{E}_{[a_1 \cdots a_{k-2}]}[k-2]) \mathbin{\subset\!\!\!\!+} \mathcal{E}_{[a_1 \cdots a_{k-1}]}[k-2]. \end{gathered}$$

In particular, *standard tractors* are 1-form-tractors:

$$\mathcal{E}_A \cong \mathcal{E}[1] \mathbin{\subset\!\!\!\!+} \mathcal{E}_a[1] \mathbin{\subset\!\!\!\!+} \mathcal{E}[-1], \tag{9.27}$$

and *adjoint tractors* are 2-form tractors:

$$\mathcal{E}_{[AB]} \cong \mathcal{E}^a \mathbin{\subset\!\!\!\!+} (\mathcal{E}_{[ab]}[2] \oplus \mathcal{E}) \mathbin{\subset\!\!\!\!+} \mathcal{E}_a\,.$$

The name "adjoint" comes from the fact that the adjoint representation of $\mathfrak{so}(n+2, \mathbf{C})$ is isomorphic to the 2-form representation of that algebra. One may also compute, for example, that trace-free symmetric p-tensor tractors, i.e. the full tractor bundle based on the trace-free symmetric representation of $\mathfrak{so}(n+2,\mathbf{C})$, look like:

$$\mathcal{E}_{(A_1\cdots A_p)_0} \cong \bigoplus{}_{q=-p}^{p} \oplus \{\mathcal{E}_{(a_1\cdots a_r)_0}[q+r] \mid (-1)^r = (-1)^p(-1)^q,\ 0 \le r \le p-|q|\}.$$

There are undoubtedly several conditions in standard Riemannian and pseudo-Riemannian geometry that have natural equivalents in tractor geometry. For example:

THEOREM 9.12. **[GN]** *A conformal manifold admits a parallel tractor which at no point is a section of a proper natural subbundle if and only if the conformal structure contains an Einstein metric.*

9.3.2. Spin-tractors. Let us return to the situation of spin-tractors. Suppose we are given just a conformal class of metrics; that is, the set of metrics of the form $\Omega^2 g$, where g is a given metric and Ω is a positive smooth function. Note that it is not really necessary to specify a "background metric" g; we can formulate things in terms of equivalence classes of metrics.

To describe spin-tractors, it is useful to alter our convention on the normalization of the Clifford symbol γ. For present purposes, we switch to a Clifford symbol $\beta = \gamma/\sqrt{2}$, so that the Clifford relation becomes

$$\beta^a\beta^b + \beta^b\beta^a = -g^{ab}\mathrm{Id}. \tag{9.28}$$

We shall call the associated Dirac operator $\not{D} := \beta^a\nabla_a$.

We start with an ordinary pair of spinors,

$$\begin{pmatrix} \psi \\ \phi \end{pmatrix}, \tag{9.29}$$

of opposite chirality. In fact, for the sake of definiteness, let us say that ψ is a positive spinor ψ^P, and ϕ is a negative spinor $\phi^{P'}$. This pair is meant to represent the spin tractor at the choice of metric g; at the conformally related choice $\hat{g} = \Omega^2 g$, we get (suppressing indices and their weights)

$$\begin{pmatrix} \hat{\psi} \\ \hat{\phi} \end{pmatrix} = \begin{pmatrix} \Omega^{w+1/2}\psi \\ \Omega^{w-1/2}(\phi + \omega_c\beta^c\psi) \end{pmatrix}. \tag{9.30}$$

Here $\omega = \log\Omega$, and ω_c is an abbreviation for $\omega_{|c} = \nabla_c\omega$. (We shall propagate this abbreviation to higher derivatives, so that $\omega_{ab} = \nabla_b\nabla_a\omega$.)

If we had a set of trivializations for the spinor bundle parameterized by some set B, what we now have is a really big set of trivializations, parameterized by $B \times \{\Omega\}$. In fact, what we have defined is $\mathcal{E}^{\mathbf{B}}[w]$, the bundle of spin-tractor-w-densities, where w is a complex number. One could also start with spinor-$(w \pm \frac{1}{2})$-densities ψ and ϕ; the effect would be to eliminate the powers of Ω from (9.30).

In order to avoid dealing with the difference between dimension $4k$ and $4k+2$ continually, it is useful to work without a chirality assumption on ψ or φ in (9.29). This means that we are really taking a direct sum of two spin-tractor bundles:

$$\begin{array}{ll} (\mathcal{E}^P \oplus\!\!\!\!+ \mathcal{E}_{P'}) \oplus (\mathcal{E}^{P'} \oplus\!\!\!\!+ \mathcal{E}_P), & n = 4k, \\ (\mathcal{E}^P \oplus\!\!\!\!+ \mathcal{E}_P) \oplus (\mathcal{E}^{P'} \oplus\!\!\!\!+ \mathcal{E}_{P'}), & n = 4k+2. \end{array} \tag{9.31}$$

In fact, with this convention, we can also handle the odd-dimensional case in unified calculations. In odd dimensions, the spin-tractor bundle is a direct sum of two copies of $\Sigma[w+\frac{1}{2}] \oplus\!\!\!\!+ \Sigma[w-\frac{1}{2}]$. A distinguished splitting comes from the action of $\mathrm{Pin}(n)$

(which doubly covers $O(n)$). This is analogous to the splitting, in the differential form bundle Λ, of $\Lambda^{(n\pm 1)/2}$, which are isomorphic under $SO(n)$, but not under $O(n)$. Using primed and unprimed indices to distinguish the two $\mathrm{Pin}(n)$-modules in question, the splitting looks typographically just like the case $n = 4k$ in (9.31).

Though it is not completely obvious, an easy calculation shows that (9.30) is an action of the group of positive functions Ω, and thus is consistent. That is, if we make the change of trivialization forced by the scale change $g \longrightarrow \Omega^2 g$, and then the change by $\Omega^2 g \longrightarrow \Xi^2\Omega^2 g$, this is the same as if we make the scale change $g \longrightarrow (\Omega\Xi)^2 g$ all at once.

Remarkably, there is a conformally invariant connection on $\mathcal{E}^{\mathbf{B}}[w]$ – this is quite unlike our experience with "conventional" tensor-spinor-density bundles. At a scale g, this is built from the spin connection by

$$\nabla_b \begin{pmatrix} \psi \\ \varphi \end{pmatrix} = \begin{pmatrix} \nabla_b\psi + \beta_b\varphi \\ \nabla_b\varphi + \mathsf{P}_{ab}\beta^a\psi \end{pmatrix}. \tag{9.32}$$

Here P is the *Schouten tensor*, the normalization of the Ricci tensor that is typically used in conformal geometry:

$$\mathsf{J} := \frac{\mathrm{Scal}}{2(n-1)}, \quad \mathsf{P} = \frac{\mathrm{Ric} - \mathsf{J}g}{n-2}.$$

In fact, the appearance of the Ricci tensor signals the presence, somewhere in the background, of the *Cartan connection* – curvature coefficients for the M-structure appear as connection coefficients for the MAN-structure.

The problem with this conformally invariant connection, or with analogous connections for other tractor bundles, is that one cannot iterate it. That is, one cannot form $\nabla_b\nabla_a\Psi^{\mathbf{B}}$ in a conformally invariant way, because of the tensor content of $\nabla_a\Psi^{\mathbf{B}}$, a section of $\mathcal{E}_a \otimes \mathcal{E}^{\mathbf{B}}[w]$. To do so would require a conformally invariant connection on the contangent bundle $\mathcal{E}_a$, and of course such a thing does not exist. This state of affairs is cured by the introduction of tractor-indexed replacements for ∇_a, notably the *tractor-D* operator.

If f_* is a tractor of weight w (where $*$ denotes any number of full tractor indices of any type, in up and/or down positions), let

$$D_A f_* = \begin{pmatrix} f_* \\ w^{-1}\nabla_a f_* \\ -\frac{1}{2}[w(w+(n-2)/2)]^{-1}(\nabla^a\nabla_a + w\mathsf{J})f_* \end{pmatrix}. \tag{9.33}$$

The vertical stack on the right side reflects the standard tractor content (recall (9.27); the rightward direction is now downward). On the left side, standard tractor content is indicated by the index A. Though this formula may not make elementary sense for some values of w, it does make perfect sense over the field of *rational functions of* w; that is, over $\mathbf{C}(w)$. Evaluation at "forbidden" values of w sometimes makes sense (and in fact is very productive), when viewed in this way. For example, a zero and a pole may neutralize each other, or it may actually be a residue that is the quantity of interest.

In fact, looking back at (9.33), the residue at $w = 0$ (if $n \neq 2$) is

$$\begin{pmatrix} 0 \\ \nabla_a f_* \\ \star \end{pmatrix}.$$

This gives the invariance of the exterior derivative d on scalar functions (or of d^∇ on full tractors of weight 0, where ∇ is an invariant tractor connection). If $n = 2$,

$$D_A f_* = w^{-2} \begin{pmatrix} 0 \\ 0 \\ -\frac{1}{2}\nabla^a \nabla_a f_* \end{pmatrix} + w^{-1} \begin{pmatrix} 0 \\ \nabla_a f_* \\ 0 \end{pmatrix} + \begin{pmatrix} f_* \\ 0 \\ 0 \end{pmatrix}$$

gives the conformal invariance of the scalar Laplacian. With this in place, we may then conclude the invariance of d (or d^∇) at the w^{-1} level, and of the identity map at the $w0$ level.

Back in the case $n \neq 2$, the residue at $w = (2-n)/2$ is

$$\frac{1}{2} \begin{pmatrix} 0 \\ 0 \\ (-\nabla^a \nabla_a + \frac{n-2}{2}\mathsf{J}) f_* \end{pmatrix},$$

showing the invariance of the Yamabe operator

$$\Box : \mathcal{E}_* \left[\frac{2-n}{2}\right] \to \mathcal{E}_* \left[\frac{-2-n}{2}\right]. \tag{9.34}$$

The "Yamabe operator" which we denoted earlier by Y is a mere special case of $\Box$, in the case where our full tractor bundle is just a scalar density bundle. A substantial use of this generalized Yamabe operator (to produce the Maxwell operator and its conformally invariant *Eastwood-Singer gauge*) is given later in this section.

It is interesting to try to do things like this for spin-tractors, and in particular to see how familiar objects like the Dirac operator appear. Consider the problem of *tractor extension.* The D_A construction, in fact, solves a certain tractor extension problem for f_*, namely to extend it to a tractor with f_* in the top slot. As indicated by formulas like (9.30), the top slot is an invariant function of the tractor stack – one can *project* onto this slot. This is not true of the other slots (again, see (9.30)), but one may invariantly *inject* into the bottom slot (by writing 0 in all slots above it). This is, of course, due to the semidirect sum nature of tractors; for example (9.26).

A computationally productive way to think of this is to define *positional sections*, so that, for example, a standard (one-form) tractor is

$$\begin{pmatrix} \sigma \\ \mu_a \\ \rho \end{pmatrix} = Y_A \sigma + Z_A{}^a \mu_a + X_A \rho.$$

Because of (9.27),

$$Y_A \in \mathcal{E}_A[-1], \quad Z_A{}^a \in \mathcal{E}_A{}^a[1], \quad X_A \in \mathcal{E}_A[1].$$

The above remarks about projecting slots translate to the statement that X_A is an invariant section, but $Z_A{}^a$ and Y_A depend on the choice of metric (within the conformal class). Positional sections make it clear how we can compute (even with a computer) with sections having several tractor indices – rather than trying to work with multidimensional arrays, we just have long formulas with several indexed positional operators. Such representations may not be very evocative, but they reduce computation (when this is necessary) to a purely mechanical matter.

Let us write a spin-tractor as

$$\begin{pmatrix} \psi \\ \varphi \end{pmatrix} = \boldsymbol{y}\psi + \boldsymbol{x}\varphi.$$

Here we have suppressed the spin-tractor index, as well as the spin indices, so that our symbols are now noncommutative. By (9.30), $\boldsymbol{y}$ has weight $-1/2$, and $\boldsymbol{x}$ has weight $1/2$. By (9.32),

$$\nabla_b \boldsymbol{y} = \boldsymbol{x} \mathsf{P}_{ab}\beta^a, \quad \nabla_b \boldsymbol{x} = \boldsymbol{y}\beta_b \,.$$

Using a prime for the conformal variation, that is, letting the metric run through a conformal curve $g_\varepsilon = e^{2\varepsilon\omega} g_0$ for ω a smooth function, and letting $' = (d/d\varepsilon)|_{\varepsilon=0}$, we have (from (9.30))

$$\boldsymbol{x}' = 0, \quad \boldsymbol{y}' = -\omega_c \boldsymbol{x}\beta^c.$$

Having $\boldsymbol{y}\psi + \boldsymbol{x}\varphi$ as a tractor extension means that $(\boldsymbol{y}\psi + \boldsymbol{x}\varphi)' = 0$; this is satisfied by

$$\varphi = \frac{2}{n+2w} \not{D}\, \psi. \tag{9.35}$$

Thus the Dirac operator has appeared, but what about its conformal invariance? This is just the observation that

$$\begin{pmatrix} \psi \\ (w+n/2)^{-1} \not{D}\, \psi \end{pmatrix}$$

has a pole at $w = -n/2$, with residue

$$\begin{pmatrix} 0 \\ \not{D}\, \psi \end{pmatrix}.$$

It's not hard to see that the first (topmost) *nonzero* slot in such an invariant expression projects out to an invariant operator; the result is the conformal invariance of

$$\not{D} : \Sigma[(1-n)/2] \to \Sigma[(-1-n)/2],$$

or with (weighted) indices attached,

$$\begin{aligned} \not{D} : &\quad \mathcal{E}^P[-n/2] \to \mathcal{E}^{P'}[-n/2-1], \\ \not{D} : &\quad \mathcal{E}^{P'}[-n/2] \to \mathcal{E}^{P}[-n/2-1]. \end{aligned}$$

In this tractor extension problem, we implicitly reasoned that the expression $\not{D} = \beta^a \nabla_a$ was, up to a constant, the only quantity with the right valence, invariance and homogeneity properties to land in the bottom slot; we put it there with an undetermined coefficient, and pinned down the coefficient by conformal variation. There are also ways to do a direct computation of the tractor extension, using fundamental tractor operators applied to tractor quantities with undetermined slots, like

$$\begin{pmatrix} \psi \\ \star \end{pmatrix}.$$

We shall not explain such calculations here, but just remark that they show that it is not necessary to input the Dirac operator with an undetermined coefficient in order to get the Dirac operator as output. In the unlikely event that someone (from another planet, say) were to discover spin-tractors before discovering the Dirac operator, he, she, or it would inexorably be led to the Dirac operator anyway.

Applying the tractor-D to our spin-tractor, we obtain

(9.36)

$$\begin{aligned} D_A \Psi \;&= Y_A(\boldsymbol{y}\psi + \boldsymbol{x}\varphi) + w^{-1} Z_A{}^a(\boldsymbol{y}(\nabla_a\psi + \beta_a\varphi) + \boldsymbol{x}(\nabla_a\varphi + \mathsf{P}_{ab}\beta^b\psi)) \\ &\quad -\tfrac{1}{2}[w(w+(n-2)/2)]^{-1} X_A(\boldsymbol{y}(\nabla^a\nabla_a\psi + 2\not{D}\,\varphi + (w-\tfrac{1}{2})\mathsf{J}\psi) \\ &\quad +\boldsymbol{x}(\nabla^a\nabla_a\varphi + 2\mathsf{P}_{ab}\beta^b\nabla^a\psi + (w-\tfrac{1}{2})\mathsf{J}\varphi + \mathsf{J}_b\beta^b\psi)). \end{aligned}$$

The residue at $w = 0$ is

$$Y_A \cdot 0 + Z_A{}^a(\boldsymbol{y}(\nabla_a \psi + \beta_a \varphi) + \boldsymbol{x} \cdot \star) + X_A \cdot \star.$$

In the obvious array notation, this is

$$\begin{pmatrix} 0 & 0 \\ \nabla_a \psi + \beta_a \varphi & \star \\ \star & \star \end{pmatrix};$$

we are allowed to project out an entry if there are no nonzero entries to its left or above it. If we take (ψ, φ) to be the tractor extension of ψ as in (9.35), we have

$$Y_A \cdot 0 + Z_A{}^a\Big(\boldsymbol{y}\underbrace{\Big(\nabla_a + \frac{2}{n}\beta_a \not{D}\psi\Big)}_{(\mathbf{T}\psi)_a} + \boldsymbol{x} \cdot \star\Big) + X_A \cdot \star,$$

showing the invariance of the *twistor operator* $\mathbf{T}_{\underline{a}} : \mathcal{E}^P \to \mathcal{E}_{\underline{a}}{}^P$ (or similarly with P replaced by P'). Here the underline means that we take only the sections τ_a annihilated by interior Clifford multiplication:

$$\beta^a \tau_a = 0.$$

REMARK 9.13. One may obtain φ from ψ by tractor extension and then take the residue at $w = (2-n)/2$, construct a third-order conformally invariant operator D_3 from $\mathcal{E}^P[(2-n)/2] \to \mathcal{E}^{P'}[(-4-n)/2]$ (or the same with P and P' interchanged). Using the Lichnerowicz formula, one may write this (up to a constant factor) in the form $D_3 = \mathcal{D}_3 +$ (lower order). In array form as above, D_3 appears in the southeast slot:

$$\begin{pmatrix} 0 & 0 \\ 0 & 0 \\ 0 & D_3\psi \end{pmatrix}.$$

A tractor analogue of the Clifford section is

$$\alpha^A = X^A \boldsymbol{y}\boldsymbol{y}^* + Z^A{}_a(\boldsymbol{x}\beta^a \boldsymbol{y}^* - \boldsymbol{y}\beta^a \boldsymbol{x}^*) - Y^A \boldsymbol{x}\boldsymbol{x}^*.$$

This satisfies

$$\alpha^A \alpha^B + \alpha^B \alpha^A = -h^{AB},$$

where h^{AB} is the tractor metric

$$h_{AB} = Z_A{}^a Z_{Ba} + X_A Y_B + Y_A X_B\,.$$

Note that h is indefinite. The tractor metric and tractor Clifford section are parallel in the connections given.

To compute $\alpha^A D_A$, we need the identities

$$\begin{gathered} Y_A Y^A = 0,\ X_A X^A = 0,\ Y_A X^A = 1,\ Z_{Aa} Z^A{}_b = g_{ab}\,, \\ \boldsymbol{y}\boldsymbol{x}^* + \boldsymbol{x}\boldsymbol{y}^* = \mathrm{Id},\ \boldsymbol{x}^*\boldsymbol{y} = \mathrm{Id},\ \boldsymbol{y}^*\boldsymbol{x} = \mathrm{Id}, \\ \boldsymbol{x}^*\boldsymbol{x} = 0,\ \boldsymbol{y}^*\boldsymbol{y} = 0. \end{gathered} \tag{9.37}$$

With this, if we abbreviate the formula (9.36) as

$$D_A \Psi = Y_A(\boldsymbol{y}\psi + \boldsymbol{x}\varphi) + Z_A{}^a(\boldsymbol{y}\eta_a + \boldsymbol{x}\theta_a) + X_A(\boldsymbol{y}\pi + \boldsymbol{x}\xi),$$

we have

$$\alpha^A D_A \Psi = \boldsymbol{y}(\varphi - \beta^a \eta_a) + \boldsymbol{x}(\beta^a \theta_a - \pi).$$

If Ψ takes the form $\boldsymbol{y} \cdot 0 + \boldsymbol{x}\varphi$, this is

$$\Big(1 + \frac{n}{2}w^{-1}\Big)\boldsymbol{y}\varphi + w^{-1}\boldsymbol{x}(1 + (w + (n-2)/2)^{-1})\not{D}\varphi,$$

or in vertical notation,

$$\begin{pmatrix} (1+\frac{n}{2}w^{-1})\varphi \\ w^{-1}(1+(w+(n-2)/2)^{-1})\not{D}\varphi \end{pmatrix}.$$

The Dirac operator emerges as the residue of this at $w = (2-n)/2$. Note that since the tractor-D lowers weights by 1, this is an operator from $\Sigma[(1-n)/2] \to \Sigma[(-1-n)/2]$.

To show how the tractor Yamabe operator leads to the Maxwell operator and Eastwood-Singer gauge, we first define the bundle of k-form tractors as the $k^{\underline{\text{th}}}$ exterior power of the bundle of standard tractors. This has a composition series which, in terms of section spaces, is given by

$$\mathcal{T}^k = \Lambda^k\mathcal{T} \cong \mathcal{E}^{k-1}[k] \mathbin{\oplus\!\!\!\!\!\!\;+} \{\mathcal{E}^k[k] \oplus \mathcal{E}^{k-2}[k-2]\} \mathbin{\oplus\!\!\!\!\!\!\;+} \mathcal{E}^{k-1}[k-2]. \tag{9.38}$$

Here $\mathcal{E}^k := \mathcal{E}_{[a_1\cdots a_k]}$. Given a choice of metric g from a given conformal class, there is a splitting of this composition series corresponding to the splitting (9.27) of the standard tractor bundle. Relative to this, a typical k-form tractor field F corresponds to a 4-tuple $(\sigma, \mu, \varphi, \rho)$ of sections of the direct sum (obtained by replacing each $\mathbin{\oplus\!\!\!\!\!\!\;+}$ with $\oplus$ in (9.38)) and we write

$$\Phi = \mathbb{Y}^k \cdot \sigma + \mathbb{Z}^k \cdot \mu + \mathbb{W}^k \cdot \varphi + \mathbb{X}^k \cdot \rho, \tag{9.39}$$

where '$\cdot$' is the usual pointwise form inner product in the tensor arguments,

$$\varphi \cdot \psi = \frac{1}{p!}\varphi^{a_1\cdots a_p}\psi_{a_1\cdots a_p} \quad \text{for } p\text{-forms},$$

and for $k > 1$, if $\underline{\wedge}$ is the wedge product in the tractor arguments,

$$\mathbb{Z}^k = Z\underline{\wedge}\mathbb{Z}^{k-1}, \quad \mathbb{X}^k = X\underline{\wedge}\mathbb{Z}^{k-1}, \quad \mathbb{Y}^k = Y\underline{\wedge}\mathbb{Z}^{k-1}, \quad \mathbb{W}^k = Y\underline{\wedge}X\underline{\wedge}\mathbb{Z}^{k-2}. \tag{9.40}$$

(By convention, $\mathbb{Z}0 = 1$ and $\mathbb{Z}^{-1} = 0$.) Note that because Z is vector valued, $Z\underline{\wedge}Z$ does not vanish, though expressions like $X\underline{\wedge}X$ and $Y\underline{\wedge}Y$ do vanish. The form tractor bundles $\mathcal{T}^k$ are non-zero for $k = 0, \ldots, n+2$; $\mathbb{Z}^k$ vanishes for $k \geq n+1$; $\mathbb{W}$ vanishes for $k \leq 1$; and $\mathbb{X}^k$, $\mathbb{Y}^k$ vanish for $k = 0, n+2$. $\mathbb{X}^k$ is an invariant section, while $\mathbb{Y}^k$ depends on a choice of scale. $\mathbb{Z}^k$ depends on a choice of scale unless $k = 0$; $\mathbb{W}$ depends on a choice of scale unless $k = n+2$.

An invariant metric on $\mathcal{T}^k$ is

$$\langle(\nu, \mu, \varphi, \rho), (\tilde{\nu}, \tilde{\mu}, \tilde{\varphi}, \tilde{\rho})\rangle = \nu \cdot \tilde{\rho} + \rho \cdot \tilde{\nu} + \mu \cdot \tilde{\mu} - \varphi \cdot \tilde{\varphi}. \tag{9.41}$$

Another way of writing this is

$$\Phi \bullet \tilde{\Phi} := \frac{1}{k!}\Phi^{A_1\cdots A_k}\tilde{\Phi}_{A_1\cdots A_k}, \tag{9.42}$$

where Φ is related to $(\sigma, \mu, \varphi, \rho)$ by (9.39), and similarly for $\tilde{\Phi}$. In fact, choosing a conformal scale, we may compute that

$$(\mathbb{Y}^k \cdot \nu + \mathbb{Z}^k \cdot \mu + \mathbb{W}^k \cdot \varphi + \mathbb{X}^k \cdot \rho) \bullet (\mathbb{Y}^k \cdot \tilde{\nu} + \mathbb{Z}^k \cdot \tilde{\mu} + \mathbb{W}^k \cdot \tilde{\varphi} + \mathbb{X}^k \cdot \tilde{\rho})$$

reduces to the expression in (9.41). This follows in turn from the formulas

$$Z^{Ab}Z_{Ac} = \delta^b{}_c, \quad X^AY_A = 1,$$

with all other quadratic contractions of X, Y, Z vanishing, together with the formula

$$(\eta\underline{\wedge}\Phi)_{A_0\cdots A_k} = (k+1)\eta_{[A_0}\Phi_{A_1\cdots A_k]}$$

for the $\underline{\wedge}$ with a standard tractor.

If α and β are one-forms and φ is a form (or if these objects are form-densities), let

$$E(\alpha \otimes \beta)\varphi := \alpha \otimes \varepsilon(\beta)\varphi, \quad I(\alpha \otimes \beta)\varphi := \alpha \otimes \iota(\beta)\varphi,$$

where ε and ι are exterior and interior multiplication respectively, and extend from simple tensors $\alpha \otimes \beta$ to arbitrary 2-tensors by linearity. The formulas for the covariant derivatives of X, Y, Z at a scale imply that

$$\begin{aligned} \nabla\mathbb{X} &= -E(g)\mathbb{W} + I(g)\mathbb{Z}, & \nabla\mathbb{Z} &= -E(\mathsf{P})\mathbb{X} - E(g)\mathbb{Y}, \\ \nabla\mathbb{W} &= I(\mathsf{P})\mathbb{X} - I(g)\mathbb{Y}, & \nabla\mathbb{Y} &= I(\mathsf{P})\mathbb{Z} + E(\mathsf{P})\mathbb{W}, \end{aligned} \tag{9.43}$$

where we have suppressed the superscript k. Under a change of scale $\hat{g} = e^{2\omega}g$, the behavior of X, Y, Z gives

$$\begin{aligned} \hat{\mathbb{X}} &= \mathbb{X}, \\ \hat{\mathbb{Z}} &= \mathbb{Z} + \varepsilon(d\omega)\mathbb{X}, \\ \hat{\mathbb{W}} &= \mathbb{W} - \iota(d\omega)\mathbb{X}, \\ \hat{\mathbb{Y}} &= \mathbb{Y} - \iota(d\omega)\mathbb{Z} - \varepsilon(d\omega)\mathbb{W} + \tfrac{1}{2}(\varepsilon(d\omega)\iota(d\omega) - \iota(d\omega)\varepsilon(d\omega))\mathbb{X}. \end{aligned}$$

We are now ready to construct the Maxwell operator and Eastwood-Singer gauge. Starting with a k-form μ, we form the (conformally invariant) tractor extension

$$\mathbb{Z}^k \cdot \mu + (n-2k)^{-1}\mathbb{X}^k \cdot \delta\mu.$$

Applying the Yamabe operator $\Box$ in the case n even, $k = (n-2)/2$, we get

$$\left\{\mathbb{Z}^k \cdot \delta d\mu + \mathbb{X}^k \cdot \delta(\tfrac{1}{2}d\delta + \mathsf{J} - 2\mathsf{P}\sharp)\mu\right\} + \left\{\mathbb{Z}^k \cdot C\sharp\sharp\mu + \mathbb{X}^k \cdot \delta(A \diamond \mu + \tfrac{1}{2}C\sharp\sharp\delta\mu)\right\}, \tag{9.44}$$

where $\mathsf{P}\sharp$ is the natural action of a 2-tensor on 1-forms, extended to the Grassmann algebra as a wedge-derivation, and $C\sharp\sharp$ arises by writing C as a sum and difference of tensors $\psi\otimes\psi$, where ψ is a 2-form, and forming the corresponding sum and difference of operators $\psi\sharp\psi\sharp$. The explanation of the A term is as follows: $A^{ab}{}_c = 2\nabla^{[a}\mathsf{P}^{b]}{}_c$, and

$$(A \diamond \mu)_{a_1\cdots a_k} := -\sum_{s=2}^{k} A^{a_1 b}{}_{a_s}\mu_{a_1\cdots b\cdots a_k}.$$

Each of the two bracketed expressions in (9.44) is conformally invariant. The first expression gives the Maxwell operator δd as the $\mathbb{Z}$ coefficient, and the Eastwood-Singer gauge operator as the $\mathbb{X}$ coefficient. Conformal invariance (as a map to form-tractors, or to sections of the $(\mathbb{Z}, \mathbb{X})$ subbundle) implies that the Eastwood-Singer operator is invariant on the null space of δd. This gives a (graded injectively) elliptic extension of the non-elliptic Maxwell operator, and shows that the Maxwell and Eastwood-Singer operators are really aspects of a single, more fundamental operator.

At the same time, this calculation signals the entry of new issues into more complex tractor calculations. One of these issues is exemplified by the second bracketed term in (9.44): there are many "versions" of high-order conformally invariant operators, differing at lower-order terms. A given tractor calculation is liable to produce something other than one's favorite version (or more precisely, a version depending only on ∇ and the Ricci tensor). On the other hand, there is some scope for variation of the fundamental objects like $\Box$, which also arrives together with the complexity of the calculation. On scalar densities, weight considerations preclude modifications of $\Box$ based on the Weyl tensor, while in situations like that just above, the possibilities for such modifications abound. The result is that in hindsight, the setup of the calculation may often be carefully modified to produce the desired version of the operator in question.

Another phenomenon which appears quite natural when viewed from a tractor viewpoint is the calculation of Baum et al. [**BFGK91**], showing that the zeros of a twistor-spinor are isolated. As a by-product, the calculation exhibits a conformal Einstein metric off the zero set.

Suppose τ is a twistor-spinor; that is, one annihilated by the twistor operator. Then

$$\nabla_a \tau = -\frac{2}{n}\beta_a \not{D}\tau,$$

and as a result,

$$\begin{aligned}\nabla_b|\tau|^2 &= -\frac{4}{n}\mathrm{Re}(\beta_b \not{D}\tau, \tau),\\ \nabla_a\nabla_b|\tau|^2 &= \frac{8}{n^2}\mathrm{Re}(\beta_b\not{D}\tau, \beta_a\not{D}\tau) - \frac{4}{n}\mathrm{Re}(\beta_b\nabla_a\not{D}\tau, \tau).\end{aligned}$$

By the skew-adjointness of the Clifford symbol together with the Clifford relation, this last equation becomes

$$\nabla_a\nabla_b|\tau|^2 = \frac{4}{n^2}g_{ab}|\not{D}\tau|^2 - \frac{4}{n}\mathrm{Re}(\beta_b\nabla_a\not{D}\tau, \tau). \tag{9.45}$$

In particular, at a zero of τ, the Hessian of $|\tau|^2$ is positive definite, or else $\not{D}\tau = 0$. If the latter, the tractor extension of τ is parallel and vanishes at a point, so vanishes on the connected component of this point. Thus on a connected spin manifold, a nonvanishing twistor-spinor has isolated zeros.

One may also compute that for a twistor-spinor τ,

$$\nabla_a\not{D}\tau = -\frac{n}{2}\mathsf{P}_{ab}\beta^b\tau. \tag{9.46}$$

This is significant in that it closes up the system defined by the jet extension of the twistor equation. That is, a general spinor ψ has independent 1-jets $\not{D}\psi$ and $\mathbf{T}\psi$; for a twistor-spinor, the second of these is of course not independent. Equation (9.46) shows that the next jet is entirely dependent on the zero-jet, so that $(\tau, \not{D}\tau)$ is the complete set of independent jets of a twistor-spinor at a point. What is really happening is that the overdetermined system $\mathbf{T}\psi = 0$ is defining a finite-dimensional quotient of the infinite jet bundle, and in fact one isomorphic to the spin-tractor bundle.

Continuing with the calculation, (9.46) reduces (9.45) to

$$\nabla_a\nabla_b|\tau|^2 = \frac{4}{n^2}g_{ab}|\not{D}\tau|^2 - \mathsf{P}_{ab}|\tau|^2.$$

This may be viewed as two equations, one on the trace-free part, and one on the trace:

$$\begin{aligned}(\mathrm{Hess}_0 + \mathsf{B})|\tau|^2 &= 0,\\ (-\Delta + \mathsf{J})|\tau|^2 &= \frac{4}{n}|\not{D}\tau|^2,\end{aligned} \tag{9.47}$$

where Hess_0 is the trace-free Hessian and $\mathsf{B} := \mathsf{P} - \mathsf{J}g/n$ is the trace-free Schouten tensor. A twistor-spinor naturally carries conformal weight $\frac{1}{2}$, so $|\tau|^2$ has weight 1. The operator appearing in the trace-free part of the equation,

$$\mathrm{Hess}_0 + \mathsf{B} : \mathcal{E}[1] \to \mathcal{E}_{(ab)_0}[1], \tag{9.48}$$

is conformally invariant. The operator appearing in the trace equation is the "Box" operator of tractor theory (the $\nabla^a\nabla_a + w\mathsf{J}$ of (9.33)), at the weight $w = 1$.

The above calculations combine to show that the application of the tractor-D to $|\tau|^2$ gives something especially simple:

$$D_A|\tau|^2 = \begin{pmatrix} |\tau|^2 \\ -4n^{-1}\mathrm{Re}(\beta_b \not{D}\tau, \tau) \\ -4n^{-2}|\not{D}\tau|^2 \end{pmatrix}.$$

A calculation with (9.43) shows that remarkably, this tractor is parallel.

On the other hand, we could work with the tractor extension of τ rather than that of $|\tau|^2$. This is, of course, the spin-tractor extension

$$\Psi := \begin{pmatrix} \tau \\ 2\not{D}\tau/n \end{pmatrix}.$$

It is easily verified that this is parallel if and only if τ is a twistor-spinor. In fact, this is a special case of a very general result relating overdetermined systems to first-order tractor equations [**BvCE**$^+$].

One may make a standard tractor out of Ψ by forming

$$\Phi := (\Psi, \alpha^A \Psi).$$

A computation with (9.37) shows that this is exactly the spin-tractor $D^A|\tau|^2$ given above. In fact, this explains why Φ should be parallel – it is because Ψ, α, and the spin-tractor inner product are. These calculations are an example of a process that has been called *helicity raising and lowering*. This process reveals a relation between two overdetermined operators; in our case, the twistor operator and the operator of (9.48). To some extent, this is analogous to what happens in the Hijazi inequality, which relates two exactly determined elliptic operators (the Dirac and Yamabe operators) acting on different bundles.

The connection to Einstein scales goes as follows. Under a conformal change $\hat{g} = e^{2\omega}g$, the trace-free Schouten tensor deforms according to

$$\hat{\mathsf{B}}_{ab} = \mathsf{B}_{ab} - \omega_{|(ab)_0}\omega + \omega_{|(a}\omega_{b)_0}. \tag{9.49}$$

The Einstein condition is $\hat{\mathsf{B}} = 0$. If $m \in \mathbf{R}$,

$$((\mathrm{Hess}_0 + \mathsf{B})e^{m\omega})_{ab} = e^{m\omega}(m\omega_{|(ab)_0} + m2\omega_{|(a}\omega_{b)_0} + \mathsf{B}_{ab}).$$

Setting $m = -1$ and $e^{-\omega} = |\tau|^2$, this is

$$0 = \hat{\mathsf{B}},$$

by (9.47) and (9.49). Thus, off the zero set of τ,

$$\hat{g} = e^{2\omega}g = |\tau|^{-4}g$$

is an Einstein metric.

[**GN**] shows how one may obtain results about Einstein scales directly from standard tractor considerations.

THOMAS BRANSON, DEPARTMENT OF MATHEMATICS, 14 MLH, UNIVERSITY OF IOWA, IOWA CITY IA 52242, USA.
E-mail address: thomas-branson@uiowa.edu

CHAPTER 10

A Brief Introduction to Particle Interactions

ALI H. CHAMSEDDINE

ABSTRACT. I give a brief introduction to particle interactions based on representations of Poincaré Lie algebra. This is later generalized to interactions based on representations of the supersymmetry Lie algebra. Globally supersymmetric models with internal symmetry and locally supersymmetric models leading to supergravity theories are presented. I also discuss higher dimensional supergravity theories and some of their applications.

Contents

10.1. Introduction

Historically, the Dirac operator was put forward as a result of making the Schrödinger equation in quantum mechanics compatible with the special theory of relativity. In modern physical theories, matter at very small distances, or equivalenty, very high-energies, is made of elementary particles whose interactions are the four fundamental forces in nature: gravity, electromagnetic theory, weak-nuclear force and strong-nuclear force.

Locally the four-dimensional space-time is Minkowskian and physical states are invariant under rotations (Lorentz transformations) and translations. The physical states fall into representations of the Poincaré algebra. Particles will be characterized by their mass and spin. Matter particles have spins 0 and $\frac{1}{2}$ obeying respectively Klein-Gordon and Dirac equations. Particles with spin-1 are the mediators of gauge interactions, e.g. photons mediate the electromagnetic interactions between electrons while all particles exchange gravitons for the gravitational interactions.

The general plan is to first study representations of the Poincaré algebra and place particles in these representations, and to form physical states obeying these symmetries and write Lorentz-invariant interactions. The Poincaré algebra is then generalized to the supersymmetry algebra. We find representations of the supersymmetry algebra and construct interactions invariant under this new symmetry.

For more details the reader may consult the following references [**Wei98**], [**Wes92**] , [**WB92**] , [**Soh85**], [**vN81**] , [**Sch79**] , [**DNP86**] , [**GSW87**].

10.2. Poincaré group

Einstein's special theory of relativity states the equivalence of certain inertial frames of reference. The coordinates x^μ in an inertial frame satisfy

$$ds^2 = dx^\mu \eta_{\mu\nu} dx^\nu = dx'^\mu \eta_{\mu\nu} dx'^\nu,$$

where x^μ and x'^μ are the coordinates of the physical states in different frames and $\eta_{\mu\nu} = diag\,(1,-1,-1,-1)$ is the Minkowski metric. The transformations $x^\mu \rightarrow x'^\mu = a^\mu + \Lambda^\mu{}_\nu x^\nu$ keep the distance invariant provided that

$$\Lambda^\mu{}_\kappa \eta_{\mu\nu} \Lambda^\nu{}_\lambda = \eta_{\kappa\lambda}. \tag{10.1}$$

Taking the determinant on both sides gives $(\det\Lambda)^2 = 1$ or $\det\Lambda = \pm 1$. In particular $\Lambda^0{}_0 \eta_{00} \Lambda^0{}_0 + \Lambda^i{}_0 \eta_{ij} \Lambda^j{}_0 = \eta_{00}$, or $\left(\Lambda^0{}_0\right)^2 = 1 + \left(\Lambda^i{}_0\right)^2$. This implies that $\Lambda^0{}_0 \geq 1$, or $\Lambda^0{}_0 \leq -1$ and transformations are divided into four categories. The entire set of Λ comprises the homogeneous Lorentz group. The orthocronous Lorentz transformations are charecterised by the condition $\Lambda^0{}_0 \geq 1$ so that every positive time-like vector transforms into positive time-like vector. If $\det\Lambda = 1$ then this gives a group of restricted homogeneous Lorentz-transformations. These are the only ones that could be obtained by a continuous transformation of the identity. This is because it is not possible to continously transform states with $\det\Lambda > 0$ to states with $\det\Lambda < 0$, or states with $\Lambda^0{}_0 \geq 1$ to states with $\Lambda^0{}_0 \leq -1$. Therefore these states correspond to an $SO(3,1)$ group. The translations a^μ form a seperate group, an abelian one, defined as $T\,(4)$. We write

$$\begin{aligned} x' &= \Lambda x + a \equiv g\,(a,\Lambda)\, x, \\ x'' &= \Lambda' x + a' = \Lambda'\,(\Lambda x + a) + a' = \Lambda'\Lambda x + (\Lambda' a + a')\,. \end{aligned}$$

But $x'' = g\,(a',\Lambda')\, x' = g\,(a',\Lambda')\, g\,(a,\Lambda)\, x \equiv g\,(a'',\Lambda'')\, x$, which implies

$$g\,(a',\Lambda')\, g\,(a,\Lambda) = g\,(\Lambda' a + a', \Lambda'\Lambda)\,.$$

The subgroup $SO(3,1)$ is defined by

$$g\,(0,\Lambda')\, g\,(0,\Lambda) = g\,(0,\Lambda'\Lambda)$$

and the abelian subgroup T^4 by

$$g\,(a',1)\, g\,(a,1) = g\,(a+a',1)\,.$$

The inverse transformation is determined by the two conditions $\Lambda' a + a' = 0$, and $\Lambda'\Lambda = 1$ which implies

$$g^{-1}\,(a,\Lambda) = g\left(-\Lambda^{-1} a, \Lambda^{-1}\right).$$

10.3. Poincaré Lie algebra

We write $\Lambda^{\mu}_{\ \nu} = \delta^{\mu}_{\nu} + \zeta^{\mu}_{\ \nu}$ where $\zeta^{\mu}_{\ \nu}$ is infinitesimal. This equation implies that $\zeta_{\mu\nu} = \eta_{\mu\lambda}\zeta^{\lambda}_{\ \nu} = -\zeta_{\nu\mu}$ is antisymmetric. Assume a^{μ} is infinitesimal then we need four essential parameters for translation and six essential parameters for rotations

$$U(a,\Lambda) = 1 + i\delta a^{\mu} P_{\mu} - \frac{i}{2}\zeta^{\mu\nu} J_{\mu\nu}.$$

The commutation relations of the Lie algebra can be obtained via the usual basis of scalar functions

$$U(a,\Lambda) f(x) = f\left(U^{-1}(a,\Lambda)\, x\right) = f\left(\Lambda^{-1}(x-a)\right).$$

By going to the infinitesimal limit we get

$$\begin{aligned}\left(1 + i\delta a^{\mu} P_{\mu} - \frac{i}{2}\zeta^{\mu\nu} J_{\mu\nu}\right) f\left(x^{\lambda}\right) &= f\left(x^{\lambda} - \delta a^{\lambda} - \zeta^{\lambda}_{\ \kappa} x^{\kappa}\right) \\ &= f\left(x^{\lambda}\right) - \left(\delta a^{\lambda} + \zeta^{\lambda}_{\ \kappa} x^{\kappa}\right) \partial_{\lambda} f(x) + \cdots\end{aligned}$$

This implies that

$$\begin{aligned}P_{\mu} &= i\frac{\partial}{\partial x^{\mu}} \equiv i\partial_{\mu}, \\ J_{\mu\nu} &= \left(x_{\mu} P_{\nu} - x_{\nu} P_{\mu}\right).\end{aligned}$$

These generators obey the Lie algegra

$$\begin{aligned}\left[P_{\mu}, P_{\nu}\right] &= 0, \\ \left[J_{\mu\nu}, P_{\lambda}\right] &= i\left(\eta_{\nu\lambda} P_{\mu} - \eta_{\mu\lambda} P_{\nu}\right), \\ \left[J_{\mu\nu}, J_{\rho\sigma}\right] &= -i\left(\eta_{\mu\rho} J_{\nu\sigma} - \eta_{\nu\rho} J_{\mu\sigma} - \eta_{\mu\sigma} J_{\nu\rho} + \eta_{\nu\sigma} J_{\mu\rho}\right).\end{aligned}$$

By exponentiating, a general finite transformation is expressible in the form

$$U(a,\Lambda) = \exp\left(i\delta a^{\mu} P_{\mu} - \frac{i}{2}\zeta^{\mu\nu} J_{\mu\nu}\right).$$

The Casimir invariants (operators that commute with the generators of the Lie algebra) are $P^2 = P^{\mu}\eta_{\mu\nu}P^{\nu}$ and $W^2 = W^{\mu}\eta_{\mu\nu}W^{\nu}$ where $W_{\mu} = -\frac{1}{2}\epsilon_{\mu\nu\rho\sigma}P^{\nu}J^{\rho\sigma}$. One can easily verify that

$$\begin{aligned}\left[P^2, P_{\mu}\right] &= 0, \quad \left[P^2, J_{\mu\nu}\right] = 0, \\ \left[W^2, P_{\mu}\right] &= 0, \quad \left[W^2, J_{\mu\nu}\right] = 0.\end{aligned}$$

If $P^2 \neq 0$ we can use the spin pseudo-vector $S_{\mu} = \frac{1}{\sqrt{P^2}} W_{\mu}$. The Casimir eigenvalues of P^2 and W^2 (or $j(j+1)$ associated with S^2) are used to specify irreducible representations:

$$\begin{aligned}P_{\mu}\left|p,j,\lambda\right\rangle &= p_{\mu}\left|p,j,\lambda\right\rangle, \\ W^2\left|p,j,\lambda\right\rangle &= W^2(j)\left|p,j,\lambda\right\rangle, \\ W_0\left|p,j,\lambda\right\rangle &= W_0(j)\left|p,j,\lambda\right\rangle.\end{aligned}$$

There are four distinct cases. i) $P^2 > 0$, ii) $P^2 = 0$, iii) $P^2 < 0$, iv) $P_{\mu} = 0$. If $P^2 \neq 0$ then we can take $P_1 = P_2 = 0$. The little group leaving this choice invariant contains $J_3 = J_{12}$, so we can associate λ with integer or $\frac{1}{2}$-integer eigenvalues of J_3 :

$$J_3\left|\widehat{p},j,\lambda\right\rangle = \lambda\left|\widehat{p},j,\lambda\right\rangle,$$

so that λ is always discrete.

If $P^2 = m^2 > 0$ we choose the rest frame

$$P_\mu = \pm (m, 0, 0, 0),$$
$$W_\mu = \pm m \left(0, \vec{J}\right), \quad J_{ij} = \epsilon_{ijk} J^k,$$
$$S_\mu = \pm \left(0, \vec{J}\right).$$

The little group is $O(3)$ generated by $[J_i, J_j] = \epsilon_{ijk} J^k$ and is described by the state vectors

$$|j, \lambda\rangle, \quad \lambda = -j, \cdots, j, \quad j = 0, \frac{1}{2}, 1, \cdots$$

where $W^2 = -m^2 j(j+1)$.

If $P^2 = 0$, we can take $P_\mu = (w, 0, 0, w)$, and $W_\mu = w(J_3, L_1, L_2, J_3)$ where $L_1 = J_1 + K_2$, $L_2 = J_2 - K_1$ and $J_{0i} = K_{0i}, \quad i = 1, 2, 3$. The generators L_1 L_2 and J_3 satisfy the commutation relations

$$[L_1, L_2] = 0, \quad [J_3, L_1] = -iL_2, \quad [J_3, L_2] = -iL_1.$$

The relations $P^2 = 0$, $W^2 = 0$ and $P^\mu W_\mu = 0$ imply that P^μ and W^μ are parallel vectors and therefore we can write

$$W_\mu = \lambda P^\mu,$$

where λ is the helicity. This implies that $W^2 = -w^2 L^2$ and unitary representations are labelled by $|l, \lambda\rangle$.

For the Lorentz group we can write

$$N_i = \frac{1}{2}(J_i + iK_i), \quad N_i^\dagger = \frac{1}{2}(J_i - iK_i),$$

where these generators satisfy the commutation relations

$$\left[N_i, N_j^\dagger\right] = 0,$$
$$[N_i, N_j] = i\epsilon_{ijk} N^k,$$
$$\left[N_i^\dagger, N_j^\dagger\right] = i\epsilon_{ijk} N^{\dagger k}.$$

The Casimir operators are given, in terms of the generators N_i and $N_i^\dagger$, by $N_i N_i$ and $N_i^\dagger N_i^\dagger$ with eigenvalues $n(n+1)$ and $m(m+1)$ respectively. The N_i and $N_i^\dagger$ obey independently Lie algebras of $SU(2)$. We can label representations with (m, n) where $m, n = 0, \frac{1}{2}, 1, \frac{3}{2}, \cdots$ are the separate $SU(2)$ numbers. The two $SU(2)$'s are not independent. The spin representation are given by $m + n$. As examples we have:

- $(0, 0)$ is spin zero and is the scalar representation.
- $(\frac{1}{2}, 0)$ or $(0, \frac{1}{2})$ is the spin-$\frac{1}{2}$ representation for Weyl spinors, while Dirac spinors are represented by $(\frac{1}{2}, 0) \oplus (0, \frac{1}{2})$.
- $(\frac{1}{2}, \frac{1}{2}) = (\frac{1}{2}, 0) \otimes (0, \frac{1}{2})$ is the spin-1 representation.

We represent $(\frac{1}{2}, 0)$ and $(0, \frac{1}{2})$ by two-component complex spinors ψ_α and $\overline{\psi}^{\dot{\alpha}}$ which transform under Lorentz transformation as

$$\psi'_\alpha = M_\alpha{}^\beta \psi_\beta, \quad \overline{\psi}^{\dot{\alpha}} = \left(M^{*-1}\right)^{\dot{\alpha}}_{\dot{\beta}} \overline{\psi}^{\dot{\beta}}$$

This can be denoted by $\psi \to M\psi$ and $\overline{\psi} \to \left(M^{\dagger}\right)^{-1}\overline{\psi}$. We identify generators J_i with $\frac{1}{2}\sigma_i$ and K_i with $-\frac{i}{2}\sigma_i$. We can therefore write

$$M = e^{\frac{i}{2}\sigma_i\left(w^i - iv^i\right)}, \quad \left(M^{\dagger}\right)^{-1} = e^{\frac{i}{2}\sigma_i\left(w^i + iv^i\right)}.$$

We also introduce the 2×2 matrices $\left(\sigma^{\mu}\right)_{\alpha\dot{\beta}}$ defined by $\left(\sigma^{0}\right)_{\alpha\dot{\beta}} = I_2$, the identity matrix and $\left(\sigma^{i}\right)_{\alpha\dot{\beta}} = -\sigma_i$ are the Pauli matrices. The matrix P is defined by $P = P_{\mu}\left(\sigma^{\mu}\right)$ and transforms under Lorentz transformations according to $P' = MPM^{\dagger}$ or equivalently $P_{\mu}\sigma^{\mu} \to P_{\mu}M\sigma^{\mu}M^{\dagger}$. We also define

$$\overline{\sigma}^{\mu\dot{\alpha}\beta} = \epsilon^{\dot{\alpha}\dot{\gamma}}\epsilon^{\beta\delta}\sigma^{\mu}{}_{\delta\dot{\gamma}}.$$

These satisfy the proporties

$$\left(\sigma^{\mu}\overline{\sigma}^{\nu} + \sigma^{\nu}\overline{\sigma}^{\mu}\right)_{\alpha}^{\ \beta} = 2\eta^{\mu\nu}\delta_{\alpha}^{\beta},$$

$$\left(\overline{\sigma}^{\mu}\sigma^{\nu} + \overline{\sigma}^{\nu}\sigma^{\mu}\right)_{\dot{\alpha}}^{\ \dot{\beta}} = 2\eta^{\mu\nu}\delta_{\dot{\alpha}}^{\dot{\beta}}.$$

From these matrices we can construct representations of the Lorentz algebra. They are given by

$$\sigma^{\mu\nu}{}_{\alpha}^{\ \beta} = \left(\sigma^{\mu}\overline{\sigma}^{\nu} - \sigma^{\nu}\overline{\sigma}^{\mu}\right)_{\alpha}^{\ \beta},$$

$$\overline{\sigma}^{\mu\nu}{}^{\dot{\beta}}{}_{\dot{\alpha}} = \left(\overline{\sigma}^{\mu}\sigma^{\nu} - \overline{\sigma}^{\nu}\sigma^{\mu}\right)_{\dot{\alpha}}^{\ \dot{\beta}}.$$

A Dirac spinor takes the form

$$\psi = \begin{pmatrix} \chi_{\alpha} \\ \overline{\lambda}^{\dot{\alpha}} \end{pmatrix},$$

and the Dirac gamma matrices are defined by

$$\gamma^{\mu} = \begin{pmatrix} 0 & \sigma^{\mu} \\ \overline{\sigma}^{\mu} & 0 \end{pmatrix},$$

which satisfy the anticommutation relations $\{\gamma^{\mu}, \gamma^{\nu}\} = 2\eta^{\mu\nu}I_4$. This defines a Clifford algebra. The representation of the Lorentz algebra in terms of the gamma matrices is given by

$$S^{\mu\nu} = \frac{i}{4}\left[\gamma^{\mu}, \gamma^{\nu}\right].$$

Let $\Lambda_{\frac{1}{2}} = \exp\left(-\frac{i}{2}\zeta_{\mu\nu}S^{\mu\nu}\right)$ then $\Lambda_{\frac{1}{2}}^{-1}\gamma^{\mu}\Lambda_{\frac{1}{2}} = \Lambda^{\mu}{}_{\nu}\gamma^{\nu}$ where $\Lambda^{\mu}{}_{\nu} = \delta_{\mu}^{\nu} + \zeta^{\mu}{}_{\nu}$. These properties will be used in the next section when we derive the Dirac equation.

10.4. Dynamical equations for free and interacting fields

10.4.1. Spin-0 : Klein-Gordon field. The Hamiltonian in relativistic dynamics has eigenvalues E which can be determined from the relation $P_{\mu}P^{\mu} = m^2$. This relation implies that $P_0^2 = \overrightarrow{P}.\overrightarrow{P} + m^2 = E^2$, where we have set the velocity of light c to 1. Using the quantum mechanical correspondence $P_{\mu} = \frac{\hbar}{i}\frac{\partial}{\partial x^{\mu}}$ and setting $\hbar = 1$) and acting with the operator $\left(P_{\mu}P^{\mu} - m^2\right)$ on the scalar field ϕ gives

$$\left(\eta^{\mu\nu}\partial_{\mu}\partial_{\nu} + m^2\right)\phi = 0,$$

which is the Klein-Gordon equation. This equation could be derived from a varriational principle using the action

$$I_0 = \int d^4xL = \frac{1}{2}\int d^4x\left(\eta^{\mu\nu}\partial_{\mu}\phi\partial_{\nu}\phi - m^2\phi^2\right).$$

The Euler-Lagrange equation obtained by demanding that $\delta I = 0$, with the field ϕ vanishing at the boundary, is given by

$$\frac{\delta L}{\delta \phi} = \partial_\mu \left(\frac{\delta L}{\delta \partial_\mu \phi} \right),$$

and this implies the Klein-Gordon equation.

10.4.2. Spin-$\frac{1}{2}$: Dirac field. The Schrödinger equation is non-relativistic and is linear in time-derivatives but quadratic in space-derivatives

$$i\hbar \frac{\partial \psi}{\partial t} = -\frac{\hbar^2}{2m} \overrightarrow{\nabla}^2 \psi.$$

The relativistic equation should be linear in time and one must find a first order differential equation of the form

$$i\hbar \frac{\partial \psi}{\partial t} = \frac{\hbar c}{i} \alpha^i \frac{\partial \psi}{\partial x^i} + \beta m \psi,$$

where α^i and β are 4×4 matrices. The square of this equation should yield an equation of the Klein-Gordon type. This is possible if α^i and β satisfy the properties

$$\alpha^i \alpha^j + \alpha^j \alpha^i = 2\delta^{ij}, \quad \beta^2 = 1, \quad \alpha^i \beta + \beta \alpha^i = 0.$$

Multiplying both sides of the equation by β and defining $\gamma^0 = \beta$, $\gamma^i = \beta \alpha^i$ gives

$$\left(i\gamma^\mu \partial_\mu - m \right) \psi = 0.$$

The matrices γ^μ satisfy the anticommutation relations $\{\gamma^\mu, \gamma^\nu\} = 2\eta^{\mu\nu}$ and can be identified with the gamma matrices defined in the previous section. This is the Dirac equation. One can easily verify that it is Lorentz covariant

$$\begin{aligned}
\left(i\gamma^\mu \partial_\mu - m \right) \psi &\to \left(i\gamma^\mu \Lambda^{-1\nu}_{\mu} \partial_\nu - m \right) \Lambda_{\frac{1}{2}} \psi \left(\Lambda^{-1} x \right) \\
&= \Lambda_{\frac{1}{2}} \Lambda_{\frac{1}{2}}^{-1} \left(i\gamma^\mu \Lambda^{-1\nu}_{\mu} \partial_\nu - m \right) \Lambda_{\frac{1}{2}} \psi \left(\Lambda^{-1} x \right) \\
&= \Lambda_{\frac{1}{2}} \left(i\gamma^\mu \partial_\mu - m \right) \psi \left(\Lambda^{-1} x \right),
\end{aligned}$$

where we have used the property $\Lambda_{\frac{1}{2}}^{-1} \gamma^\mu \Lambda_{\frac{1}{2}} = \Lambda^\mu{}_\nu \gamma^\nu$ derived in the last section.

The action

$$I_{\frac{1}{2}} = \int d^4 x \overline{\psi}(x) \left(i\gamma^\mu \partial_\mu - m \right) \psi(x),$$

is Lorentz invariant because $\overline{\psi} \to \overline{\psi} \Lambda_{\frac{1}{2}}^{-1}$. In the special case when $m = 0$, the Dirac equation simplifies to

$$\begin{pmatrix} 0 & i\sigma^\mu \partial_\mu \\ i\overline{\sigma}^\mu \partial_\mu & 0 \end{pmatrix} \begin{pmatrix} \psi_L \\ \psi_R \end{pmatrix} = 0,$$

and this splits into two independent equations, one for the left-handed component ψ_L and the other for the right handed component ψ_R

$$i\sigma^\mu \partial_\mu \psi_L = 0, \quad i\overline{\sigma}^\mu \partial_\mu \psi_R = 0.$$

10.4.3. Spin-1: Maxwell field. We are interested in massless vectors, in particular, the photon which obeys Maxwell's equations

$$\vec{\nabla} \cdot \vec{E} = \rho, \quad \vec{\nabla} \times \vec{E} = -\frac{\partial \vec{B}}{\partial t},$$

$$\vec{\nabla} \cdot \vec{B} = 0, \quad \vec{\nabla} \times \vec{B} = \frac{\partial \vec{E}}{\partial t} + \vec{J}.$$

The second and fourth equations are solved by

$$\vec{E} = -\vec{\nabla} A^0 - \frac{\partial \vec{A}}{\partial t}, \quad \vec{B} = \vec{\nabla} \times \vec{A},$$

and are invariant under the gauge transformations

$$\vec{A}' = \vec{A} - \vec{\nabla} \Lambda, \quad A_0' = A_0 + \frac{\partial \Lambda}{\partial t}.$$

Therefore one can impose a gauge choice, the Lorentz gauge, on the potential $\vec{A}$ and A_0

$$\nabla \cdot A + \frac{\partial A^0}{\partial t} = 0.$$

In this gauge, Maxwell equations simplify to

$$\eta^{\mu\nu}\partial_\mu \partial_\nu A^0 = -\rho, \quad \eta^{\mu\nu}\partial_\mu \partial_\nu \vec{A} = \vec{J},$$

which is a standard wave equation with a source. Maxwell equations could be written in a Lorentz covariant manner if we define the curvature

$$F_{\mu\nu} = \partial_\mu A_\nu - \partial_\nu A_\mu$$

so that

$$\partial_\nu F^{\mu\nu} = J^\nu.$$

We can replace the definition of $F_{\mu\nu}$ by the requirement

$$\partial_\rho F_{\mu\nu} + \partial_\mu F_{\nu\rho} + \partial_\nu F_{\rho\mu} = 0,$$

which, locally, can be solved by the above equation defining A_μ. The curvature $F_{\mu\nu}$ is invariant under the gauge transformations $A'_\mu = A_\mu + \partial_\mu \Lambda$. Maxwell equations could be obtained from the action

$$I_1 = \int d^4x \left(-\frac{1}{4} F_{\mu\nu} F_{\rho\sigma} \eta^{\mu\rho} \eta^{\nu\sigma} + A_\mu J^\mu \right).$$

According to Noether's theorem, there is a conserved charged associated to every symmetry of the action. The symmetry in this case gives electric charge conservation. To find the interactions of electrons with photons, we note that the action for electrons $I_{\frac{1}{2}}$ is invariant under the global symmetry $\psi(x) \to e^{-ie\Lambda}\psi(x)$ where Λ is a constant phase. This global symmetry is lost if the symmetry is promoted to a local symmetry by allowing Λ to depend on x. In this case

$$\overline{\psi}(x) i\gamma^\mu \partial_\mu \psi(x) \to \overline{\psi}(x) i\gamma^\mu \partial_\mu \psi(x) + e\partial_\mu \Lambda \overline{\psi}(x)\gamma^\mu \psi(x).$$

To restore the symmetry, we add to the Dirac-action, the photon-electron interaction term

$$-e \int d^4x \overline{\psi}(x) \gamma^\mu \psi(x) A_\mu,$$

which transforms under the local symmetry according to

$$-e\overline{\psi}(x)\gamma^\mu \psi(x) A_\mu \to -e\overline{\psi}(x)\gamma^\mu \psi(x) A_\mu - e\partial_\mu \Lambda \overline{\psi}(x) \gamma^\mu \psi(x).$$

We deduce that the action for the spin-$\frac{1}{2}$, spin-1 system governing the interaction of photons and electrons and invariant under the local gauge transformations $\psi(x) \rightarrow e^{-ie\Lambda}\psi(x)$ and $A_\mu \rightarrow A_\mu + \partial_\mu \Lambda$, is given by

$$I_{(\frac{1}{2},1)} = \int d^4x \left(-\frac{1}{4} F_{\mu\nu} F^{\mu\nu} + \overline{\psi}(x) \left(i\gamma^\mu D_\mu - m \right) \psi(x) \right),$$

where $D_\mu = \partial_\mu + ieA_\mu$. Therefore, interaction terms are obtained by replacing ordinary derivatives ∂_μ in the free Dirac action by covariant derivatives D_μ. These covariant derivatives satisfy the commutation relations

$$[D_\mu, D_\nu] = ieF_{\mu\nu}.$$

10.4.4. Spin-$\frac{3}{2}$: Rarita-Schwinger field. The field that describes a spin-$\frac{3}{2}$ particle is a vector-spinor $\psi_{\mu\alpha}$ where μ is a space-time vector index and α is a spinor index. The free field satisfies the Rarita-Schwinger equation

$$\gamma^{\mu\nu\rho}\psi_{\nu\rho} = 0$$

where $\psi_{\mu\nu} = \partial_\mu \psi_\nu - \partial_\nu \psi_\mu$, and

$$\gamma^{\mu\nu\rho} = \frac{1}{3!}\left(\gamma^\mu\gamma^\nu\gamma^\rho - \gamma^\nu\gamma^\mu\gamma^\rho + \gamma^\nu\gamma^\rho\gamma^\mu - \gamma^\rho\gamma^\nu\gamma^\mu + \gamma^\rho\gamma^\mu\gamma^\nu - \gamma^\mu\gamma^\rho\gamma^\nu \right)$$

is completely antisymmetric in the indices $\mu\nu\rho$. The product of spin-1 and spin-$\frac{1}{2}$ representations gives spin-$\frac{3}{2}$ and spin-$\frac{1}{2}$ representations

$$\left(\frac{1}{2}, \frac{1}{2} \right) \otimes \left(\frac{1}{2}, 0 \right) = \left(1, \frac{1}{2} \right) \oplus \left(0, \frac{1}{2} \right).$$

It will be seen that the Rarita-Schwinger equation is necessary to eliminate the spin-$\frac{1}{2}$ component from $\psi_{\mu\alpha}$.

10.4.5. Spin-2 : Gravitational field. The massless spin-2 field is the graviton which is a symmetric traceless tensor $h_{\mu\nu}$ with gauge invariance

$$h_{\mu\nu} \rightarrow h_{\mu\nu} + \partial_\mu \xi_\nu + \partial_\nu \xi_\mu.$$

To see that this is none other than the linearized form of the metric tensor of the underlying space-time manifold, let us assume that we would like to promote the global Lorentz invariance of spinors to a local one:

$$\psi_\alpha(x) \rightarrow \exp\left(\frac{i}{4} \Lambda^{ab}(x)\, \gamma_{ab} \right)_\alpha^{\ \beta} \psi_\beta(x),$$

where $\gamma_{ab} = \frac{1}{2}\left(\gamma^a\gamma^b - \gamma^b\gamma^a \right)$. To restore the invariance of the Dirac equation under local Lorentz transformations, and in analogy with the electromagnetic case, we introduce the covariant derivative

$$D_\mu = \partial_\mu + \frac{1}{4}\omega_\mu^{\ ab}\gamma_{ab}.$$

The term $\overline{\psi}(x) \left(i\gamma^\mu D_\mu - m \right) \psi(x)$ becomes invariant provided that

$$\omega_\mu^{\prime ab} = \partial_\mu \Lambda^{ab} + \omega_\mu^{\ ac}\Lambda_c^{\ b} - \omega_\mu^{\ bc}\Lambda_c^{\ a}.$$

The curvature tensor is defined by

$$[D_\mu, D_\nu] = \frac{1}{4} R_{\mu\nu}^{\ \ ab}\gamma_{ab},$$

which can be easily evaluated to be

$$R_{\mu\nu}^{\ \ ab} = \partial_\mu \omega_\nu^{\ ab} - \partial_\nu \omega_\mu^{\ ab} + \omega_\mu^{\ ac}\omega_{\nu c}^{\ \ b} - \omega_\nu^{\ ac}\omega_{\mu c}^{\ \ b},$$

and is covariant under local Lorentz transformations. To make contact with the Riemann curvature tensor, we introduce the soldering form e^a_μ satisfying

$$\nabla_\mu e^a_\nu = \partial_\mu e^a_\nu - \Gamma^\rho_{\mu\nu} e^a_\rho + \omega_\mu{}^{ab} e_{\nu b} = 0,$$

where $\Gamma^\rho_{\mu\nu} = \frac{1}{2} g^{\rho\sigma} \left(g_{\mu\sigma,\nu} + g_{\nu\sigma,\mu} - g_{\mu\nu,\sigma} \right)$ is the Christoffel connection and $g_{\mu\nu} = e^a_\mu \eta_{ab} e^b_\nu$. This equation implies that the covariant derivative of the metric under coordinate transformations vanishes. The constraint equation can be solved for $\omega_\mu{}^{ab}$ in terms of e^a_μ. Substituting the solution into $R_{\mu\nu}{}^{ab}(\omega)$ one can show that

$$R_{\mu\nu}{}^{ab}(\omega)\, e_{\rho a} e_{\sigma b} = R_{\mu\nu\rho\sigma}(g),$$

is identical to the Riemann tensor as a function of the metric g.

Equivalently, we can introduce gauge fields corresponding to both the Lorentz generators J_{ab} and translations P_a

$$D_\mu = \partial_\mu + \omega_\mu{}^{ab} J_{ab} + e^a_\mu P_a,$$

then the curvature is

$$[D_\mu, D_\nu] = R_{\mu\nu}{}^{ab} J_{ab} + T^a_{\mu\nu} P_a,$$

where $R_{\mu\nu}{}^{ab}$ is the same as before while

$$T^a_{\mu\nu} = \partial_\mu e^a_\nu - \partial_\nu e^a_\mu + \omega_\mu{}^{ab} e_{\nu b} - \omega_\nu{}^{ab} e_{\mu b}.$$

By setting the torsion $T^a_{\mu\nu}$ to zero we can uniquely solve for $\omega_\mu{}^{ab}$ in terms of e^a_μ and substitute this expression into $R_{\mu\nu}{}^{ab}$ to obtain the Riemann tensor.

This last formulation is related to the Cartan structure equations

$$\begin{aligned} T^a &= de^a + \omega^{ab} \wedge e_b, \\ R^{ab} &= d\omega^{ab} + \omega^{ac} \wedge \omega_c{}^b. \end{aligned}$$

By writing $e^a = e^a_\mu dx^\mu$, $\omega^{ab} = \omega^{ab}_\mu dx^\mu$, $R^{ab} = \frac{1}{2} R_{\mu\nu}{}^{ab} dx^\mu \wedge dx^\nu$ and $T^a = \frac{1}{2} T^a_{\mu\nu} dx^\mu \wedge dx^\nu$ we can recover all the previous formulas.

The Einstein-Hilbert action is given by

$$I_2 = -\int d^4x \frac{e}{4} e^\mu_a e^\nu_b R_{\mu\nu}{}^{ab},$$

where $e = \det\left(e^a_\mu\right)$. The dynamical equations governing the interaction of electrons and photons with gravity can be read from the equation

$$\begin{aligned} I_{(\frac{1}{2},1,2)} = &\int d^4x e \left(\overline{\psi} \left(i\gamma^\mu \left(\partial_\mu + eA_\mu + \frac{1}{4} \omega_\mu{}^{ab} \gamma_{ab} \right) - m \right) \psi \right) \\ &- \int d^4x e \left(\frac{1}{4} F_{\mu\nu} F_{\rho\sigma} g^{\mu\rho} g^{\nu\sigma} + \frac{1}{4} e^\mu_a e^\nu_b R_{\mu\nu}{}^{ab} \right). \end{aligned}$$

10.4.6. The standard model. The principle of gauge invariance plays a fundamental role in physics. The Poincaré symmetry is a space-time symmetry effecting local space-time coordinates. On the other hand the known elementary particles in nature, fall into group representations, e.g. fermions have the local gauge invariance under the transformations $\psi_{\alpha I} \to \left(e^{ig_a \Lambda^a(x) T^a} \right)_I^J \psi_{\alpha J}$ where $(T^a)_I^J$ are the representations of the gauge group. As in the Maxwell case, connections are introduced to insure local gauge invariance. Let

$$D = d + igA, \quad A = A^a_\mu T^a dx^\mu,$$

so that the curvature of the connection A will be given by

$$D^2 = igF = ig\left(dA + igA^2\right),$$

$$F = \frac{1}{2} F^a_{\mu\nu} T^a dx^\mu \wedge dx^\nu,$$

so that

$$F^a_{\mu\nu} = \partial_\mu A^a_\nu - \partial_\nu A^a_\mu + g f^{abc} A^b_\mu A^c_\nu,$$

where $\left[T^a, T^b\right] = if^{abc}T^c$. Let us denote the quarks by $q = \begin{pmatrix} u_L \\ d_L \end{pmatrix}$, u_R, d_R which are in the representations $\left(3, 2, -\frac{1}{3}\right)$, $\left(\overline{3}, 1, -\frac{4}{3}\right)$ and $\left(\overline{3}, 1, \frac{2}{3}\right)$ of $SU(3)_c \times SU(2)_w \times U(1)_Y$. The leptons are denoted by $l = \begin{pmatrix} e^-_L \\ \nu_L \end{pmatrix}$, e^+_R which are in the $(1, 2, 1)$ and $(1, 1, 2)$ of $SU(3)_c \times SU(2)_w \times U(1)_Y$.. The Lagrangian that govern all known interactions is given by

$$\begin{aligned}
e^{-1}L = & -\frac{1}{4\kappa^2}R - \frac{1}{4}\left(F^i_{\mu\nu}F^{\mu\nu i} + F^\alpha_{\mu\nu}F^{\mu\nu\alpha} + B_{\mu\nu}B^{\mu\nu}\right) \\
& + D_\mu H^\dagger D^\mu H - \mu^2 H^\dagger H + \lambda\left(H^\dagger H\right)^2 \\
& + i\overline{q}\gamma^\mu\left(\partial_\mu + \frac{1}{4}\omega_\mu^{\ ab}\gamma_{ab} - \frac{i}{2}g_2 A^\alpha_\mu \sigma^\alpha - \frac{i}{6}g_1 B_\mu - \frac{i}{2}g_3 V^i_\mu \lambda^i\right) q \\
& + i\overline{d}_R\gamma^\mu\left(\partial_\mu + \frac{1}{4}\omega_\mu^{\ ab}\gamma_{ab} + \frac{i}{3}g_1 B_\mu - \frac{i}{2}g_3 V^i_\mu \lambda^i\right) d_R \\
& + i\overline{u}_R\gamma^\mu\left(\partial_\mu + \frac{1}{4}\omega_\mu^{\ ab}\gamma_{ab} - \frac{2i}{3}g_1 B_\mu - \frac{i}{2}g_3 V^i_\mu \lambda^i\right) u_R \\
& + i\overline{l}\gamma^\mu\left(\partial_\mu + \frac{1}{4}\omega_\mu^{\ ab}\gamma_{ab} - \frac{i}{2}g_2 A^\alpha_\mu \sigma^\alpha + \frac{i}{2}g_1 B_\mu\right) l \\
& + i\overline{e}_R\gamma^\mu\left(\partial_\mu + \frac{1}{4}\omega_\mu^{\ ab}\gamma_{ab} + ig_1 B_\mu\right) e_R \\
& + \left(k^d\overline{q}Hd_R + k^u\overline{q}\tau^2 Hu_R + k^e\overline{l}He_R + h.c.\right),
\end{aligned}$$

where the Higgs field H is in the $(1, 2, 1)$ representation so that $D_\mu H = \partial_\mu H - \frac{i}{2}g_2 A^\alpha_\mu \sigma^\alpha - \frac{i}{2}g_1 B_\mu H$ and V^i_μ, A^α_μ and B_μ are the gauge fields for the gauge Lie algebras $SU(3)_c$, $SU(2)_w$ and $U(1)_Y$. The k^d, k^u and k^e are 3×3 matrices mixing the three generations of particles. Minimizing the potential of the Higgs fields H gives a vacuum

$$\langle H \rangle = \begin{pmatrix} 0 \\ v \end{pmatrix},$$

which generates masses for the quarks and leptons, e.g. the leptons will acquire the term

$$k^e\overline{l}He_R = k^e v\overline{e}_L e_R.$$

Similarly the term $D_\mu H^\dagger D^\mu H$ will yield the mass terms

$$\frac{1}{4}v^2 g_2^2\left(W^+_\mu W^{\mu -} + \frac{1}{\cos^2\theta}Z_\mu Z^\mu\right),$$

where $W^\pm_\mu = \left(A_\mu \pm iA^2_\mu\right)$, $Z_\mu = \frac{1}{\sqrt{g_1^2+g_2^2}}\left(g_2 A^3_\mu - g_1 B_\mu\right)$ and $\sin^2\theta = \frac{g_1^2}{g_1^2+g_2^2}$. The symmetry of the Lagrangian is diffeomorphism$\times$ local internal symmetry. All fermions are chiral (Weyl fermions) and acquire mass only after the symmetry is broken spontaneously from $SU(3)_c \times SU(2)_w \times U(1)_Y$ to $SU(3)_c \times U(1)_{em}$. At the quantum level chiral fermions break gauge invariance. The chirality condition is $\gamma_5\psi_\pm = \pm\psi_\pm$ where $\gamma_5 = i\gamma_0\gamma_1\gamma_2\gamma_3$ so that $\gamma_5^2 = 1$. A Dirac spinor is decomposed according to $\psi = \psi_+ + \psi_-$. The eigenvalue equation $D\psi = \lambda\psi$ implies $\gamma_5 D\psi_\pm = \mp\lambda D\psi_\pm$ so that the eigenvalue problem could not be set for chiral fermions alone without their partners except for zero eigevalues. After quantization one must sum over all states which would involve computing the determinat

$$\int D\psi D\overline{\psi} e^{\overline{\psi} D\psi} \rightarrow \det D,$$

and this involve the step of taking the trace over all states, which must now be restricted to the chiral ones

$$\langle\psi_{n-}|\, D\, |\psi_{n+}\rangle\,.$$

This is not invariant under the chiral rotation $\psi_{n+} \rightarrow e^{i\theta\gamma_5}\psi_{n+}$. To see this explicitely we have

$$Tr(F(D)) = \sum_n \langle\psi_{n-}|\, F(D)\, |\psi_{n+}\rangle \rightarrow \sum_n \left\langle e^{-i\theta^a T^a}\psi_{n-}\right|\, F(D) \left|e^{i\theta^a T^a}\psi_{n+}\right\rangle.$$

Since $F(D)$ depends only on geometrical quantities, one can use the heat kernel expansion to find the lowest terms

$$\begin{aligned} Tr(F(D)) &\rightarrow 2i\theta^a Tr\left(\gamma_5 T^a \gamma^{\mu\nu} F^b_{\mu\nu} T^b \gamma^{\rho\sigma} F^c_{\rho\sigma} T^c\right) \\ &= 2i\theta^a \epsilon^{\mu\nu\rho\sigma} F^b_{\mu\nu} F^c_{\rho\sigma} Tr\left(T^a \left\{T^b, T^c\right\}\right). \end{aligned}$$

We deduce that gauge invariance is not destroyed by chirality if

$$Tr\left(T^a \left\{T^b, T^c\right\}\right) = 0.$$

If we include gravitational terms then the lowest order term is

$$2i\theta^a Tr\left(\gamma_5 T^a \gamma^{\mu\nu} R_{\mu\nu}{}^{cd} \gamma^{\rho\sigma} R_{\rho\sigma cd}\right),$$

which vanishes if

$$Tr\left(T^a\right) = 0,$$

insuring the absence of gravitational anomalies. Both conditions on the representations of the chiral fermions in the standard model are satisfied. These are very strong constraints on possible representations of the fermions in a realistic model and lead to the conclusion that these are not satisfied by accident but are derivable from higher principles. One such explanation is unification where the chiral fermionic representations

$$\left(3, 2, -\frac{1}{3}\right) \oplus \left(\overline{3}, 1, \frac{2}{3}\right) \oplus \left(\overline{3}, 1, -\frac{4}{3}\right) \oplus (1, 2, 1) \oplus (1, 1, 2)$$

could be obtained from the $\overline{5} \oplus 10$ representations of $SU(5)$ to be broken spontaneously into two stages

$$SU(5) \rightarrow SU(3)_c \times SU(2)_w \times U(1)_Y \rightarrow SU(3)_c \times U(1)_{em}.$$

This is not the unique possibility but the simplest.

10.5. Supersymmetry

In nature there are particles of different spins, in particular integral and half-integral spins. In the ultimate theory there must be a symmetry that places particles with different spin in the same multiplet. Until 1974 all internal symmetries were studied and there was a theorem by Coleman and Mandula stating that, under certain physical assumptions, it is impossible to have a space-time symmetry larger than the Poincaré symmetry. This obstruction was overcome by the simple extension of considering $\mathbb{Z}_2$ graded algebras whose generators are classified into two classes, even (bosonic) and odd (fermionic) and obey

$$[even, even] = even, \quad [even, odd] = odd, \quad \{odd, odd\} = even.$$

With every generator A in the graded Lie algebra we associate a number $a = \begin{cases} 0 \text{ if } A \text{ is even} \\ 1 \text{ if } A \text{ is odd} \end{cases}$ and define the graded commutator by

$$[A, B\} = AB - (-1)^{ab} BA,$$

which satisfies the graded Jacobi identity

$$[A, [B, C\}\} + (-1)^{a(b+c)} [B, [C, A\}\} + (-1)^{c(a+b)} [C, [A, B\}\} = 0.$$

Denote by Q^i_α the fermionic generators where the index α transforms under the Lorentz group and the index i transforms under an internal algebra with generators T_r

$$[T_r, T_s] = f_{rst} T_t, \quad [P_\mu, T_r] = 0, \quad [J_{\mu\nu}, T_r] = 0,$$
$$\left[Q^i_\alpha, J_{\mu\nu}\right] = (b_{\mu\nu})^\beta_\alpha Q^i_\beta, \quad \left[Q^i_\alpha, P_\mu\right] = (c_\mu)^\beta_\alpha Q^i_\beta,$$
$$\left\{Q^i_\alpha, Q^j_\beta\right\} = r\,(\gamma^\mu C)_{\alpha\beta} P_\mu \delta^{ij} + s\,(\gamma^{\mu\nu} C)_{\alpha\beta} J_{\mu\nu} \delta^{ij}$$
$$+ C_{\alpha\beta} U^{ij} + (\gamma_5 C)_{\alpha\beta} V^{ij} + (\gamma^\mu \gamma_5 C)_{\alpha\beta} L^{ij}_\mu.$$

By the Coleman-Mandula theorem we can set L^{ij}_μ to zero because it carries a Lorentz vector index. Without any loss in generality we can assume that Q^i_α satisfy the Majorana condition $Q^i_\alpha = C_{\alpha\beta} \overline{Q}^{\beta i}$.

Using the Jacobi identity

$$\left[J_{\mu\nu}, \left[J_{\rho\sigma}, Q^i_\alpha\right]\right] + \left[J_{\rho\sigma}, \left[Q^i_\alpha\right]\right] + \left[Q^i_\alpha, \left[J_{\mu\nu}, J_{\rho\sigma}\right]\right] = 0,$$

we deduce that $(b_{\mu\nu})^\beta_\alpha$ should form a representation of the Lorentz group and must be identified with

$$(b_{\mu\nu})^\beta_\alpha = \frac{i}{2} (\gamma_{\mu\nu})^\beta_\alpha$$

From the identity

$$\left[P_\mu, \left[P_\nu, Q^i_\alpha\right]\right] + \left[P_\nu, \left[Q^i_\alpha, P_\mu\right]\right] + \left[Q^i_\alpha, \left[P_\mu, P_\nu\right]\right] = 0,$$

one deduces that $[c_\mu, c_\nu] = 0$. But it is always possible to change $(c_\mu)^\beta_\alpha$ to the form $c_\mu = c\,(\gamma_\mu)^\beta_\alpha$ and this implies that $c = 0$. From the identity

$$\left[Q^i_\alpha, \left\{Q^j_\beta, Q^k_\gamma\right\}\right] + \left[Q^j_\beta, \{Q^k_\gamma, Q^i_\alpha\}\right] + \left[Q^k_\gamma, \left\{Q^i_\alpha, Q^j_\beta\right\}\right] = 0,$$

we deduce that U^{ij} and V^{ij} commute with all the generators. Finally from the identity

$$\left[P_\mu, \left\{Q^i_\alpha, Q^j_\beta\right\}\right] + \left\{Q^i_\alpha, \left[Q^j_\beta, P_\mu\right]\right\} - \left\{Q^j_\beta, \left[P_\mu, Q^i_\alpha\right]\right\} = 0,$$

we deduce that $s = 0$. We shall rescale the parameter r to 2. Summarizing the results we have the most general graded algebra which is an extension of the Poincaré symmetery and consistent with unitarity:

$$\left[Q^i_\alpha, J_{\mu\nu}\right] = -\frac{i}{2}\left(\gamma_{\mu\nu}\right)^\beta_\alpha Q^i_\beta, \quad \left[Q^i_\alpha, P_\mu\right] = 0,$$
$$\left\{Q^i_\alpha, Q^j_\beta\right\} = -2\left(\gamma^\mu C\right)_{\alpha\beta} P_\mu \delta^{ij} + C_{\alpha\beta} U^{ij} + \left(\gamma_5 C\right)_{\alpha\beta} V^{ij}.$$

10.5.1. Irreducible representations of supersymmetry. The Casimir operators are P^2 and $\widehat{W}^2$ where

$$\widehat{W}_\mu = \epsilon_{\mu\nu\rho\sigma} P^\nu J^{\rho\sigma} + \overline{Q}^i \gamma_\mu \gamma_5 Q^i.$$

Since P^2 is a Casimir, all particles belonging to the same multiplet must have the same mass. Now introduce the fermion number $(-1)^{N_F}$ which acts on bosonic and fermionic states as follows

$$(-1)^{N_F} |boson\rangle = |boson\rangle,$$
$$(-1)^{N_F} |fermion\rangle = -|fermion\rangle,$$

which implies that

$$(-1)^{N_F} Q^i_\alpha |.\rangle = -Q^i_\alpha (-1)^{N_F} |.\rangle.$$

For any finite dimensional representation we have

$$Tr\left((-1)^{N_F}\left\{Q^i_\alpha, \overline{Q}^{i\beta}\right\}\right) = Tr\left((-1)^{N_F}(Q^i_\alpha \overline{Q}^{i\beta} + \overline{Q}^{i\beta} Q^i_\alpha\right)$$
$$= Tr\left((-1)^{N_F}(\gamma^\mu)^\beta_\alpha P_\mu\right)$$
$$= Tr\left(-Q^i_\alpha (-1)^{N_F} \overline{Q}^{i\beta} + Q^i_\alpha (-1)^{N_F} \overline{Q}^{i\beta}\right) = 0.$$

From this we deduce that

$$Tr\left((-1)^{N_F}\right) = 0,$$

implying that the number of fermionic and bosonic degrees of freedom are identical.

The energy in supersymmetric theories is non-negative. To see this consider

$$0 \leq \sum_i \left(Q^i_\alpha (Q^i_\alpha)^\dagger + (Q^i_\alpha)^\dagger Q^i_\alpha\right) = \sum_i \left(Q^i_\alpha \left(\overline{Q}^i\gamma_0\right)^\alpha + \left(\overline{Q}^i\gamma_0\right)^\alpha Q^i_\alpha\right)$$
$$= Tr \sum_i \left\{Q^i_\alpha, \left(\overline{Q}^i\gamma_0\right)^\alpha\right\} = 2N Tr\left(\gamma^\mu P_\mu \gamma_0\right) = 8NP_0.$$

Therefore $E \geq 0$.

We would like to characterize the massless representations of supersymmetry: $P^2 = 0$, and we can choose $P_\mu = (w, 0, 0, w)$. In this frame, and after setting the central charges U^{ij} and V^{ij} to zero, we get

$$\left\{Q^i_\alpha, Q^j_\beta\right\} = 2\delta^{ij}\left(\gamma^\mu C\right)_{\alpha\beta} P_\mu = 2w\delta^{ij}\left(\left(\gamma_0 + \gamma_3\right) C\right)$$
$$= -4w\delta^{ij}\begin{pmatrix} 0 & 0 & 0 & 1 \\ 0 & 0 & 0 & 0 \\ 0 & 0 & 0 & 0 \\ 1 & 0 & 0 & 0 \end{pmatrix}.$$

We therefore have one independent relation $\left\{Q_1^i, Q_4^j\right\} = -4E\delta^{ij}$. From the relation $Q_\alpha^i = C_{\alpha\beta}\overline{Q}^{i\beta}$ we have

$$Q_3^i = Q_2^{i\dagger}, \quad Q_4^i = -Q_1^{i\dagger}$$

so we can write $\left\{Q_1^i, Q_1^{j\dagger}\right\} = 4w\delta^{ij}$. Normalizing the Q_1^i we can finally write

$$\left\{Q_1^i, Q_1^{j\dagger}\right\} = \delta^{ij}.$$

The subgroup of the Lorentz group leaving the state $(w, 0, 0, w)$ invariant is $\Lambda_{03} = 0$, $\Lambda_{10} = -\Lambda_{13}$, $\Lambda_{20} = -\Lambda_{23}$ implying that the generators are

$$T_1 = J_{10} + J_{13}, \quad T_2 = J_{20} + J_{23}, \quad J = J_{12}.$$

These obey the commutation relations

$$[J, T_1] = T_2, \quad [J, T_2] = -T_1, \quad [T_1, T_2] = 0,$$
$$\left[J, Q_1^i\right] = \frac{1}{2}Q_1^i, \quad \left[J, \overline{Q}_1^i\right] = -\frac{1}{2}\overline{Q}_1^i, \quad \left\{Q_1^i, Q_1^{j\dagger}\right\} = \delta^{ij}.$$

Since Q_1^i and $Q_1^{i\dagger}$ form a Clifford algebra, they correspond to a set of N fermi creation operators $Q_1^{i\dagger}$ and N fermi annihilation operators Q_1^i. The vacuum state is defined by $|\lambda\rangle$

$$Q_1^i\,|\lambda\rangle = 0, \quad J\,|\lambda\rangle = \lambda\,|\lambda\rangle\,,$$
$$Q_1^{i\dagger}\,|\lambda\rangle = \left|\lambda - \frac{1}{2}, i\right\rangle, \quad Q_1^{i\dagger}Q_1^{j\dagger}\,|\lambda\rangle = |\lambda - 1, [ij]\rangle\,,$$
$$Q_1^{1\dagger}Q_1^{2\dagger}\cdots Q_1^{N\dagger}\,|\lambda\rangle = \left|\lambda - \frac{N}{2}\right\rangle.$$

These form 2^N states falling into two classes. Each class has 2^{N-1} states obtained from $|\lambda\rangle$ by acting with odd and even number of $Q_1^{i\dagger}$ respectively. The 2^N states have helicities ranging from $\lambda_{\max}$ to $\lambda_{\max} - \frac{N}{2}$. By the CPT theorem physical states must have both helicities λ and $-\lambda$ present which implies that $\lambda_{\max} - \frac{N}{2} = -\lambda_{\max}$ and therefore $\lambda_{\max} = \frac{N}{4}$. If this is not satisfied one must add the conjugate states where $\lambda'_{\max} = \frac{N}{2} - \lambda_{\max}$ and act on this state with $Q_1^{i\dagger}$. A massless non CPT conjugate multiplet has 2^{N+1} states. For gauge and matter theories $\lambda_{\max} \leq 1$ which implies that $N \leq 4$. We can construct the following table:

	I	II	III	IV	V	VI
λ	$N=1$	$N=2$	$N=1$	$N=2$	$N=3$	$N=4$
1		1		1	1	1
$\frac{1}{2}$	1	1	$1+1'$	2	$3+1$	4
0	$1+1$		$2+2'$	$1+1'$	$\overline{3}+3$	6
$-\frac{1}{2}$	1	$1'$	$1+1'$	2	$1+\overline{3}$	$\overline{4}$
-1		1		1	1	1

The multiplets in cases I and II have $\lambda_{\max} = \frac{1}{2}$, while in all the other cases $\lambda_{\max} = 1$. Case I corresponds to an $N = 1$ chiral multiplet with $2^{1+1} = 4$ states. Case II is $N = 1$ vector multiplet also with $2^{1+1} = 4$ states. Case III is an $N = 2$ hypermultiplet with $2^{2+1} = 8$ states. Case IV is an $N = 2$ vector multiplet with $2^{2+1} = 8$ states. Finally case VI give the $N = 4$ vector multiplet with $2^4 = 16$ states as it is CPT self conjugate. The $N = 3$ in case V is identical to $N = 4$ because one must add the CPT conjugate states.

Next we consider the situation where $\lambda_{\max} = 2$ which implies that $N \leq 8$. All these multiplets will contain the graviton and will correspond to supergravity multiplets:

λ	$N=1$	$N=2$	$N=3$	$N=4$	$N=5$	$N=6$	$N=7$	$N=8$
2	1	1	1	1	1	1	1	1
$\frac{3}{2}$	1	2	3	4	5	6	$7+1$	8
1		1	$\overline{3}$	6	$10+1$	15	$21+7$	28
$\frac{1}{2}$			1	$\overline{4}$	$\overline{10}+1$	$20+1$	$35+21$	56
0				$1+1$	$\overline{5}+5$	$\overline{6}+6$	$\overline{35}+35$	70
$-\frac{1}{2}$			1	4	$1+10$	$1+\overline{20}$	$\overline{21}+\overline{35}$	$\overline{56}$
-1		1	3	6	$\overline{10}+1$	$\overline{15}$	$\overline{21}+\overline{7}$	$\overline{28}$
$-\frac{3}{2}$	1	2	$\overline{3}$	$\overline{4}$	$\overline{5}$	$\overline{6}$	$\overline{7}+1$	$\overline{8}$
-2	1	1	1	1	1	1	1	1

Notice that $N = 7$ supergravity is identical to $N = 8$ supergravity, so there are seven different supergravity theories. Only the field content of the first three supergravity theories is fixed uniquely. The simplest is $N = 1$ supergravity where only two fields are needed, the metric (graviton) and the gravitino (Rarita-Schwinger).

10.5.2. Supersymmetric field theories.

10.5.2.1. *Lagrangian for chiral multiplet.* The Lagrangian for $N = 1$ chiral multiplet (A, B, ψ) is given by

$$L = \frac{1}{2}\left(\partial_\mu A \partial^\mu A + \partial_\mu B \partial^\mu B + i\overline{\psi}\gamma^\mu \partial_\mu \psi\right).$$

The action is invariant under the supersymmetry transformations

$$\delta A = \overline{\epsilon}\psi, \quad \delta B = \overline{\epsilon}\gamma_5\psi,$$
$$\delta\psi = \left(-i\partial_\mu A + \partial_\mu B\gamma_5\right)\gamma^\mu\epsilon,$$

as can be easily verified. The supersymmetry parameter ϵ is independent of the coordinates x^μ and supersymmetry is a global symmetry.

10.5.2.2. *Lagrangian for vector multiplet.* The vecor multiplet is given by (A^μ, λ) where $A_\mu = A_\mu^a T^a$ and $\lambda = \lambda^a T^a$ which are Lie algebra valued: $\left[T^a, T^b\right] = if^{abc}T^c$. The Lagrangian which is valid in dimensions $D = 4, 6$ and 10 is given by

$$L = Tr\left(-\frac{1}{4}F_{\mu\nu}F^{\mu\nu} + \frac{i}{2}\lambda\Gamma^\mu D_\mu\lambda\right),$$

where $F = dA + A^2 = \frac{1}{2}F_{\mu\nu}dx^\mu \wedge dx^\nu$, $F_{\mu\nu} = F_{\mu\nu}^a T^a$ and $D_\mu\lambda = \partial_\mu\lambda + ig\left[A_\mu, \lambda\right] = D_\mu\lambda^a T^a$. The action is invariant under the supersymmetry transformations

$$\delta A_\mu = i\epsilon\Gamma_\mu\lambda,$$
$$\delta\lambda = \frac{1}{2}F_{\mu\nu}\Gamma^{\mu\nu}\epsilon.$$

To verify supersymmetry one has to use Fierz identities for the fermions which are valid only in dimensions $4, 6$ and 10.

One can couple vector multiplets to chiral multiplets and have a generalization of the standard model. In the supersymmetric standard model it turned out that all known fields must have partners, e.g., the electron must have a bosonic partner, the s-electron, while gauge fields A_μ will have as partners the gauginos λ. As we have seen, the fermions in the standard model are chiral. The chirality condition cannot be satisfied for theories with more than $N = 1$ supersymmetry. In other words, for supersymmetry to be of any relevance, the Lagrangian at low energies must have $N = 1$ supersymmetry. However, in the real world supersymmetry is not present because no partners for the observed particles with the same mass are found, so it must be broken. One of the main advantages of supersymmetry is that it has good quantum behaviour, where divergencies in the renormalization of the bosonic fields are cancelled by those coming from the fermions. To preserve good quantum behaviour, the symmetry must be broken spontaneously. The Higgs mechanism which is essential in the bosonic case must also be present here. The Goldstone phenomena where the scalar fields associated with the broken generators are absorbed by the massless vector fields to become massive must be generalized. In this case the field associated with the broken supersymmetry generator, the Goldstino, must be absorbed by a massless vector-spinor field. In the discussion in the last section we have seen that this is the Rartia-Schwinger field. This implies that the supergravity multiplet must be coupled to the matter Lagrangian of the standard model in such a way that supersymmetry is broken spontaneously, so that the gravitino absorb one fermionic field to become massive. In supergravity the gravitational multiplet plays a role in the breaking of supersymmetry and this can be used to induce the electroweak breaking in the standard model. This is a novel phenomena as this implies that physics of the extremely weak gravitational field influences the low-energy sector.

10.5.2.3. *N=1 supergravity Lagrangian.* The easiest way to construct the supergravity Lagrangian is to generalize the construction of the Einstein-Hilbert action based on gauging the Poincaré algebra. Here we gauge the supersymmetry algebra instead. Let

$$D_\mu = \partial_\mu + \omega_\mu^{\ ab} J_{ab} + e_\mu^a P_a + \psi_\mu^\alpha Q_\alpha,$$

be the connection associated with the supersymmetry algebra. The curvature is easily computed to be

$$[D_\mu, D_\nu] = R_{\mu\nu}^{\ \ ab} J_{ab} + T_{\mu\nu}^a P_a + \psi_{\mu\nu}^\alpha Q_\alpha,$$

where

$$\begin{aligned} R_{\mu\nu}^{\ \ ab} &= \partial_\mu \omega_\nu^{\ ab} - \partial_\nu \omega_\mu^{\ ab} + \omega_\mu^{\ ac} \omega_{\nu c}^{\ \ \ b} - \omega_\nu^{\ ac} \omega_{\mu c}^{\ \ \ b}, \\ T_{\mu\nu}^a &= \partial_\mu e_\nu^a - \partial_\nu e_\mu^a + \omega_\mu^{\ ab} e_{\nu b} - \omega_\nu^{\ ab} e_{\mu b} + \overline{\psi}_\mu \gamma^a \psi_\nu, \\ \psi_{\mu\nu} &= \left(\partial_\mu + \frac{1}{4}\omega_\mu^{\ ab}\gamma_{ab}\right)\psi_\nu - \left(\partial_\nu + \frac{1}{4}\omega_\nu^{\ ab}\gamma_{ab}\right)\psi_\mu, \end{aligned}$$

and the action of J_{ab} on the spinor ψ_μ is represented by $\frac{1}{4}\gamma_{ab}$. To proceed one first impose the torsion free constraint

$$T_{\mu\nu}^a = 0,$$

which can be solved to express $\omega_\mu^{\ ab}$ in terms of e_μ^a and ψ_μ to obtain

$$\omega_\mu^{\ ab} = \omega_\mu^{\ ab}(e) + \frac{1}{4}\left(\overline{\psi}_\mu \gamma^a \psi^b - \overline{\psi}_\mu \gamma^b \psi^a + \overline{\psi}^a \gamma_\mu \psi^b\right)$$

where $\omega_\mu^{\ ab}(e)$ is the same expression we had in the non-supersymmetric case, and indices are changed from flat to curved by using the vierbein e_μ^a and its inverse e_a^μ.

The supersymmetry transformations are given by

$$\begin{aligned}\delta e_\mu^a &= \bar{\epsilon}\gamma^a\psi_\mu,\\ \delta\omega_\mu{}^{ab} &= 0,\\ \delta\psi_\mu &= \left(\partial_\mu + \frac{1}{4}\omega_\mu{}^{ab}\gamma_{ab}\right)\epsilon.\end{aligned}$$

The torsion constraint is not preserved under these transformations. We can preserve the torsion constraint by modifying the $\delta\omega_\mu{}^{ab}$ accordingly. The result is

$$\delta'\omega_{\mu ab} = -\frac{1}{2}e_a^\nu e_b^\rho\left(\bar{\epsilon}\gamma_\rho\psi_{\mu\nu} - \bar{\epsilon}\gamma_\mu\psi_{\nu\rho} + \bar{\epsilon}\gamma_\nu\psi_{\rho\mu}\right).$$

The Lagrangian invariant under the supersymmetry transformations is given by

$$e^{-1}L_{SG} = -\frac{1}{4}R + \overline{\psi}_\mu\gamma^{\mu\nu\rho}\left(\partial_\nu + \frac{1}{4}\omega_\nu{}^{ab}\gamma_{ab}\right)\psi_\rho$$

The proof of invariance is simplified by using that $eR = \frac{1}{2}\epsilon^{\mu\nu\rho\sigma}\epsilon_{abcd}e_\mu^a e_\nu^b R_{\rho\sigma}{}^{cd}$ and by writing the varriation with respect to $\omega_\mu{}^{ab}$ in the form $\frac{\delta L}{\delta\omega_\mu{}^{ab}}\delta'\omega_\mu{}^{ab}$. Since $\omega_\mu{}^{ab}$ appears linearly and quadratically one can easily see that, after integrating by parts, its varriation $\frac{\delta L}{\delta\omega_\mu{}^{ab}}$ vanishes if the torsion vanishes and therefore there is no need to substitute the very complicated expression for $\delta'\omega_\mu{}^{ab}$. This method is known as the 1.5 formalism and was used widely in the early formulations of supergravity because it drastically simplifies the proof of the supersymmetry invariance of the proposed actions.

10.6. Higher dimensional theories

By matching the bosonic and fermionic degrees of freedom we shall determine what are the possible supersymmetric multiplets in all possible dimensions. First a massless vector field in D dimensions have $D-2$ degrees of freedom as the A_0 component does not propagate because it has no time derivatives in the action and one of the transverse components of A_i can be gauged away. For fermions a Dirac spinor has $2^{\left[\frac{D}{2}\right]}$ components, but in certain cases could be subjected to either the Weyl chirality condition or to the Majorana condition or both. The first step is to define the Clifford algebra in dimensions D. The Dirac gammas matrices Γ^M where $M = 0, 1, \cdots, D-1$ will be $2^{\left[\frac{D}{2}\right]} \times 2^{\left[\frac{D}{2}\right]}$ matricess. In even dimensions we can define the gamma matrix $\Gamma_{D+1} = \eta\Gamma^0\Gamma^1\cdots\Gamma^{D-1}$ where $(\Gamma_{D+1})^2 = 1$ so that $\eta^2 = (-1)^{s-t} = 1$ $(mod8)$, where s is the number of space-like coordinates and t the number of time like coordinates. For our considerations we will always take $s = D-1$ and $t = 1$. Therefore $(\Gamma_{D+1})^2 = 1$ and one can always impose the Weyl condition $\Gamma_{D+1}\psi_\pm = \pm\psi_\pm$. No Weyl fermions can exist in odd dimensions. The Majorana condition $\psi = C\overline{\psi}$ requires the existence of imaginary representations of the gamma matrices in D dimensions so that $\Gamma^{M^*} = -C\Gamma^M C^{-1}$. This is only possible in $D = 2, 3, 4$ $(mod8)$. It is also possible to impose both the Weyl condition and the Majorana condition simultaneously only in $D = 2$ $(mod8)$. Therefore a spinor λ have $2^2 \times \frac{1}{2} = 2$ components in 4 dimensions as one imposes the Majorana condition. (Supersymmetric multiplets are always defined with respect to Majorana spinors so the Weyl condition can only be imposed in dimensions 2 $(mod8)$. The number of components for λ in 5 dimensions is 4 while in 6 dimensions it is $2^3 \times \frac{1}{2} = 4$ components. In dimensions 7 and 8 it is 8 components. In dimension 9 it is 16 (although in this case it is possible to modify the Majorana condition to reduce it to 8 components). In dimension 10 one can impose simultaneously the Weyl and Majorana conditions so that the number of independent components of λ is 8. We notice that the number of independent components of A_M and λ match in $D = 3, 4, 6$ and 10 and this explains why the super Yang-Mills action is invariant under supersymmetry transformations only in those dimensions. In dimensions higher than 10 the number of independent fermionic components increase much faster than the number of independent components for the vectors A_M and supersymmetric actions for the pure vector multiplet do not exist. This is to be expected because when these higher dimensional theories are analyzed as viewd in four-dimensions, they will correspond to vector multiplets with supersymmetry higher than $N = 4$ which is not possible from our analysis of the supersymmetry representations.

To analyse supergravity theories in higher dimensions we must first determine the number of independent degrees of freedom for the graviton. This is essentially a symmetric metric with $\frac{D(D+1)}{2}$ components where the components g_{0i} do not propagate because they do not acquire second time derivatives. From the other components g_{ij} the D diffeomorphism parameters could be used to reduce the components to $\frac{D(D-3)}{2} = \frac{(D-2)(D-1)}{2} - 1$ so essentially the physical degrees are those of a symmetric traceless metric in $(D-2)$ dimensions. From the Rarita-Schwinger equation for a free field, $\Gamma^{MNP}\partial_N\psi_P = 0$, we can easily show by first contracting this equation with Γ^M and making the gauge choice $\Gamma^M\psi_M = 0$ that $\Gamma^N\partial_N\psi_M = 0$ and $\partial^M\psi_M = 0$. Therefore ψ_M behaves like a massless vector with $D-2$ spinor components and the condition $\Gamma^N\partial_N\psi_M = 0$ eliminates one more spinor component to give finally $(D-3)\,2^{\left[\frac{D}{2}\right]} \times r$ where the reduction factor r is $\frac{1}{2}$ if the Majorana condition is imposed, r is $\frac{1}{4}$ if the Majorana-Weyl condition is imposed and $r = 1$ if no conditions are imposed. From this analysis it is easy to see that the supergravity multiplet in $D = 4$ is the metric (or vierbein) e_μ^a and the gravitino ψ_μ as we have seen before. In five dimensions, the metric has 5 degrees of freedom while the gravitino has 8. A vector field has 3 degrees of freedom so the number of fermionic and bosonic degrees match if we take the multiplet $\left(e_\mu^a, A_\mu, \psi_\mu\right)$. We can classify all possible theories in various dimensions, but it is easier to start with maximal supergravity which is $N = 8$.

10.6.1. N=1 eleven dimensional supergravity. A spinor in 11 dimensions

have $2^{\left[\frac{11}{2}\right]} \times \frac{1}{2} = 16$ degrees of freedom because the Majorana condition could be imposed. As viewed from 4 dimensions where a spinor have 2 degrees of freedom, this corresponds to a spinor with $SO(8)$ symmetry, and according to the classification discussed before, this corresponds to a theory with $N = 8$ supersymmetry in four-dimensions. Therefore $D = 11$ is the highest dimensions where a supergravity theory could be constructed. Also in higher dimensions the number of fermionic degrees increase much faster than those of the metric and other bosonic fields. To determine the field content of the supergravity multiplet in $D = 1$ we note that the metric has 44 degrees of freedom while the gravitino has 128. The difference is 84 degrees which must be present in other bosonic configuration. An antisymmetric tensor of rank p (a p-form) has $\begin{pmatrix} D-2 \\ p \end{pmatrix}$ degrees of freedon, which is equal to 84 when $p = 3$. We deduce that in the multiplet $\left(e_M^A, A_{MNP}, \psi_M\right)$ the bosonic and fermionic degrees of freedom match. This is a necessary but not sufficient condition to have a supersymmetric multiplet and construct a supersymmetric action. This is indeed possible and the full supersymmetric Lagrangian is

$$
\begin{aligned}
e^{-1}L = & -\frac{1}{4}R - \frac{i}{2}\overline{\psi}_M\Gamma^{MNP}D_N\left(\frac{\omega+\widehat{\omega}}{2}\right)\psi_P - \frac{1}{48}F_{MNPQ}F^{MNPQ} \\
& + \frac{2e^{-1}}{12^4}\epsilon^{M_1\cdots M_8PQR}F_{M_1\cdots M_4}F_{M_5\cdots M_8}A_{PQR} \\
& + \frac{1}{192}\left(\overline{\psi}_M\Gamma^{MNPQRS}\psi_N + 12\overline{\psi}^P\Gamma^{QR}\psi^S\right)\left(F_{PQRS}+\widehat{F}_{PQRS}\right),
\end{aligned}
$$

where

$$
\begin{aligned}
F_{MNPQ} &= \frac{1}{4}\left(\partial_M A_{NPQ} - \partial_N A_{PQM} + \partial_P A_{QNM} - \partial_Q A_{MNP}\right), \\
\widehat{\omega}_M^{\ AB} &= \omega_M^{\ AB}(e) + \frac{1}{4}\left(\overline{\psi}_M\Gamma^A\psi^B - \overline{\psi}_M\Gamma^B\psi^A + \overline{\psi}^A\Gamma_M\psi^B\right), \\
\widehat{F}_{PQRS} &= F_{PQRS} - 3\overline{\psi}_{[P}\Gamma_{QR}\psi_{S]}.
\end{aligned}
$$

The action is invariant under the supersymmetry transformations

$$
\begin{aligned}
\delta e_M^A &= -i\overline{\epsilon}\Gamma^A\psi_M, \\
\delta\psi_M &= \left(\partial_M + \frac{1}{4}\widehat{\omega}_M^{\ AB}\Gamma_{AB} + \frac{i}{144}\left(\Gamma_M^{\ PQRS} - 8\delta_M^P\Gamma^{QRS}\right)\widehat{F}_{PQRS}\right)\epsilon, \\
\delta A_{MNP} &= \frac{3}{2}\overline{\epsilon}\Gamma_{[MN}\psi_{P]}.
\end{aligned}
$$

A theory in higher dimensions can produce more complicated content in lower dimensions either by compactification or by dimensional reduction. In dimensional reduction one assumes that the fields are independent of the coordinates in a certain number of internal directions. The action then reduces to that of a lower dimensional theory with more fields. As an exmaple a metric in D dimensions reduces in d dimensions to a metric, $(D-d)$ vectors and $\frac{(D-d)(D-d+1)}{2}$ scalars. A vector in D dimensions reduces in d dimensions to a vector and $(D-d)$ scalars.

10.6.2. N=1 ten-dimensional Supergravity and Super Yang-Mills. The super Yang-Mills action can be reduced from 10 dimensions to 4. The vector fields A_M^a reduce in 4 dimensions to the vector A_μ^a and the 6 scalars A_i^a, $B_i^a = A_{i+3}^a$, $i = 4,5,6$. The spinors $\lambda_{\widehat{\alpha}}^a$ in 10 dimensions reduce to four spinors in 4 dimensions $\lambda_{\alpha K}$, $K = 1,\cdots,4$. The reduced action takes the form

$$
\begin{aligned}
I = \int d^4x Tr\Big(& -\frac{1}{4}F_{\mu\nu}^a F^{\mu\nu a} + \frac{1}{2}D_\mu A_i^a D^\mu A_i^a + \frac{1}{2}D_\mu B_i^a D^\mu B_i^a \\
& + \frac{g}{2}\overline{\lambda}_K\left[\alpha_{KL}^i A_i + i\gamma_5\beta_{KL}^i B_i, \lambda_L\right] \\
& + \frac{g^2}{4}\left([A_i,A_j]^2 + [B_i,B_j]^2 + 2[A_i,B_j]^2\right)\Big),
\end{aligned}
$$

where α_{KL}^i and β_{KL}^i are 6 real antisymmetric 4×4 matrices obeying $SU(2)\times SU(2)$ algebra.

Of particular importance is $N=1$ supergravity in 10 dimensions. The supermultiplet consists of the vielbein e_M^A, an antisymmetric field B_{MN}, a dilaton field ϕ, a gravitino ψ_M and a spinor λ. The supergravity action in ten dimensions can be obtained from the eleven dimensional action by dimensional reduction and consistent truncation. The eleven dimensional vierbein gives in ten dimensions a vielbein, a vector and a scalar dilaton field. The antisymmetric field A_{MNP} gives in ten dimensions antisymmetric tensors of ranks three and two. The gravitino in eleven dimensions give in ten dimensions two gravitinos and two spinors. A consistent truncation is to set the vector, the rank three antisymmetric tensor, one of the gravitinos and one of the spinors to zero. One can show that one of the two supersymmetries will be preserved by this truncation leaving a theory with $N = 1$ supersymmetry. The untruncated theory has $N = 2$ supersymmetry in ten dimensions. The $N = 1$ supergravity Lagrangian is given by

$$\begin{aligned} e^{-1}L = & -\frac{1}{4}R + \frac{1}{12}e^{-2\phi}H_{MNP}H^{MNP} + \frac{1}{2}\partial_M\phi\partial^M\phi \\ & -\frac{i}{2}\overline{\psi}_M\Gamma^{MNP}\left(D_N + \widehat{D}_N\right)\psi_P + \frac{i}{2}\overline{\chi}\Gamma^M\left(D_N + \widehat{D}_N\right)\chi \\ & + \frac{1}{2\sqrt{2}}\overline{\psi}_M\Gamma^N\Gamma^M\chi\left(\partial_N\phi + \widehat{D}_N\phi\right) \\ & + \frac{i}{48}\left(\overline{\psi}_M\Gamma^{MNPQR}\psi_R + 6\overline{\psi}^N\Gamma^P\psi^Q - i\sqrt{2}\overline{\psi}_M\Gamma^{NPQ}\Gamma^M\chi\right)\left(H_{NPQ} + \widehat{H}_{NPQ}\right), \end{aligned}$$

where $H_{MNP} = 3\partial_{[M}B_{NP]}$. This action is invariant under the supersymmetry transformations

$$\begin{aligned} \delta e_M^A &= -i\overline{\epsilon}\Gamma^A\psi_M, \\ \delta\phi &= -\frac{1}{\sqrt{2}}\overline{\epsilon}\chi, \\ \delta B_{MN} &= e^{\phi}\left(i\epsilon\Gamma_{[M}\psi_{N]}\right), \\ \delta\psi_M &= \widehat{D}_M\epsilon - \frac{\sqrt{2}}{48}e^{-\phi}\left(\Gamma_M^{\ NPQ} - 9\delta_M^N\Gamma^{PQ}\right)\widehat{H}_{NPQ}, \\ \delta\chi &= \frac{i}{\sqrt{2}}\Gamma^M\epsilon\widehat{D}_M\phi + \frac{i}{12\sqrt{2}}e^{-\phi}\Gamma^{NPQ}\widehat{F}_{NPQ}. \end{aligned}$$

The main advantage of the $N = 1$ supergravity Lagrangian in ten dimensions is that it can be coupled to the ten-dimensional super Yang-Mills theory with an arbitrary gauge group. The main modification to the pure supergravity Lagrangian is that the field H_{MNP} is replaced with

$$\begin{aligned} \widehat{H}_{MNP} &= H_{MNp} - \omega_{MNP}, \\ \omega_{MNP} &= Tr\left(A_{[M}\partial_N A_{P]} + \frac{g}{3}A_{[M}A_N A_{P]}\right), \end{aligned}$$

where ω_{MNP} is the Chern-Simons form over the gauge Lie algebra. When higher derivative terms are allowed in the supergravity Lagrangian then the modifications to H_{MNP} will also include the Chern-Simons gravitational term. The field $\widehat{H}_{MNP}$ satisfies the equation

$$d\widehat{H} = tr\left(-F \wedge F + R \wedge R\right),$$

where $\widehat{H}$ is a three-form, F is the gauge field strength two-form and R is the curvature tensor two-form.

10.7. Further directions

The topics considered so far in these lectures would enable us to study some very interesting questions in physics using results from mathematics. For lack of space, these directions will only be described briefly. The interested reader can pursue these problems in the literature.

10.7.1. Index of Dirac operators. Consider Dirac equation for a massless spinor in ten-dimensions

$$iD_{10}\psi = 0 = i\Gamma^M D_M \psi = 0,$$
$$= i\left(D_4 + D_K\right)\psi,$$

where K is the internal 6 dimensional manifold. Let $\Gamma^{(4)} = i\Gamma_0\Gamma_1\Gamma_2\Gamma_3$ and $\Gamma^{(K)} = -i\Gamma_4\Gamma_5\cdots\Gamma_9$ and $\Gamma^{(10)} = \Gamma_0\Gamma_1\cdots\Gamma_9$. This implies that $\left(\Gamma^{(4)}\right)^2 = \left(\Gamma^{(K)}\right)^2 = \left(\Gamma^{(10)}\right)^2 = 1$ and $\Gamma^{(10)} = \Gamma^{(4)}\Gamma^{(K)}$. The Weyl condition on fermions in ten-dimensions $\Gamma^{(10)}\psi = \psi$ is equivalent to $\Gamma^{(4)}\psi = \Gamma^{(K)}\psi$. Defining $\widehat{D}_4 = \Gamma^{(4)}D_4$ and $\widehat{D}_K = \Gamma^{(4)}D_K$ it is easy to see that one can diagonalize simultaneously $\widehat{D}_4$ and $\widehat{D}_K$. Let $H = \left(iD_K\right)^2$ and let ψ be an eigenstate with energy E, $H\psi = E\psi$. As H commutes with $\Gamma^{(K)}$ we have

$$H\left(iD_K\psi\right) = E\left(iD_K\psi\right),$$

so that ψ and $D_K\psi$ are degenerate in energy. On the other hand

$$D_K\left(\Gamma^{(K)}\psi\right) = -\Gamma^{(K)}D_K\psi,$$

so that ψ and $D_K\psi$ have opposite chiralities with respect to $\Gamma^{(K)}$ and are therefore linearly independent, unless $D_K\psi = 0$. For every state with $\Gamma^{(K)} = 1$ there is a state with $\Gamma^{(K)} = -1$ except for zero eigenvalues which do not have to be paired. The index of the Dirac operator is then defined by

$$index\left(iD_K\right) = n_+ - n_-$$

where n_+ and n_- are, respectively, the number of zero eigenvalues with eigenvalues $\Gamma^{(K)} = 1$ and $\Gamma^{(K)} = -1$. In the absence of gauge fields the Dirac operator satisfies the properties

$$\left(D_K\psi\right)^* = D_K\psi^*, \qquad \left(\Gamma^{(K)}D_K\psi\right)^* = -\Gamma^{(K)}D_K\psi^*.$$

This is so because in six-dimensions the gamma matrices are real and $\Gamma^{(K)}$ is purely imaginary , and therefore complex conjugation exchanges n_+ and n_-, so that $n_+ = n_-$ and the index is zero. In presence of gauge fields, and for spinors in some representation Q of the gauge group we have

$$index_Q\left(iD_K\right) = -index_{Q^*}\left(iD_K\right)$$

because complex conjugation exchanges the representation Q with its complex conjugate Q^*, and this implies that the index is zero if the representation Q is real or pseudoreal. If Q is complex then the index is not necessarily zero. In terms of the gauge fields and the spin-connection of the underlying manifold the index of the Dirac operator can be computed to be

$$index_Q\left(iD_K\right) = \frac{1}{48\left(2\pi\right)^3}\int\left(tr_Q\left(F\wedge F\wedge F\right) - \frac{1}{8}tr_Q\left(F\right)\wedge tr\left(R\wedge R\right)\right).$$

One can apply this property to explain physical phenomena. We have seen that in the standard model of particle physics, there are three families of particles. In higher dimensional theories it is assumed that particles are in some large representation of a gauge group which is spontaneously broken and the four dimensional theory is obtained by compactification from the higher dimensional one. From the higher dimensional point of view all these particles are massless as they acquire their small masses (compared with the high energy scale of compactification) at a later stage.

The number of massless families is then linked to the index of the Dirac operator on the internal manifold. Therefore we can write

$$index_Q\left(iD_K\right) = \frac{1}{2}\chi\left(K\right)$$

where $\chi\left(K\right)$ is the Euler number of the internal manifold. To build a realistic model with three families one must choose the Euler number of the internal manifold to be 6.

10.7.2. Holonomy and N=1 supersymmetry. At low-energies it is desirable to have a field theory with $N = 1$ supersymmetry in four-dimensions. To have some unbroken supersymmetry, let Q be the supersymmetry generator annihilating the vacuum $|0\rangle$. This implies that

$$\langle 0|\left\{Q, A\right\}|0\rangle = 0,$$

for all operators A. If A is bosonic then $\{Q, A\}$ is fermionic, but $\langle 0|$ fermionic $|0\rangle = 0$ as a fermionic state has no expectation value. If A is fermionic then $\{Q, A\} = \delta A$ and the above relation implies that $\delta A = 0$. In supergravity theories we have seen that the supersymmetry transformations of the gravitino and the other fermions are proportional to the other bosonic fields and the supersymmetry parmeter. For example, assuming that bosonic fields, except for the metric, are set to zero, the gravitino transformation is given by

$$\delta\psi_M = D_M\epsilon = 0.$$

But this is a Killing spinor equation and ϵ is a covariantly constant spinor. This can be satisfied provided Killing spinors exist on the manifold. This also imply consistency conditions

$$\left[D_M, D_N\right]\epsilon = 0,$$

or equivalently $R_{MNPQ}\Gamma^{PQ}\epsilon = 0$. Multiplying by Γ^N and using the property

$$\Gamma^N\Gamma^{PQ} = \Gamma^{NPQ} + g^{NP}\Gamma^Q - g^{NQ}\Gamma^P,$$

where Γ^{NPQ} is completely antisymmetric in the three indices, and using the Riemann identity $R_{M[NPQ]} = 0$, we deduce that

$$R_{MQ}\Gamma^Q\epsilon = 0.$$

Let us impose the constraint that the vacuum state of the ten-dimensional space is of the form $M^4 \times K$ where M^4 is a maximally symmetric four-dimensional space and K is a compact six-dimensional manifold. In this case the consistency condition on the Riemann tensor does not admit deSitter or anti deSitter spaces as solutions because in this case

$$R_{\mu\nu\rho\sigma} = \frac{r}{12}\left(g_{\mu\rho}g_{\nu\sigma} - g_{\mu\sigma}g_{\nu\rho}\right)$$

which togother with the consistency condition implies that $r = 0$.

Upon parallel transport around a contractible closed curve γ, a field f is transformed into Uf where

$$U = P\exp_\gamma \int \omega dx$$

where ω is the spin-connection of the n-dimensional internal manifold K. The $SO(n)$ matrices U obtained this way always form a group H called the holonomy group. We can then ask : Under what conditions the manifold K admits a spinor field ϵ obeying $D_i\epsilon = 0$? A covariantly constant spinor ϵ keeps its original value under parallel transport: $U\epsilon = \epsilon$. As an example, let us consider compactifying ten-dimensional supergrravity to four dimensions. In this case the internal manifold K is 6 dimensional. The subgroup of the $SO(6)$ is isomorphic to $SU(4)$ and the left-handed spinor obeying the above condition will be in the fundamental 4-representation. By an $SU(4)$ rotation we can always transform ϵ to the form

$$\epsilon = \begin{pmatrix} 0 \\ 0 \\ 0 \\ \epsilon_0 \end{pmatrix}$$

and this is left invariant by the $SU(3)$ subgroup. We can write the following decomposition

$$\begin{aligned} SO(1,9) &\rightarrow SO(1,3) \times SO(6) \sim SO(1,3) \times SU(4), \\ 16 &= (2,4) \oplus (2',\overline{4}) \end{aligned}$$

where we have decomposed a 16 dimensional spinor of $SO(1,9)$ (the Lorentz group in ten dimensions) in terms of the four-dimensional spinors of $SO(1,3)$ and the 4 and $\overline{4}$ spinor representations of $SO(6)$. From these considerations we conclude that by requiring the manifold K to have $SU(3)$ holonomy, the compactified theory will have $N = 1$ supersymmetry in four-dimensions. Therefore the question to ask now is: What sort of manifold K admits a metric with $SU(3)$ holonomy?

If ϵ is a covariantly constant spinor, then so would be the Kählerian form $k_{ij} = \bar{\epsilon}\Gamma_{ij}\epsilon$, the complex structure $J^i_j = g^{il}k_{lj}$ and the volume form $\omega_{ijk} = \epsilon^T\Gamma_{ijk}\epsilon$. In six-dimensions, if the holonomy group is not $SO(6)$ but $U(3)$ then at any point in K tangent vectors which transform as vectors of $SO(6)$ decompose as $3+\overline{3}$ under $U(3)$. There is a unique matrix J^i_j acting on tangent vectors in the 3 and $\overline{3}$ representations. J defines an almost complex structure and is invariant under the holonomy group. Therefore it is covariantly constant and its derivatives vanish. A manifold of $U(N)$ holonomy is complex and is also Kähler. If we ask whether it is possible to find a Kähler manifold with $SU(N)$ holonomy instead of $U(N)$ holonomy, then the answer lies in the Calabi-Yau conjecture. This states that a Kähler manifold of vanishing first Chern class admits a Kähler metric of $SU(N)$ holonomy and this metric is unique.

10.7.3. Solutions of Einstein equations. There are advantages to deal with a supersymmetric system when solving Einstein equations resulting from a coupled gravitational system. This is done by embedding the bosonic system in a supergravity theory, whenever this is possible. One first starts with an ansatz for the metric dictated by the symmetry of the problem. The supersymmetry transformations depend on first order derivatives of the fields. Setting the supersymmetric transformations to zero and requiring that the solution preserve all or part of the supersymmetry, results in a system of first order differential equations, in contrast to the second order differential equations which one have in the original bosonic system. This method has been very successful in finding new solutions for Einstein-coupled systems. To illustrate the power of this method we consider the follwing model.

Consider the coupling of gravity to a dilaton field ϕ and an $SU(2)$ Yang-Mills field A^a_μ. described by the action

$$S = \int \left(-\frac{1}{4} R + \frac{1}{2} \partial_\mu\phi \, \partial^\mu\phi - \frac{1}{4} e^{2\phi} F^a_{\mu\nu}F^{a\mu\nu} + \frac{1}{8} e^{-2\phi} \right) \sqrt{-\mathbf{g}} \, d^4x,$$

where $F^a_{\mu\nu} = \partial_\mu A^a_\nu - \partial_\nu A^a_\mu + \varepsilon_{abc} A^b_\mu A^c_\nu$, and the dilaton potential can be viewed as an effective negative, position-dependent cosmological term $\Lambda(\phi) = -\frac{1}{4}e^{-2\phi}$.

We shall consider static, spherically symmetric, purely magnetic configurations of the bosonic fields, and for this we parameterize the fields as follows:

$$ds^2 = N\sigma^2 dt^2 - \frac{dr^2}{N} - r^2(d\theta^2 + \sin^2\theta\, d\varphi^2),$$

$$\alpha^a A^a_\mu dx^\mu = w\,(-\alpha^2\, d\theta + \alpha^1\, \sin\theta\, d\varphi) + \alpha^3\, \cos\theta\, d\varphi,$$

where N, σ, w, as well as the dilaton ϕ, are functions of the radial coordinate r. The field equations, following from the action, read

$$(rN)' + r^2 N\phi'^{\,2} + U + r^2\Lambda(\phi) = 1,$$

$$\left(\sigma N r^2 \phi'\right)' = \sigma\,(U - r^2\Lambda(\phi)),$$

$$r^2\left(N\sigma e^{2\phi}\, w'\right)' = \sigma e^{2\phi}\, w(w^2 - 1),$$

$$\sigma' = \sigma\,(r\phi'^{\,2} + 2e^{2\phi}\, w'^{\,2}/r),$$

where $U = 2e^{2\phi}\left(Nw'^{\,2} + (w^2-1)^2/2r^2\right)$. This is a system of second order differential equations which could not be solved. We now adopt the strategy of embedding this action into a supersymmetric action and use proporties of supersymmetry to find a solution.

The action of the N=4 gauged SU(2)×SU(2) supergravity includes a vierbein e^α_μ, four Majorana spin-3/2 fields $\psi_\mu \equiv \psi^{\mathrm{I}}_\mu$ ($\mathrm{I} = 1, \ldots 4$), vector and pseudovector non-Abelian gauge fields $A^{(1)\,a}_\mu$ and $A^{(2)\,a}_\mu$ with independent gauge coupling constants g_1 and g_2, respectively, four Majorana spin-1/2 fields $\chi \equiv \chi^{\mathrm{I}}$, the axion $\mathbf{a}$ and the dilaton ϕ . The bosonic part of the action reads

$$S = \int\Bigg(-\frac{1}{4}R + \frac{1}{2}\,\partial_\mu\phi\,\partial^\mu\phi + \frac{1}{2}\,e^{-4\phi}\partial_\mu\mathbf{a}\,\partial^\mu\mathbf{a} - \frac{1}{4}\,e^{2\phi}\sum_{s=1}^{2} F^{(s)\,a}_{\mu\nu}F^{(s)\,a\mu\nu}$$

$$-\frac{1}{2}\,\mathbf{a}\sum_{s=1}^{2} F^{(s)\,a}_{\mu\nu} * F^{(s)\,a\mu\nu} + \frac{g^2}{8}\,e^{-2\phi}\Bigg)\sqrt{-\mathbf{g}}\,d^4x.$$

Here $g^2 = g_1^2 + g_2^2$, the gauge field tensor $F^{(s)\,a}_{\mu\nu} = \partial_\mu A^{(s)\,a}_\nu - \partial_\nu A^{(s)\,a}_\mu + g_C\,\varepsilon_{abc}\,A^{(s)\,b}_\mu\,A^{(s)\,c}_\nu$ (there is no summation over $s = 1, 2$), and $*F^{(s)\,a}_{\mu\nu}$ is the dual tensor. The dilaton potential can be viewed as an effective negative, position-dependent cosmological term $\Lambda(\phi) = -\frac{1}{8}\,g^2\,e^{-2\phi}$. The ungauged version of the theory corresponds to the case where $g_1 = g_2 = 0$. We consider the truncated theory specified by the conditions $g_2 = A^{a(2)}_\nu = 0$. In addition, we require the vector field A^a_μ to be purely magnetic, which allows us to set the axion to zero.

For a purely bosonic configuration, the supersymmetry transformation laws are

$$\delta\overline{\chi} = -\frac{i}{\sqrt{2}}\,\overline{\epsilon}\,\gamma^\mu\partial_\mu\phi - \frac{1}{2}e^\phi\,\overline{\epsilon}\,\alpha^a F^a_{\mu\nu}\,\sigma^{\mu\nu} + \frac{1}{4}\,e^{-\phi}\,\overline{\epsilon},$$

$$\delta\overline{\psi}_\rho = \overline{\epsilon}\left(\overleftarrow{\partial}_\rho - \frac{1}{2}\,\omega_{\rho mn}\,\sigma^{mn} + \frac{1}{2}\,\alpha^a A^a_\rho\right) - \frac{1}{2\sqrt{2}}\,e^\phi\,\overline{\epsilon}\,\alpha^a F^a_{\mu\nu}\,\gamma_\rho\,\sigma^{\mu\nu} + \frac{i}{4\sqrt{2}}\,e^{-\phi}\,\overline{\epsilon}\,\gamma_\rho,$$

the variations of the bosonic fields being zero. In these formulas, $\epsilon \equiv \epsilon^{\mathrm{I}}$ are four Majorana spinor supersymmetry parameters, $\alpha^a \equiv \alpha^a_{\mathrm{IJ}}$ are the SU(2) gauge group generators, and $\omega_{\rho mn}$ is the tetrad connection.

The field configuration is supersymmetric, provided that there are non-trivial supersymmetry Killing spinors ϵ for which the variations of the fermion fields vanish. Specifying to the above configuration and putting $\delta\overline{\chi} = \delta\overline{\psi}_\mu = 0$, the supersymmetry constraints become a system of equations for the four spinors ϵ^I. The procedure which solves these equations is rather involved. The consistency of the algebraic constraints requires that the determinants of the corresponding coefficient matrices vanish and that the matrices commute with each other. These consistency conditions can be expressed by the following relations for the background:

$$
\begin{aligned}
N\sigma^2 &= e^{2(\phi-\phi_0)}, \\
N &= \frac{1+w^2}{2} + e^{2\phi}\frac{(w^2-1)^2}{2r^2} + \frac{r^2}{8}e^{-2\phi}, \\
r\phi' &= \frac{r^2}{8N}e^{-2\phi}\left(1 - 4e^{4\phi}\frac{(w^2-1)^2}{r^4}\right), \\
rw' &= -2w\frac{r^2}{8N}e^{-2\phi}\left(1 + 2e^{2\phi}\frac{w^2-1}{r^2}\right),
\end{aligned}
$$

with constant ϕ_0. Under these conditions, the solution of the algebraic constraints yields ϵ in terms of only two independent functions of r. The remaining differential constraint then uniquely specify these two functions up to two integration constants, which finally corresponds to two unbroken supersymmetries. Introducing the new variables $x = w^2$ and $R^2 = \frac{1}{2}r^2 e^{-2\phi}$, the above equations become equivalent to one first order differential equation

$$
2xR\left(R^2 + x - 1\right)\frac{dR}{dx} + (x+1)\,R^2 + (x-1)^2 = 0.
$$

If $R(x)$ is known, the radial dependence of the functions, $x(r)$ and $R(r)$, can be determined. Define the following substitution:

$$
x = \rho^2\, e^{\xi(\rho)}, \qquad R^2 = -\rho\frac{d\xi(\rho)}{d\rho} - \rho^2\, e^{\xi(\rho)} - 1,
$$

where $\xi(\rho)$ is obtained from

$$
\frac{d^2\xi(\rho)}{d\rho^2} = 2\, e^{\xi(\rho)}.
$$

The most general (up to reparametrizations) solution of this equation which ensures that $R^2 > 0$ is $\xi(\rho) = -2\ln\sinh(\rho - \rho_0)$. This gives us the general solution. The metric is non-singular at the origin if only $\rho_0 = 0$, in which case

$$
R^2(\rho) = 2\rho\coth\rho - \frac{\rho^2}{\sinh^2\rho} - 1,
$$

one has $R^2(\rho) = \rho^2 + O(\rho^4)$ as $\rho \to 0$, and $R^2(\rho) = 2\rho + O(1)$ as $\rho \to \infty$. The last step is to obtain $r(s)$, which finally gives us a family of completely regular solutions of the Bogomolny equations:

$$
ds^2 = a^2\,\frac{\sinh\rho}{R(\rho)}\left\{dt^2 - d\rho^2 - R^2(\rho)(d\vartheta^2 + \sin^2\vartheta d\varphi^2)\right\},
$$

$$
w = \pm\frac{\rho}{\sinh\rho}, \qquad e^{2\phi} = a^2\,\frac{\sinh\rho}{2\,R(\rho)},
$$

where $0 \le \rho < \infty$, and we have chosen $2\phi_0 = -\ln 2$. The geometry described by the above line element is everywhere regular, the coordinates covering the whole space whose topology is R^4. The geometry becomes flat at the origin, but asymptotically it is not flat, even though the cosmological term $\Lambda(\phi)$ vanishes at infinity. In the asymptotic region all curvature invariants tend to zero, however, not fast enough. The Schwarzschild metric functions for $r \to \infty$ are $N \propto \ln r$ and $N\sigma^2 \propto r^2/4 \ln r$, the non-vanishing Weyl tensor invariant being $\Psi_2 \propto -1/6r^2$.

Ali H. Chamseddine, Center for Advanced Mathematical Sciences (CAMS) *and* Physics Department, American University of Beirut, Lebanon.
E-mail address: chams@aub.edu.lb

Part 2

Workshop

CHAPTER 11

Ambient Dirac Eigenvalue Estimates and the Willmore Functional

Bernd Ammann[1]

Abstract. We prove a lower bound for the Willmore functional with the help spinors and the Dirac operator. As a corollary we verify the Willmore conjecture in a certain region of the spin-conformal moduli space for tori with sufficiently small L^p-norms of the Gauss curvature.

Contents

11.1. Introduction

Willmore conjectured [**Wil65**] that for any immersion F of the 2-torus T^2 into Euclidean space $\mathbb{R}^3$, the Willmore functional

$$W(F) := \int_{T^2} H^2$$

is bounded from below by $2\pi^2$. The conjecture has been verified for a large class of manifolds (see e.g. [**Top98**] or [**Amm00b**] for an overview over the known cases). However, the general case remains still open until today.

Any such immersion induces a metric and a spin structure on T^2, hence for any immersion F we obtain an element of the spin-conformal moduli space, that we will denote by $[F]$. Li and Yau [**LY82**] proved that there is a compact subset A of positive measure in the (spin-)conformal moduli space such that the conjecture

[1]Partially supported by The European Contract Human Potential Programme, Research Training Networks HPRN-CT-2000-00101 and HPRN-CT-1999-00118.

holds, if $[F] \in A$ (see Figure 11.1). Their methods have been extended in [**MR86a**] to a larger compact subset of the (spin-)conformal moduli space.

In this paper we study the functional near one of the ends of the moduli space. The main theorem we prove is

THEOREM 11.1. *Let $F : T^2 \to S^3$ be an immersion of the 2-dimensional torus in S^3. We endow S^3 with the standard metric g_{S^3}. Let F be regularly homotopic to an embedding. We set $g := F^* g_{S^3}$, and choose $(x, y) \in \mathcal{M}^{\mathrm{spin}}$ such that $T^2_{(x,y)}$ is spin-conformally equivalent to (T^2, g). Then*

$$\mathcal{W}(F) \geq \frac{\pi^2}{y} - \frac{1}{8} (\operatorname{osc} u) \, \|K_g\|_{L^1(T^2,g)} \, .$$

In particular, if $\|K_g\|_{L^1(T^2,g)} < 4\pi$, then for any $p > 1$

$$\mathcal{W}(F) \geq \frac{\pi^2}{y} - \frac{1}{8} \mathcal{S}_p \, \|K_g\|_{L^1(T^2,g)}$$

where $\mathcal{S}_p := \mathcal{S}_p \left(\|K_g\|_{L^1(T^2,g)} \, , \|K_g\|_{L^p(T^2,g)} \, \mathrm{area}(T^2, g)^{1-(1/p)}, \frac{\mathrm{area}(T^2,g)}{\mathrm{sys}_1(T^2,g)^2} \right)$ is an expression explicitely defined in Definition 11.3.

As a corollary we obtain the Willmore conjecture in a special case.

Another corollary states if we have a sequence of immersions F_i such that $[F_i]$ converges to a certain end of the spin-conformal moduli space in such a way that the L^p-norm of the Gaussian curvature is bounded by a sufficiently small number, then $W(F_i)$ tends to ∞ (Corollary 11.13).

In view of Theorem 11.8 it is tempting to conjecture that Theorem 11.9 could be generalized to the statement:

Naive Guess. *Let $F : (T^2, g, \sigma) \to (S^3, g_{S^3})$ be a conformal embedding. We denote by σ the spin structure on T^2 induced by F. Then the first positive eigenvalue μ_1 of the square of the Dirac operator D^2 on (T^2, g, σ) satisfies*

$$\mathcal{W}(F) \geq \mu_1 \mathrm{area}(T^2, g).$$

An example at the end of the article shows, that this naive guess does not hold. The example also shows that some of our curvature bounds are indeed necessary.

This article is the report of a talk given at the American University of Beirut in summer 2001. Many thanks to the organizers for this beautiful conference.

What we present here differs slightly from the talk. In this article, we omit several results published in previous preprints and articles. On the other hand, we include a detailed proof of Theorem 11.1. These estimates have not been published in this form previously. However, a weaker version already appeared in [**Amm98**].

The proof of our main statements rely on the fact that on a surface immersed into $\mathbb{R}^3$ a spinor is induced, a fact which plays an important role in several talks at the conference.

11.2. Uniformization theorem for the torus

Let us recall the famous uniformization theorem.

THEOREM 11.2 (Uniformization Theorem). *Let g be any metric on the two-dimensional torus. Then there is a conformal metric g_0 that is flat.*

We write $g = e^{2u} g_0$. We will recall an upper estimate for the oscillation $\operatorname{osc} u = \max u - \min u$.

DEFINITION 11.3. For any $p > 1$, let $\mathcal{S}_p$ be the function given by the expression

$$\mathcal{S}_p(\mathcal{K}, \mathcal{K}', \mathcal{V}) := \frac{p}{p-1}\left[\frac{\mathcal{K}'}{4\pi} + \frac{1}{2}\left|\log\left(1 - \frac{\mathcal{K}}{4\pi}\right)\right| + \frac{\mathcal{K}}{8\pi - 2\mathcal{K}} \log\left(\frac{2\mathcal{K}'}{\mathcal{K}}\right)\right] + \frac{\mathcal{K}\mathcal{V}}{8}$$

for $\mathcal{K} \in (0, 4\pi)$, $\mathcal{K}' \in [\mathcal{K}, \infty)$ and $\mathcal{V} \in [0, \infty)$.

We extend continuously by setting

$$\mathcal{S}_p(0, \mathcal{K}', \mathcal{V}) := \frac{p}{p-1}\frac{\mathcal{K}'}{4\pi}.$$

THEOREM 11.4 ([**Amm00a**]). *Let g be any metric on the two-dimensional torus such that the Gauss curvature K_g satisfies $\|K_g\|_{L^1(T^2,g)} < 4\pi$. Then for any $p > 1$ we obtain a bound for the oscillation of u*

$$\operatorname{osc} u \leq \mathcal{S}_p\Big(\|K_g\|_{L^1(T^2,g)}, \|K_g\|_{L^p(T^2,g)} \left(\operatorname{area}(T^2, g)\right)^{1-(1/p)}, \frac{\operatorname{area}(T^2, g)}{\operatorname{sys}_1(T^2, g)^2}\Big).$$

Equality is obtained if and only if g is flat.

11.3. Spin structures on 2-tori

We recall some facts about spin structures on 2-tori.

PROPOSITION 11.5. *For any spin structure on the two-dimensional torus the following properties are equivalent:*

(1) the associated Dirac operator has no kernel,
(2) the torus is spin corbordant zero,
(3) the spin structure is induced from an embedding into $\mathbb{R}^3$,

Obviously, the last property can be equivalently reformulated as "The spin structure is induced from an immersion into $\mathbb{R}^3$ that is regularly homotopic to an embedding."

We say that a spin structure is *nontrivial* if it satisfies any of these conditions.

11.4. The spin-conformal moduli space

Two manifolds with Riemannian metrics and spin structures are said to be *spin-conformally equivalent*, if there is a conformal diffeomorphism between them preserving the spin structure.

The *spin-conformal moduli space* $\mathcal{M}^{\text{spin}}$ (see Figure 11.1) is defined to be the set of all $(x, y) \in \mathbb{R}^2$ satisfying

$$(11.1) \qquad 0 \leq x \leq \frac{1}{2}, \qquad \left(x - \frac{1}{2}\right)^2 + y^2 \geq \frac{1}{4}, \qquad y > 0.$$

For such a pair (x, y), let $T^2_{(x,y)} = \mathbb{R}^2/\langle(1, 0), (x, y)\rangle$ be the 2-torus equipped with the (flat) metric induced from $\mathbb{R}^2$ and equipped with the spin structure that admits a parallel section along the generator $(1,0)$ of the fundamental group, and that does not admit a parallel section along the generator (x,y).

LEMMA 11.6. *For any 2-torus T^2 with a Riemannian metric g and a non-trivial spin structure σ, there is a unique $(x,y) \in \mathcal{M}^{\mathrm{spin}}$ such that (T^2, g, σ) is spin-conformally equivalent to $T^2_{(x,y)}$.*

Hence, $\mathcal{M}^{\mathrm{spin}}$ can be identified with the equivalence classes of Riemannian 2-tori with spin structures under spin-conformal equivalence.

11.5. The Willmore functional

In this section S^3 always carries the metric g_{S^3} of constant sectional curvature 1. For any immersion $F : T^2 \to S^3$ we define the Willmore functional

$$\mathcal{W}(F) := \int_{(T^2, F^* g_{S^3})} \left(|H_{T^2 \to S^3}|^2 + 1\right)$$

where $H_{T^2 \to S^3}$ is the relative mean curvature of $F(T^2)$ in S^3 and integration is the usual integration of functions $T^2 \to \mathbb{R}$ over the Riemannian manifold $(T^2, F^* g_{S^3})$. Note that the mean curvature H of $F(T^2)$ in $\mathbb{R}^4$ satisfies

$$|H|^2 = |H_{T^2 \to S^3}|^2 + 1.$$

The following proposition says that the Willmore functional $\mathcal{W}$ is essentially the same as the functional W from the introduction.

PROPOSITION 11.7 ([**Tho23**],[**Wei78**]). *Let $S : S^3 \setminus \{N\} \to \mathbb{R}^3$ denote stereographic projection. Then for any immersion $F : T^2 \to S^3 \setminus \{N\}$*

$$W(S \circ F) = \mathcal{W}(F).$$

Willmore has conjectured that for any immersion $\mathcal{W}(F) \geq 2\pi^2$. This conjecture is still open in general, although it has been verified in many special cases. Li and Yau [**LY82**, Fact 3] proved that the conjecture holds, if F is not an embedding. They also proved:

THEOREM 11.8 ([**LY82**, Theorem 1]). *Let $F : (T^2, g) \to (S^3, g_{S^3})$ be a conformal embedding. Let λ_1 be the first positive eigenvalue of the Laplacian Δ on (T^2, g) then*

$$\mathcal{W}(F) \geq \frac{1}{2} \lambda_1 \mathrm{area}(T^2, g).$$

From this theorem the conjectured inequality $\mathcal{W}(F) \geq 2\pi^2$ follows, if the spin-conformal equivalence class $[F]$ lies in a compact subset of $\mathcal{M}^{\mathrm{spin}}$ with positive measure (see Figure 11.1).

A similar lower bound for $\mathcal{W}(F)$ in terms of Dirac eigenvalues has been given by Bär.

THEOREM 11.9 ([**Bär98**]). *Let $F : (T^2, g, \varphi) \to (S^3, g_{S^3})$ be an* ***isometric*** *immersion preserving the spin structure. Then for the first eigenvalue μ_1 of the square of the Dirac operator on (T^2, g, φ) the inequality*

$$\mathcal{W}(F) \geq \mu_1 \mathrm{area}(T^2, g)$$

holds.

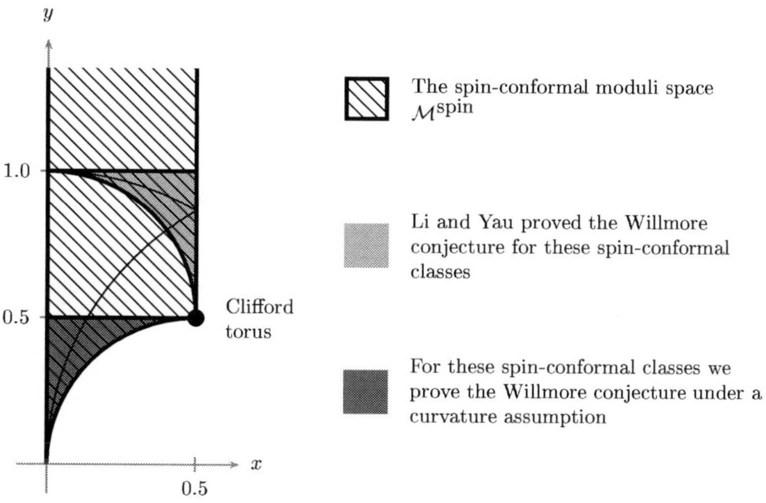

FIGURE 11.1. The spin-conformal moduli space

Note that this estimate is non-trivial only if F is regularly homotopic to an embedding.

REMARK 11.10. At the end of this section we will show by example that in general "isometric" can not be replaced be "conformal" in this theorem.

In this article, our approach is to modify the techniques of the proof of Theorem 11.9. This yields together with Theorem 11.4 new results about the Willmore functional.

THEOREM 11.11. *Let $F : T^2 \to S^3$ be an immersion of the 2-dimensional torus in S^3. We endow S^3 with the standard metric g_{S^3}. Let F be regularly homotopic to an embedding. We set $g := F^* g_{S^3}$, and choose $(x, y) \in \mathcal{M}^{\text{spin}}$ such that $T^2_{(x,y)}$ is spin-conformally equivalent to (T^2, g). Then*

$$\mathcal{W}(F) \geq \frac{\pi^2}{y} - \frac{1}{8} (\operatorname{osc} u) \, \|K_g\|_{L^1(T^2,g)} .$$

In particular, if $\|K_g\|_{L^1(T^2,g)} < 4\pi$, then for any $p > 1$

$$\mathcal{W}(F) \geq \frac{\pi^2}{y} - \frac{1}{8} \mathcal{S}_p \, \|K_g\|_{L^1(T^2,g)}$$

with $\mathcal{S}_p := \mathcal{S}_p \left(\|K_g\|_{L^1(T^2,g)}, \|K_g\|_{L^p(T^2,g)} \operatorname{area}(T^2, g)^{1-(1/p)}, \frac{\operatorname{area}(T^2,g)}{\operatorname{sys}_1(T^2,g)^2} \right).$

PROOF. We write the induced metric g on T^2 in the form $g = e^{2u} g_0$ with g_0 flat. Any Killing spinor on S^3 with the Killing constant $\alpha = (1/2)$ induces a spinor field ψ on (T^2, g) satisfying

$$D_g \psi = H\psi + \nu\psi,$$

where

$$\nu = \gamma(e_1)\gamma(e_2) = \begin{pmatrix} -i & 0 \\ 0 & i \end{pmatrix} \in \operatorname{End}\left(\Sigma^+ T^2 \oplus \Sigma^- T^2\right)$$

(see e.g. [**Bär98**]). There is an isomorphism of vector bundles [**Hit74**],[**Hij86a**, 4.3.1] from the spinor bundle ΣT^2 associated to g to the spinor bundle $\Sigma_0 T^2$ associated to g_0

$$\begin{array}{ccc} \Sigma T^2 & \to & \Sigma_0 T^2 \\ \Psi & \mapsto & \widehat{\Psi} \end{array}$$

with

$$e^u \widehat{D_g \Psi} = D_{g_0} \widehat{\Psi} + \frac{1}{2} \gamma_{g_0}(\operatorname{grad}_{g_0} u) \widehat{\Psi}$$

and

$$|\widehat{\Psi}| = |\Psi|.$$

Here γ_{g_0} means Clifford multiplication corresponding to the metric g_0.

We apply this transformation to the induced spinor ψ from above. We obtain

$$D_{g_0} \widehat{\psi} = -\frac{1}{2} \gamma_{g_0}(\operatorname{grad}_{g_0} u) \widehat{\psi} + e^u H \widehat{\psi} + e^u \nu \widehat{\psi}.$$

As ν, $\gamma(V)$ and $\nu\gamma(V)$ are skew-hermitian for any vector V, this yields

$$\begin{aligned} \left| D_{g_0} \widehat{\psi} \right|^2 &= \frac{1}{4} \left| \gamma_{g_0}(\operatorname{grad}_{g_0} u) \widehat{\psi} \right|^2 + e^{2u} H^2 \left| \widehat{\psi} \right|^2 + e^{2u} \left| \nu \widehat{\psi} \right|^2 \\ &= \frac{1}{4} |du|_{g_0}^2 + e^{2u} H^2 + e^{2u}. \end{aligned}$$

Integration over (T^2, g_0) provides

$$\tilde{\lambda}_1 \operatorname{area}(T^2, g_0) \leq \frac{1}{4} \int_{T^2} |du|_{g_0}^2 \, \mathrm{dvol}_{g_0} + \mathcal{W}(F), \tag{11.2}$$

where $\tilde{\lambda}_1$ denotes the smallest eigenvalue of the square of the Dirac operator on (T^2, g_0).

On the other hand

$$\begin{aligned} \int_{T^2} |du|_{g_0}^2 \, \mathrm{dvol}_{g_0} &= \int_{T^2} u \Delta_{g_0} u \, \mathrm{dvol}_{g_0} \\ &= \int_{T^2} e^{2u} u K_g \, \mathrm{dvol}_{g_0} \\ &= \int_{T^2} u K_g \, \mathrm{dvol}_g \\ &\leq \frac{1}{2} (\operatorname{osc} u) \, \|K_g\|_{L^1(T^2, g)}. \end{aligned}$$

Together with Theorem 11.4 we obtain the statement. □

COROLLARY 11.12. *For any $\kappa_1 \in \,]0, 4\pi[$, any $p > 1$ and any $\kappa_p > 0$ there is a neighborhood U of the $(y \to 0)$-end of $\mathcal{M}^{\mathrm{spin}}$ with the following property:*
If $F : T^2 \to S^3$ is an immersion such that $[F] \in U$ and if the curvature conditions

$$\|K_g\|_{L^1(T^2,g)} < \kappa_1 \quad \text{and} \quad \|K_g\|_{L^p(T^2,g)} \operatorname{area}(T^2, g)^{1-(1/p)} < \kappa_p$$

are satisfied, then the Willmore conjecture

$$W(F) \geq 2\pi^2$$

holds.

COROLLARY 11.13. *Let $F_i : T^2 \to S^3$ be a sequence of immersions. The induced metrics $g_i := F_i^* g_{S^3}$ together with the induced spin structures define a sequence (x_i, y_i) in the spin-moduli space $\mathcal{M}^{\mathrm{spin}}$. Assume that $y_i \to 0$ and that the curvature conditions*

$$\|K_{g_i}\|_{L^1(T^2,g_i)} < \kappa_1 < 4\pi \quad \text{and} \quad \|K_g\|_{L^p(T^2,g)} \operatorname{area}(T^2,g)^{1-(1/p)} < \kappa_p$$

are satisfied for some $p > 1$ and $\kappa_p < \infty$. Then

$$\mathcal{W}(F_i) \to \infty.$$

The corollaries follow directly from the theorem.

11.6. An example showing the necessity of the curvature condition

The conclusion of the Corollary 11.13 is false if we drop the curvature conditions. To see this we construct a sequence of immersions with $y_i \to 0$ and $\mathcal{W}(F_i) < const.$ We start with an embedding $F : T^2 \to S^3$ which looks in a neighborhood of some point like a cylinder. Now we "strangle" the torus as in the picture below:

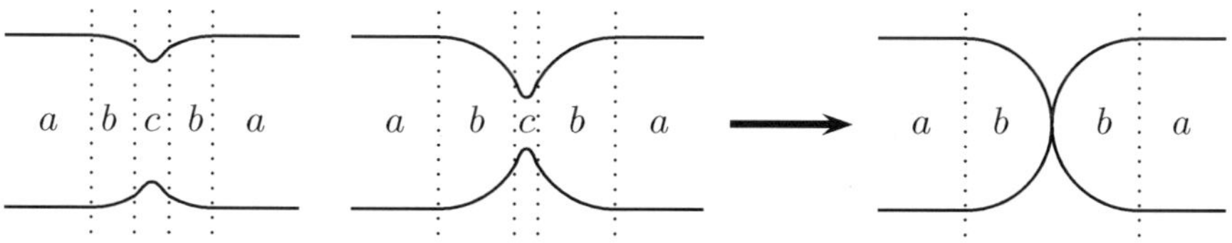

We get a sequence $F_i : T^2 \to S^3$ of C^1-embeddings with the following properties:

(i) $F_i(T^2)$ coincides with $F(T^2)$ in region a
(ii) $F_i(T^2)$ coincides with a part of a half-sphere in region b,
(iii) $F_i(T^2)$ coincides with a minimal surface in region c.

Note that the regions a, b and c depend on i. In the limit $i \to \infty$, region c disappears. After smoothing we get a family of smooth embeddings satisfying both $y_i \to 0$ and $\mathcal{W}(F_i) < const$ and $\operatorname{area}(T^2, F_i^* g_{S^3}) \to const.$

Hence, the first eigenvalue of D^2 is bounded from above. But the first eigenvalue of the spin-conformally equivalent flat torus with unit volume converges to ∞. This implies that there are spin-conformal classes in which the optimal constants in Lott's inequality [**Lot86**, Prop. 1] are not attained by flat metrics.

From this example we can also conclude that Theorem 11.9 does no longer hold, if we replace the condition "isometric spin immersion" by "conformal spin immersion".

11.7. Latest news about the conjecture

After submission of this conference proceeding, Martin U. Schmidt has annouced a proof of the Willmore conjecture in full generality [**Sch02**]. The induced spinor on the immersed surface (as described above) is the starting point of Schmidt's considerations.

Bernd Ammann, Institut Élie Cartan, Université Henri Poincaré, Nancy I, B.P. 239, 54506 Vandœuvre-Lès-Nancy Cedex, France.
E-mail address: ammann@iecn.u-nancy.fr

CHAPTER 12

The Geometric Structure of Lorentzian Manifolds with Twistor Spinors in low Dimension

HELGA BAUM AND FELIPE LEITNER[1]

ABSTRACT. In this chapter we discuss the twistor equation in Lorentzian spin geometry. In particular, we explain the local conformal structure of Lorentzian manifolds, which admit twistor spinors inducing lightlike Killing fields. Furthermore, we derive all geometries with singularity free twistor spinors that occur up to dimension 6.

Contents

12.1. Introduction

A classical object in differential geometry are conformal Killing fields. These are by definition infinitesimal conformal symmetries i.e. the flow of such vector fields preserves the conformal class of the metric. The number of linearely independing conformal Killing fields measures the degree of conformal symmetry of the manifold. This number is bounded by $\frac{1}{2}(n+1)(n+2)$, where n is the dimension of the manifold. It is the maximal one if the manifold is conformally flat and simply connected. S. Tachibana and T. Kashiwada (cf. [**TK69**], [**Kas68**]) introduced a generalisation of conformal Killing fields, the conformal Killing forms (or twistor forms). Conformal Killing forms are solutions of a conformally invariant twistor type equation on differential forms. They were studied in General Relativity mainly from the local viewpoint in order to integrate the equation of motion (e.g. [**PW70**]), furthermore they were used to obtain symmetries of field equations ([**BC97**], [**BCK97**]). Recently, U. Semmelmann ([**Sem01**]) started to discuss global properties of conformal

[1]The second author was partially supported by the SFB 288 "'Differential Geometry and Quantum Physics"'.

Killing forms in Riemannian geometry. Another generalisation of conformal Killing vectors is that of conformal Killing spinors (or twistor spinors), which are solutions of the conformally invariant twistor equation on spinors introduced by R. Penrose in General Relativity (cf. [**PR86**]). Whereas conformal Killing fields are classical symmetries, conformal Killing spinors define infinitesimal symmetries on supermanifolds (cf. [**ACDS98**]). Special kinds of such spinors, parallel and special Killing spinors, occur in supergravity and string theories. In 1989 A. Lichnerowicz and T. Friedrich started a systematic study of twistor spinors in conformal Riemannian geometry. Whereas the global structure of Riemannian manifolds admitting conformal Killing spinors is quite well understood (cf. e.g. [**Lic88b**], [**Lic88a**], [**Lic89**], [**Fri89**] [**Lic90b**], [**BFGK91**], [**Hab90**], [**Hab93**], [**Hab94**], [**Hab96**], [**KR94**], [**KR96**], [**KR97**], [**KR98**]), the state of art in its origin, Lorentzian geometry, is far from being satisfactory. We are mainly interested in the following problems:

(1) *Which Lorentzian geometries admit twistor spinors?*
(2) *How the properties of twistor spinors are related to the geometric structures where they can occur?*

J. Lewandowski ([**Lew91**]) described the local normal forms of 4-dimensional spacetimes with zero free twistor spinors. His results indicated that there are interesting relations between conformal Killing spinors, different global contact structures and Lorentzian geometry, that should be discovered. We discuss here some of these structures which appear up to dimension 6. We consider only the case of zero free twistor spinors. Results for twistor spinors with zeros can be found in [**Lei01b**].
After recalling the definition of twistor spinors we discuss in Section 12.3 Brinkman spaces, Lorentzian Einstein Sasaki structures and Fefferman spaces and their relation to the problem in question. In Section 12.4, we study the local conformal structure of Lorentzian manifolds that admit twistor spinors inducing lightlike Killing fields. In Section 12.5, we derive all local conformal structures of Lorentzian manifolds admitting singularity free solutions of the twistor equation in low dimensions $n \leq 6$ (see Theorem 12.14).
In this paper we give only a short sketch of the proofs. A detailed version will appear in [**BL**].

12.2. The twistor equation on spinors

In this section we recall the definion of twistor spinors and fix the notations. For more details we refer to [**PR86**] or [**BFGK91**].
Let (M^n, g) be a semi-Riemannian spin manifold of dimension $n \geq 3$. We denote by S the spinor bundle and by $\mu : T^*M \otimes S \longrightarrow S$ the Clifford multiplication. The 1-forms with values in the spinor bundle decompose into two subbundles

$$T^*M \otimes S = V \oplus Tw,$$

where V, being the orthogonal complement to the "twistor bundle" $Tw := Ker\mu$, is isomorphic to S. Usually, we identify TM and T^*M using the metric g.
We obtain two differential operators of first order by composing the spinor derivative ∇^S with the orthogonal projections onto each of these subbundles,
the *Dirac operator* D

$$D : \Gamma(S) \xrightarrow{\nabla^S} \Gamma(T^*M \otimes S) = \Gamma(S \oplus Tw) \xrightarrow{pr_S} \Gamma(S)$$

and the *twistor operator* P

$$P : \Gamma(S) \xrightarrow{\nabla^S} \Gamma(T^*M \otimes S) = \Gamma(S \oplus Tw) \xrightarrow{pr_{Tw}} \Gamma(Tw).$$

Locally, these operators are given by the following formulas

$$\begin{aligned} D\varphi &= \sum_{i=1}^{n} \sigma^i \cdot \nabla^S_{s_i}\varphi \\ P\varphi &= \sum_{i=1}^{n} \sigma^i \otimes (\nabla^S_{s_i}\varphi + \frac{1}{n} s_i \cdot D\varphi), \end{aligned}$$

where $(s_1, \dots, s_n)$ is a local orthonormal basis, $(\sigma^1, \dots, \sigma^n)$ its dual and $\cdot$ denotes the Clifford multiplication. Both operators are conformally covariant. More exactly, if $\tilde{g} = e^{2\sigma} g$ is a conformal change of the metric, the Dirac and the twistor operator satisfy

$$\begin{aligned} D_{\tilde{g}} &= e^{-\frac{n+1}{2}\sigma} D_g e^{\frac{n-1}{2}\sigma} \\ P_{\tilde{g}} &= e^{-\frac{\sigma}{2}} P_g e^{-\frac{\sigma}{2}}. \end{aligned}$$

A spinor field is called *twistor spinor* or *conformal Killing spinor*, if it lies in the kernel of the twistor operator P. Using the local formula for the twistor operator one obtains the following characterisation of a twistor spinor: A spinor field $\varphi \in \Gamma(S)$ is a twistor spinor if and only if

$$\nabla^S_X \varphi + \frac{1}{n} X \cdot D\varphi = 0 \qquad \text{for all vector fields } X.$$

It is a well-known fact, that – as in the case of conformal vector fields – the dimension of the space of twistor spinors is bounded and the maximal possible dimension is attained only for conformally flat manifolds. More exactly, it holds

PROPOSITION 12.1. *(1) The dimension of the space of twistor spinors is a conformal invariant and bounded by*

$$\dim \ker P \leq 2\,\mathrm{rank} S = 2^{[\frac{n}{2}]+1} =: d_n.$$

(2) If $\dim \ker P = d_n$, *then* (M^n, g) *is conformally flat.*
(3) If (M^n, g) *is simply connected and conformally flat, then* $\dim \ker P = d_n$.

Hence, for example, all simply connected space forms $\mathbb{R}^n_k, \tilde{\mathbb{H}}^n_k, \tilde{S}^n_k$ admit the maximal number of linearly independent twistor spinors.

12.3. Twistor spinors on Lorentzian spin manifolds - Examples

Now, we restrict our attention to the case of Lorentzian signature $(- + \dots +)$. In this section we will explain the special geometries that occur in the Theorem 12.14. Let (M^n, g) be an oriented and time-oriented Lorentzian spin manifold. On the spinor bundle S there exists an indefinite non-degenerate inner product $\langle \cdot, \cdot \rangle$ such that

$$\begin{aligned} \langle X \cdot \varphi, \psi \rangle &= \langle \varphi, X \cdot \psi \rangle \qquad \text{and} \\ X(\langle \varphi, \psi \rangle) &= \langle \nabla^S_X \varphi, \psi \rangle + \langle \varphi, \nabla^S_X \psi \rangle, \end{aligned}$$

for all vector fields X and all spinor fields φ, ψ. Each spinor field $\varphi \in \Gamma(S)$ defines a vector field V_φ on M, the so-called Dirac current, by

$$g(V_\varphi, X) := -\langle X \cdot \varphi, \varphi \rangle. \tag{12.1}$$

A direct calculation shows the following properties of the Dirac current

PROPOSITION 12.2. *Let φ be a spinor field on a Lorentzian spin manifold (M^n, g) with Dirac current V_φ. Then*

(1) V_φ is causal and future-directed.
(2) The zero sets of φ and V_φ coincide.
(3) If φ is a twistor spinor, V_φ is a conformal Killing field.

Now, let us discuss 3 types of special Lorentzian geometries that admit twistor spinors.

12.3.1. Brinkman spaces with parallel spinors. A Lorentzian manifold is called *Brinkman space* if it admits a non-trivial lightlike parallel vector field. Let us consider two examples of such spaces.

EXAMPLE 12.3 (pp-manifolds.). A *pp-manifold* is a Brinkman space with the following additional property of the curvature tensor $\mathcal{R}$

$$\mathrm{Trace}_{(3,5),(4,6)} \mathcal{R} \otimes \mathcal{R} = 0.$$

Equivalently, pp-manifolds can be characterized as those Lorentzian manifolds (M^n, g), where the metric has the following local normal form depending only on one function f of $(n-1)$ variables

$$g = dt\, ds + f(s, x_1, \ldots, x_{n-2}) ds^2 + \sum_{i=1}^{n-2} dx_i^2.$$

(cf. [**Sch74**]). In terms of holonomy, pp-manifolds can be characterised as those Lorentzian manifolds for which the holonomy algebra is contained in the abelian ideal $\mathbb{R}^{n-2}$ of the algebra $(\mathbb{R} \oplus \mathfrak{so}(n-2)) \times_s \mathbb{R}^{n-2} \subset \mathfrak{so}(1, n-1)$, cf. [**Lei01a**]. Using the latter fact one can easily prove that for each simply connected pp-manifold

$$\dim \ker P \geq \frac{d_n}{4}.$$

Furthermore, on non-conformally flat pp-manifolds each twistor spinor is parallel. An important example of geodesically complete pp-manifolds are the Lorentzian symmetric spaces with solvable transvection group, which occur also as a special model for a certain string equation in 10+1-dimension (cf. [**FOP01**], [**BFOHP01**]).

EXAMPLE 12.4 (Brinkman spaces with special Kähler flag.). Let (M^n, g) be a Brinkman space with the lightlike parallel vector field V. V defines a flag $\mathbb{R}V \subset V^\perp \subset TM$ in TM, where $V^\perp = \{Y \in TM \mid g(V, Y) = 0\}$. We equip the bundle $E := V^\perp / \mathbb{R}V$ with the positive definite inner product $\tilde{g}$ induced by g and the metric connection $\tilde{\nabla}$ induced by the Levi-Civita connection of g. We call the flag $\mathbb{R}V \subset V^\perp \subset TM$ a *special Kähler flag*, if the bundle E admits an orthogonal almost complex structure $J : E \to E$, $J^2 = -id$, such that $\mathrm{Trace}\,(J \circ \mathcal{R}^{\tilde{\nabla}}(X, Y)) = 0$ for all $X, Y \in TM$. It was proved by I. Kath in [**Kat99**] that (M^{2m}, g) is a Brinkman space with special Kähler flag if and only if (M^{2m}, g) has pure parallel spinors.

12.3.2. Twistor spinors on Lorentzian Einstein–Sasaki manifolds. An odd-dimensional Lorentzian manifold $(M^{2m+1}, g; \xi)$ equipped with a vector field ξ is called *Lorentzian Sasaki manifold* if

(1) ξ is a timelike Killing field with $g(\xi, \xi) = -1$.
(2) The map $J := -\nabla\xi : TM \to TM$ satisfies

$$\begin{array}{rcll} J^2 X &=& -X - g(X,\xi)\xi & \text{and} \\ (\nabla_X J)(Y) &=& -g(X,Y)\xi + g(Y,\xi)X. & \end{array}$$

Let us consider the metric cone $C_-(M) := (\mathbb{R}^+ \times M, -dt^2 + t^2 g)$ with timelike cone axis over (M, g). The cone metric has signature $(2, 2m)$. Then the following relations between properties of M and those of its cone are easy to verify

$(M^{2m+1}, g; \xi)$		cone $C_-(M)$
Lorentzian Sasaki	$\Longleftrightarrow$	(pseudo) Kähler
Lorentzian Einstein–Sasaki ($R < 0$)	$\Longleftrightarrow$	Ricci-flat and (pseudo) Kähler
Lorentzian Einstein–Sasaki ($R < 0$)	$\Longleftrightarrow$	$\mathrm{Hol}_0(C_-(M)) \subset SU(1, m)$

The standard example for regular Lorentzian Einstein–Sasaki manifolds are S^1-bundles over Riemannian Kähler Einstein spaces of negative scalar curvature: Let (X^{2m}, h) be a Riemannian Kähler Einstein spin manifold of scalar curvature $R_X < 0$ and let $(M^{2m+1}, \pi, X; S^1)$ denote the S^1-principal bundle associated to the square root $\sqrt{\Lambda^{m,0}X}$ of the canonical bundle of X given by the spin structure. Furthermore, let A be the connection on M induced by the Levi-Civita connection of (X, h). Then

$$g := \pi^* h - \frac{16m}{(m+1)R_X} A \odot A$$

defines a Lorentzian Einstein–Sasaki metric on the spin manifold M^{2m+1}. Lorentzian Einstein–Sasaki manifolds admit a special kind of twistor spinors.

PROPOSITION 12.5. *(*[**Kat99**], [**Boh03**]*) Let (M, g) be a simply connected Lorentzian Einstein–Sasaki manifold. Then M is spin and admits a twistor spinor φ on M such that*

a) *V_φ is a timelike Killing field with $g(V_\varphi, V_\varphi) = -1$*
b) *$V_\varphi \cdot \varphi = -\varphi$*
c) *$\nabla_{V_\varphi}\varphi = -\frac{1}{2} i\varphi$.*

Converselly, if (M, g) is a Lorentzian spin manifold with a twistor spinor satisfying a), b) and c). Then $\xi := V_\varphi$ is a Lorentzian Einstein–Sasaki structure on (M, g).

12.3.3. Twistor spinors on Fefferman spaces. Fefferman spaces are Lorentzian manifolds which appear in the frame work of CR geometry. Let us first explain the necessary notations from CR geometry. Let N^{2m+1} be a smooth oriented manifold of odd dimension $2m + 1$. A *CR structure on* N is a pair (H, J) where

1. $H \subset TM$ is a real $2m$-dimensional subbundle.
2. $J : H \to H$ is an almost complex structure on H, $J^2 = -Id$.

3. If $X, Y \in \Gamma(H)$, then $[JX, Y] + [X, JY] \in \Gamma(H)$ and
$J([JX, Y] + [X, JY]) - [JX, JY] + [X, Y] \equiv 0$ (integrability condition).

Standard examples of CR manifolds are the following:

- Real hypersurfaces N of a complex manifold (M, J_M).
  The CR structure is given by $H := TN \cap J_M(TN)$, $J := J_M\big|_H$.
- Riemannian Sasaki manifolds (N, g, ξ).
  The CR structure is given by $H := \xi^\perp$ and $J := -\nabla\xi$.
- Heisenberg manifolds $\mathcal{H}e(m) = He(m)/\Gamma$, where $He(m)$ is the Heisenberg group of matrices

$$He(m) = \left\{ \begin{pmatrix} 1 & x^t & z \\ 0 & I_m & y \\ 0 & 0 & 1 \end{pmatrix} \middle|\, x, y \in \mathbb{R}^m, z \in \mathbb{R} \right\}$$

  and Γ is a discrete lattice in $He(m)$.

Let (N, H, J) be a CR manifold. In order to define Fefferman spaces we fix a contact form $\theta \in \Omega^1(N)$ on N such that $\theta|_H = 0$. Let us denote by T the Reeb vector field of θ which is defined by the conditions $\theta(T) = 1$ and $T \lrcorner\, d\theta = 0$. In the following we suppose that the Levi form $L_\theta : H \times H \to \mathbb{R}$

$$L_\theta(X, Y) := d\theta(X, JY)$$

is positive definite. Then (N, H, J, θ) is called a *strictly pseudoconvex manifold.* The tensor field $g_\theta := L_\theta + \theta \odot \theta$ defines a Riemannian metric on N. There is a special metric covariant derivative on a strictly pseudoconvex manifold, the *Tanaka-Webster connection* $\nabla^W : \Gamma(TN) \to \Gamma(TN^* \otimes TN)$ uniquely defined by the conditions

$$\begin{aligned} \nabla^W g_\theta &= 0 \\ Tor^W(X, Y) &= L_\theta(JX, Y) \cdot T \\ Tor^W(T, X) &= -\frac{1}{2}([T, X] + J[T, JX]) \end{aligned}$$

for $X, Y \in \Gamma(H)$. This connection preserves the subbundle H and satisfies $\nabla^W J = 0$ and $\nabla^W T = 0$ (cf. [**Tan75**], [**Web78**]). Let us denote by $T_{10} \subset TN^{\mathbb{C}}$ the eigenspace of the complex extension of J on $H^{\mathbb{C}}$ to the eigenvalue i. Then L_θ extends to a hermitian form on T_{10} by

$$L_\theta(U, V) := -i\, d\theta(U, \overline{V}), \quad U, V \in T_{10}.$$

For a complex 2-form $\omega \in \Lambda^2 N^{\mathbb{C}}$ we denote by $\mathrm{trace}_\theta \omega$ the θ-trace of ω:

$$\mathrm{trace}_\theta \omega := \sum_{\alpha=1}^{m} \omega(Z_\alpha, \overline{Z}_\alpha),$$

where $(Z_1, \ldots, Z_m)$ is a unitary basis of (T_{10}, L_θ). Let $\mathfrak{R}^W$ be the $(4,0)$-curvature tensor of the Tanaka-Webster connection ∇^W on the complexified tangent bundle of N

$$\mathfrak{R}^W(X, Y, Z, V) := g_\theta(([\nabla^W_X, \nabla^W_Y] - \nabla^W_{[X,Y]})Z, \overline{V}).$$

and let us denote by

$$\mathrm{Ric}^W := \mathrm{trace}_\theta^{(3,4)} := \sum_{\alpha=1}^{m} \mathfrak{R}^W(\cdot, \cdot, Z_\alpha, \overline{Z}_\alpha)$$

the *Tanaka-Webster Ricci curvature* and by $R^W := \mathrm{trace}_\theta \mathrm{Ric}^W$ the *Tanaka-Webster scalar curvature.* The Ricci curvature Ric^W is a $(1,1)$-form on N with $\mathrm{Ric}^W(X,Y) \in i\mathbb{R}$ for real vectors $X, Y \in TN$. The scalar curvature R^W is a real function.
Now, let us suppose that (N^{2m+1}, H, J, θ) is a strictly pseudoconvex spin manifold. The spin structure of (N, g_θ) defines a square root $\sqrt{\Lambda^{m+1,0}N}$ of the canonical line bundle

$$\Lambda^{m+1,0}N := \{\omega \in \Lambda^{m+1}N^C \mid V \lrcorner \omega = 0 \ \forall V \in \overline{T}_{10}\}.$$

We denote by (F, π, N) the S^1-principal bundle associated to $\sqrt{\Lambda^{m+1,0}N}$. We fix a connection form A on F and the corresponding decomposition of the tangent bundle $TF = ThF \oplus TvF = H^* \oplus \mathbb{R}T^* \oplus TvF$ into the horizontal and vertical part, where H^* and T^* denote the horizontal lifts of H and T. Then a Lorentzian metric h is defined by

$$h := \pi^* L_\theta - i\, c\, \pi^* \theta \odot A,$$

where c is a non-zero real number. The Fefferman metric is a metric of the latter type, where the choice of A and c is done in such a way that the conformal class of h does not depend on the pseudohermitian form θ. Such a choice can be made with the connection

$$A_\theta := A^W - \frac{i}{4(m+1)} R^W \cdot \theta,$$

where A^W is the connection form of F defined by the Tanaka-Webster connection ∇^W. The curvature form of A^W is $\Omega^{A^W} = -\frac{1}{2}\mathrm{Ric}^W$. Then

$$h_\theta := \pi^* L_\theta - i\, \frac{8}{m+2} \pi^* \theta \odot A_\theta$$

is a Lorentzian metric such that the conformal class $[h_\theta]$ is an invariant of the CR-structure (N, H, J). The metric h_θ is S^1-invariant, the fibres of the S^1-bundle are lightlike. We call (F^{2m+2}, h_θ) with its canonically induced spin structure *Fefferman space of the strictly pseudoconvex spin manifold* (N, H, J, θ).
The Fefferman metric was first discovered by C. Fefferman for the case of strictly pseudoconvex hypersurfaces $N \subset \mathbb{C}^{m+1}$ ([**Fef76**]). The considerations of Fefferman were extended by Burns, Diederich and Snider ([**BDS77**]) and by Lee ([**Lee86**]) to the case of abstract (not nessarily embedded) CR-manifolds. A geometric characterization of Fefferman metrics was given by Sparling (cf. [**Spa85**], [**Gra87**]).
Fefferman spaces admit a special kind of twistor spinors.

PROPOSITION 12.6. ([**Bau99**]) *Let (N, H, J, θ) be a strictly pseudoconvex spin manifold with the Fefferman space (F, h_θ). Then there exist two linearly independent twistor spinors φ on (F, h_θ) such that*

a) *V_φ is a regular lightlike Killing field*
b) $V_\varphi \cdot \varphi = 0$
c) $\nabla_{V_\varphi} \varphi = i\, c\, \varphi$ *where* $c \in \mathbb{R} \backslash \{0\}$.

Conversely, if (M, g) is an even dimensional Lorentzian spin manifold with a twistor spinor satisfying a), b) and c), then there exists a strictly pseudoconvex spin manifold (N, H, J, θ) such that its Fefferman space is locally isometric to (M, g).

12.4. Twistor spinors inducing lightlike Killing fields

In this section we want to discuss special solutions of the Lorentzian twistor equation, those with lightlike Dirac current. As we noticed in Proposition 12.2, each twistor spinor φ induces a causal conformal vector field V_φ. Now we study the situation where V_φ is lightlike and Killing.
Let us start with some notations. Let $W : \Lambda^2 M \to \Lambda^2 M$ denote the Weyl tensor of (M^n, g) considered as selfadjoint maps on the space of 2-forms, Ric denotes the Ricci tensor of (M^n, g) considered here as $(1,1)$-tensor or as $(2,0)$-tensor as it is needed. In conformal geometry there are 2 further curvature tensors that play an important role, the Rho tensor K

$$K(X) := \frac{1}{n-2}\Big(\frac{R}{2(n-1)}X - \mathrm{Ric}(X)\Big) \quad , X \in TM$$

and the Cotton-York tensor

$$C(X,Y) := (\nabla_X K)(Y) - (\nabla_Y K)(X), \quad X, Y \in TM.$$

For twistor spinors we have the following properties of the Cotton-York and the Weyl tensor

PROPOSITION 12.7. *Let φ be an arbitrary twistor spinor. Then the Dirac current V_φ annihilates the Cotton-York tensor: $V_\varphi \lrcorner C = 0$.*

PROOF. We use the following well known integrability conditions for the twistor spinor φ (cf. [**BFGK91**])

$$W(\eta) \cdot \varphi = 0 \quad \text{for all 2-forms } \eta \tag{12.2}$$

$$W(X \wedge Y) \cdot D\varphi = n\, C(X,Y) \cdot \varphi . \tag{12.3}$$

Using these conditions we obtain

$$\begin{aligned} C(V_\varphi, X, Y) &= g(V_\varphi, C(X,Y)) = -\langle C(X,Y) \cdot \varphi, \varphi \rangle \\ &= -\frac{1}{n}\langle W(X \wedge Y) \cdot D\varphi, \varphi \rangle = \frac{1}{n}\langle D\varphi, W(X \wedge Y) \cdot \varphi \rangle = 0. \end{aligned}$$

□

Now, let us mention an algebraic property of spinor fields with lightlike Dirac current:

LEMMA 12.8. *(cf. [**Lei01b**]) Let φ be a spinor field on a Lorentzian manifold with lightlike Dirac current V_φ. Then*

(1) $V_\varphi \cdot \varphi = 0$.
(2) $\langle \varphi, \varphi \rangle = 0$.
(3) If $\eta \in \Omega^2(M)$ is a 2-form such that $\eta \cdot \varphi = 0$. then $V_\varphi \lrcorner \eta = 0$.

As consequence we obtain for the Weyl tensor

PROPOSITION 12.9. *Let φ be a Lorentzian twistor spinor with lightlike Dirac current V_φ, then $V_\varphi \lrcorner W = 0$.*

PROOF. According to 12.2 we have $W(\eta) \cdot \varphi = 0$ for all 2-forms η on M. Since $W(\eta)$ is a 2–form, Lemma 12.8 yields $V_\varphi \lrcorner W(\eta) = 0$ for all 2-forms η on M. Hence, $V_\varphi \lrcorner W = 0$. □

Let V be a vector field and let θ denote the dual 1-form $\theta(X) = g(X, V)$. Then V is called *twisting* if the 3-form $d\theta \wedge \theta$ nowhere vanishes and *non-twisting* if $d\theta \wedge \theta = 0$. By the Frobenius Theorem the latter means that the distribution $V^{\perp} \subset TM$ is integrable. For twistor spinors with lightlike non-twisting Dirac current one has the following

PROPOSITION 12.10. *(cf. [**Lei01b**]) Let φ be a twistor spinor with lightlike, non-twisting Dirac current. Then (M, g) is locally conformal equivalent to a Brinkman space with parallel spinor.*

Using these conditions one is able to characterize – at least locally – the geometries that admit a zero free twistor spinor with lightlike Dirac current. In particular, the following Proposition explains the role, Fefferman spaces and Brinkman spaces are playing among all Lorentzian geometries that admit twistor spinors.

PROPOSITION 12.11. *Let (M, g) be a Lorentzian spin manifold admitting a twistor spinor φ such that V_φ is lightlike and Killing. Then the function $\mathrm{Ric}(V_\varphi, V_\varphi)$ is constant and non-negative on M. Furthermore,*

(1) $\mathrm{Ric}(V_\varphi, V_\varphi) > 0$ if and only if (M, g) is locally isometric to a Fefferman space. In this case the Dirac current ist twisting and the dimension of M is even.
(2) $\mathrm{Ric}(V_\varphi, V_\varphi) = 0$ if and only if (M, g) is locally conformal equivalent to a Brinkman space with parallel spinors. In this case the Dirac current is non-twisting .

PROOF. We give a short sketch of the proof. A more detailed version can be found in ([**BL**]): If V is a lightlike Killing field such that $V \lrcorner W = 0$ and $V \lrcorner C = 0$, then $\mathrm{Ric}(V, V) = \mathrm{const}$ (see [**Gra87**]). Now, let us consider $V = V_\varphi$. Then $\mathrm{Ric}(V_\varphi, V_\varphi) = \mathrm{const}$ follows from Propositions 12.7 and 12.9. We denote by η and θ the following 1-forms

$$\eta(X) := K(V, X), \qquad \theta(X) = g(V, X).$$

Furthermore, let T be the vector field dual to η. For the following we normalize the spinor φ so, that $\eta(V) = -(n-2)\,\mathrm{Ric}(V, V) =: \varepsilon \in \{0, 1, -1\}$. Now, let us consider the endomorphism

$$J : TM \longrightarrow TM \qquad J(X) := \nabla_X V$$

Then one can prove that J is a skew-adjoint map with $JV = JT = 0$ and

$$J^2(X) = \varepsilon X - \theta(X)T - \eta(X)V \tag{12.4}$$

$$g(T, V) = \varepsilon\,, \qquad g(V, V) = g(T, T) = 0 \tag{12.5}$$

$$d\theta(X, Y) = 2g(JX, Y) \tag{12.6}$$

Graham considered the case $\varepsilon = -1$ and proved that the manifold (M, g) is locally isometric to a Fefferman space. The case $\varepsilon = 1$ can not occur, since by (12.4) and (12.5) J would be a skew-adjoint involution on the positive definite subbundle $\mathrm{span}(V, T)^{\perp}$ of TM. In case $\varepsilon = 0$, T is parallel to V and using (12.6) one can prove, that θ is non-twisting. Hence, the second assertion of the Theorem follows from Proposition 12.10. ∎

12.5. Twistor spinors in dimension $n \leq 6$

In this section we discuss the twistor equation on Lorentzian spin manifolds in the dimensions $n = 3, 4, 5$ and 6. For solutions of the twistor equation without singularities we derive a complete list of possible underlying Lorentzian geometries in these low dimensions. For dimension 4 this was already proved by J. Lewandowski. To begin with we state some special properties for spinors on low dimensional Lorentzian spin manifolds.

LEMMA 12.12. (Properties of spinors on low dimensional Lorentzian manifolds)

1. *Let $n = 3, 5$. Then each spinor field $\varphi \in \Gamma(S)$ satisfies $V_\varphi \cdot \varphi = \langle\varphi, \varphi\rangle\varphi$ and $g(V_\varphi, V_\varphi) = -\langle\varphi, \varphi\rangle^2$.*
2. *Let $n = 2, 4, 6$. Then each spinor field $\varphi \in \Gamma(S^\pm)$ satisfies $V_\varphi \cdot \varphi = 0$ and $g(V_\varphi, V_\varphi) = 0$.*
3. *Let $n = 3$. Then there exists a real structure $\tau : S \longrightarrow S$ on the spinor bundle such that $\tau(X \cdot \varphi) = -X \cdot \tau(\varphi)$ and $\langle\tau\varphi, \tau\psi\rangle = -\langle\psi, \varphi\rangle$.*
   If φ is a real spinor field, i.e. $\tau(\varphi) = \varphi$, then the Dirac current V_φ is lightlike.
4. *Let $n = 5$. Then there exists a quaternionic structure $J : S \longrightarrow S$ such that $J(X \cdot \varphi) = X \cdot J(\varphi)$ and $\langle J\varphi, J\psi\rangle = -\langle\varphi, \psi\rangle$.*

PROPOSITION 12.13. *The dimension of the space of twistor spinors on a non-conformally flat Lorentzian manifold (M^n, g) is bounded by*

(1) $n = 3$: $\dim_{\mathbb{C}} \operatorname{Ker} P \leq 1$.
(2) $n = 4$: $\dim_{\mathbb{C}} \operatorname{Ker} P \leq 2$.
(3) $n = 5$: $\dim_{\mathbb{C}} \operatorname{Ker} P \leq 2$.

PROOF. Let us show this in dimension 5. By the preceeding Lemma there is a quaternionic structure J on the spinor bundle. The space of twistor spinors is invariant under J. Assume, that $\dim_{\mathbb{C}} \operatorname{Ker} P > 2$. Then there is a dense subset $A \subset M$ such that $\dim_{\mathbb{H}}\{\varphi(x) \mid \varphi \in \operatorname{Ker} P\} = 2$ for all $x \in A$. Since the Clifford multiplication is $\mathbb{H}$-linear and $\operatorname{rank}_{\mathbb{H}} S = 2$, the integrability condition $W(\eta)\cdot\varphi = 0$ for each twistor spinor φ and each 2-form η yields $W = 0$. □

We call a twistor spinor φ *singularity free* if it has no zeros and if the zero set of the length function $\langle\varphi, \varphi\rangle$ is empty or the whole manifold M. There are the following geometric structures of Lorentzian manifolds with singularity free twistor spinors in dimension $n \leq 6$.

THEOREM 12.14. *(*[**Lew91**], [**Lei01b**]*) Let (M^n, g) be a Lorentzian manifold with a singularity free twistor spinor. Then (M^n, g) is locally conformal equivalent to one of the following kinds of Lorentzian structures.*

$n = 3:$ • *pp-manifold*

$n = 4:$ • *pp-manifold*
• *Fefferman space*

$n = 5:$ • *pp-manifold*
• *Lorentzian Einstein Sasaki manifold*
• $(\mathbb{R}, -dt^2) \times (N^4, h)$, *where* (N^4, h) *is a Riemannian Ricci-flat Kähler manifold*

$n = 6:$ • *pp-manifold*
• *Fefferman space*
• $\mathbb{R}^{1,1} \times (N^4, h)$, *where* (N^4, h) *is a Ricci-flat Kähler manifold*
• *Brinkman space with special Kähler flag*

PROOF. In dimension $n = 4$ and $n = 6$ we may assume, that φ is a zero free half-spinor in $S^{\pm}$, since we are only interested in local considerations. Since V_φ has no zero (cf. Proposition 12.2(2)) and is lightlike (cf. Lemma 12.12) there is a local conformal transformation of the metric such that V_φ is Killing and lightlike with respect to the new metric. Hence, according to Proposition 12.11, (M, g) is locally conformal equivalent to a Fefferman space or a Brinkman space with parallel spinors. The low-dimensional Lorentzian manifolds with parallel spinors are discussed for example in a paper of R. Bryant ([**Bry00**]). Each 4-dimensional Lorentzian manifold with a parallel spinor is a pp-manifold. Studying the orbit types of the spin representation in dimension 6 for signature $(-+++++)$, one obtains the types of geometries listed in the Theorem for dimension $n = 6$. Now, let $n = 3$. If φ is a zero free twistor spinor, then locally the real part $\Re\varphi := \frac{1}{2}(\varphi + \tau(\varphi))$ or the imaginary part $\Im\varphi := \frac{i}{2}(\tau(\varphi) - \varphi)$ is a zero free real spinor (here τ denotes the real structure on S). Hence, according to Lemma 12.12, we have a zero free twistor spinor with lightlike Dirac current, which becomes a Killing field after a conformal change of the metric. Since the dimension is odd, (M, g) is local conformally equivalent to a manifold with parallel spinors (Proposition 12.11). 3-dimensional manifolds with parallel spinors are known to be pp-manifolds.
Finally, let us sketch the proof in case of dimension $n = 5$. Let φ be a zero free twistor spinor. In case that $\langle\varphi, \varphi\rangle = 0$, the Dirac current V_φ is lightlike by Lemma 12.12. Since we are only interested in the local conformal structure we may assume that it is Killing as well. The dimension is odd, therefore Proposition 12.11 shows that (M, g) has to have parallel spinors. Hence it is locally conformal equivalent to a pp-manifold. Now, suppose that the lenght function of φ has no zeros. Consider the conformally changed metric $\tilde{g} := \langle\varphi, \varphi\rangle^{-2} g$. The spinor field $\psi := |\langle\varphi, \varphi\rangle|^{-1/2}\varphi$ is a twistor spinor of constant length ± 1 for $\tilde{g}$. Hence, we may suppose that we have a twistor spinor φ of constant length ± 1 on (M, g). In that case $\langle D\varphi, D\varphi\rangle$ is constant as well and (M, g) is an Einstein space of scalar curvature

$$R = -\frac{4(n-1)}{n} \frac{\langle D\varphi, D\varphi\rangle}{\langle\varphi, \varphi\rangle}.$$

If $R = 0$, φ or $D\varphi$ is a non-vanishing parallel spinor with timelike Dirac current. Hence, (M, g) is a product of a timelike line and a Riemannian spin manifold with parallel spinor, which is known to be Ricci-flat and Kaehler. If R is non-zero, either φ is a Killing spinor or the spinors

$$\psi_{\pm} = \frac{1}{2}\varphi \pm \sqrt{\frac{n-1}{nR}} D\varphi$$

are Killing spinors to different Killing numbers. In the first case the Killing number of φ has to be imaginary since the length is constant. Using Lemma 12.12 one can show that (after normation and a change of the time-orientation if the length of the spinor is $+1$) the conditions of Proposition 12.5 are satisfied. Hence, (M, g) has to be Einstein Sasaki. In the second case, if $\psi_\pm$ are imaginary Killing spinors we can proceed in the same way. If $\psi_\pm$ are real Killing spinors, then $\psi_\pm$, $J\psi_\pm$ are 4 $\mathbb{C}$-linearely independent Killing spinors (here J denotes the quaternionic structure on the spinor bundle). Hence, by Proposition 12.13, (M, g) is conformally flat. □

HELGA BAUM, HUMBOLDT-UNIVERSITÄT ZU BERLIN, INSTITUT FÜR MATHEMATIK, SITZ: RUDOWER CHAUSSEE 25,D-10099 BERLIN, GERMANY.
E-mail address: baum@mathematik.hu-berlin.de

FELIPE LEITNER, UNIVERSITY OF EDINGBURGH, DEPARTMENT OF MATHEMATICS AND STATISTICS, JCMB–KING'S BUILDINGS, EH9 3JZ EDINGBURGH, SCOTTLAND.
E-mail address: felipe@maths.ed.ac.uk

CHAPTER 13

Geometric Spectrum of Isospectral Spherical Space Forms

FIDA EL CHAMI

ABSTRACT. In this paper, we are interested in the relation between the Laplace spectrum on the differential forms and the geometric spectrum. We consider examples of isospectral non-isometric spherical space forms constructed by A. Ikeda: lens spaces isospectral up to a fixed order and not above, and spherical space forms with fundamental groups of type 1 isospectral on the forms. We compute and compare their geometric spectra. We obtain the same geometric spectrum for the first family but not for the second.

Contents

Introduction

Two compact connected Riemannian manifolds M_1 and M_2 are said to be p-isospectral if the spectrum of the Laplacian acting on the space of differential p-forms is the same that in M_1 and M_2. If $p = 0$, we say simply that M_1 and M_2 are isospectral. We know that the Laplace spectrum on the functions determines the dimension, volume and total scalar curvature. Under certain conditions, it determines the collection of lengths of closed geodesics. It is not known whether this spectrum always determines the length spectrum (definition 13.1). It can be considered that the collection of all p-spectrum contains more geometric information of a manifold than does the p-spectrum for a fixed p. We will show that the collection of all p-spectrum does not determine the length spectrum.

Many methods can be employed to establish relations between the Laplace spectrum and the geometric spectrum. A general result obtained by T. Sunada [**Sun85**] shows that under some conditions, two manifolds are p-isospectral for all p and have the same length of closed geodesics. Another method is the trace formula ([**dV73**] for example). We can also calculate explicitly the Laplace spectrum or the geometric spectrum and then compare. The last method is used by C. S. Gordon [**Gor86**] for some Heisenberg manifolds.

A compact connected Riemannian manifold of constant curvature 1 is said to be a spherical space form. If the fundamental group of a spherical space form is cyclic then the spherical space form is called a lens space. For the classification of the spherical space forms see [**Wol84**].

In this paper, we calculate the length spectrum and sometimes the volume spectrum for the isospectral and non-isometric spherical space forms constructed by A. Ikeda ([**Ike89**], [**Ike83**]). Next, we compare the spectra of two isospectral and non-isometric manifolds.

We begin with lens spaces p-isospectral but not $(p+1)$-isospectral. We calculate their length spectra and we find they have the same. This spaces do not verify the T. Sunada theorem [**Sun85**].

We can not determine all the minimal submanifolds of this spaces. We calculate the volumes of the totally geodesic submanifolds and of another family of minimal submanifolds. The list of the obtained volumes is the same for two isospectral lens spaces.

We note that there is isospectral and non-isometric lens spaces of the same family which are not homotopy equivalent (A. Ikeda [**Ike80**]). These spaces don't have the same marked length spectrum (R. Gornet [**Gor96**]).

Last, we consider the spherical space forms with finite fixed point free non-cyclic fundamental group of type 1 which are p-isospectral for all p and non-isometric ([**Ike83**]). This spaces verify the T. Sunada's theorem conditions. For a such space, we calculate the length spectrum. The length spectra of two spaces in this family are not always identical.

In [**Gor94**], R. Gornet gave an example of a pair of compact three-dimensional Riemannian nilmanifolds verifying the Sunada conditions but having different length spectra. The fundamental groups of her spaces are discrete but not finite.

These results are obtained in my thesis. For more details on the calculus, see [**Cha00**].

13.1. Preliminaries

In this section we give definitions and results which we need in the following sections.

Let M be a compact Riemannian manifold. If M has fundamental group Γ, the free homotopy classes of loops in M correspond to the conjugacy classes $[\gamma]_\Gamma$ in Γ. Each free homotopy class contains at least one closed geodesic.

There is a number of different definitions of length spectrum in the literature. For example, we can define the multiplicity of a length l as the number of closed geodesic of length l. In the examples considered, there is sufficient symmetry so that all lengths which occur have infinite multiplicity under this definition. Hence we will use a more informative definition of the length spectrum.

DEFINITION 13.1. We call length spectrum of M the collection of lengths of closed geodesics of M with the multiplicity of the length l defined to be the total number of free homotopy classes containing a geodesic of length l.

Now, we describe the lengths of closed geodesics in a quotient space $M = \Gamma\backslash X$. We say that a geodesic γ of X is invariant by an element σ of the group Γ if there is a real number $\omega_\sigma > 0$ such that $\sigma.\gamma(t) = \gamma(t + \omega_\sigma)$ for all $t \in \mathbb{R}$. The lowest ω_σ verifying this relation is the period of γ relatively to σ. The lowest ω_σ where σ belongs to Γ, is the length ω of the prime geodesic of M, associated to the projection of γ. The iterates of this prime geodesic have lengths $k\omega$, $k \geq 1$. The lengths of closed geodesics (without multiplicity) are simply the positive numbers ω verifying $\sigma.\gamma(t) = \gamma(t + \omega)$, γ is a geodesic in X and $\sigma \in \Gamma$.

DEFINITION 13.2. We call volume p-spectrum of M the collection of volumes of compact minimal p-dimensional submanifolds of M. The multiplicity of the volume v is defined to be the total number of minimal submanifolds of volume v.

Let γ the geodesic of the sphere S^n given by:

$$\gamma(t) = {}^t(\cos t, \sin t, 0, \ldots, 0).$$

Every geodesic of S^n is of the form $h.\gamma$ with $h \in O(n + 1)$. The geodesics of a spherical space form $\Gamma\backslash S^n$ ($\Gamma \subset \mathrm{O}(n + 1)$) are precisely the projections of the geodesics of S^n. The next lemma shows that the lengths of closed geodesics of $\Gamma\backslash S^n$ are obtained by calculating the eigenvalues of the matrices of Γ.

LEMMA 13.3. *The length of closed geodesics of $\Gamma\backslash S^n$ are the positive numbers ω such that $2\cos\omega$ is a multiple eigenvalue of order 2 of $\sigma + {}^t\sigma$, for a $\sigma \in \Gamma$.*

PROOF. Let $\omega > 0$ be the length of a closed geodesic of $\Gamma\backslash S^n$, projection of $h.\gamma$, $h \in \mathrm{O}(n + 1)$. Then there is $\sigma \in \Gamma$ such that $\sigma h\gamma(t) = h\gamma(t + \omega)$ for all $t \in \mathbb{R}$. Then we obtain:

$$h^{-1}\sigma h = \begin{pmatrix} \cos\omega & -\sin\omega & (0) \\ \sin\omega & \cos\omega & (0) \\ (0) & (0) & A \end{pmatrix}$$

where $A \in \mathrm{O}(n-1)$. Hence, $2\cos\omega$ is an eigenvalue of $\sigma + {}^t\sigma$ of order 2. Conversely, if ω is a positive real number such that $2\cos\omega$ is a eigenvalue of order 2 of the matrix $\sigma + {}^t\sigma$ ($\sigma \in \Gamma$), we take h_1 and h_2 the two corresponding orthonormal eigenvectors and $h \in \mathrm{O}(n)$ a matrix having h_1 and h_2 the first and second columns. The geodesic $h\gamma$ of S^n is invariant by σ and has a projection on $\Gamma\backslash S^n$ of length ω. □

DEFINITION 13.4. Let G be a group. Two subgroups Γ_1 and Γ_2 of G are said to be almost conjugate if there is a bijection φ of Γ_1 on Γ_2 satisfying that $\varphi(g)$ is conjugate to g in G for each g in Γ_1.

THEOREM 13.5. *(*[**Bér93**], [**BP97**], [**DG89**]*) Let (M, g) be a compact Riemannian manifold where g is the corresponding metric, G an isometry compact group of M. Let Γ_1 and Γ_2 be two finite fixed point free subgroups of G such that $\Gamma_1\backslash M$ and $\Gamma_2\backslash M$ are compact with the induced metric. Assume $\sharp([g]_G \cap \Gamma_1) = \sharp([g]_G \cap \Gamma_2)$ for all $g \in G$, $[g]_G$ being the conjugacy class in G. Then:*

(1) $\Gamma_1\backslash M$ and $\Gamma_2\backslash M$ are p-isospectral for all p.
(2) There is a length preserving bijection between the set of closed geodesics of $\Gamma_1\backslash M$ and the set of closed geodesics of $\Gamma_2\backslash M$.

REMARK 13.6. The condition $\sharp([g]_G \cap \Gamma_1) = \sharp([g]_G \cap \Gamma_2)$ for all $g \in G$ in theorem 13.5 is equivalent to the fact that Γ_1 and Γ_2 are almost conjugate in G.

13.2. Lens spaces

Let $S^{2n-1} = \mathrm{SO}(2n)/\mathrm{SO}(2n-1)$ be the sphere of dimension $2n-1$ and Γ a finite fixed point free group of isometries of S^{2n-1}. In this section, we calculate the length spectrum and a part of the volume spectrum of the lens space $\Gamma\backslash S^{2n-1}$ which not verify the conditions of T. Sunada theorem 13.5. Finally, we compare the obtained spectra.

Let q be the order of Γ. We can write $\Gamma = \langle\sigma\rangle$ where σ is the orthogonal matrix given by:

$$\sigma = \mathrm{diag}\,(R(\theta_1), R(\theta_2), \ldots, R(\theta_n)) \tag{13.1}$$

where $\theta_i = 2\pi p_i/q$ for $1 \le i \le n$ and

$$R(\theta) = \begin{pmatrix} \cos\theta & \sin\theta \\ -\sin\theta & \cos\theta \end{pmatrix},$$

$p_1, p_2, \ldots, p_n$ are integers prime with q. We denote by $L(q : p_1, ..., p_n)$ the lens space $\Gamma\backslash S^{2n-1}$. We define:

$$\tilde{\mathfrak{L}}_0(q,n) = \{L(q : p_1, ..., p_n);\ p_i \not\equiv \pm p_j \pmod q,\ 1 \le i < j \le n\}.$$

and $\mathfrak{L}_0(q,n)$ the set of isometry classes of $\tilde{\mathfrak{L}}_0(q,n)$.

In the rest, we take q prime verifying $\frac{q-1}{2} = n+2$. The set K_q of equivalent classes (modulo q) of integers prime with q is a cyclic group of order $q-1$. There is a bijection ω between $\mathfrak{L}_0(q,n)$ and $\mathfrak{L}_0(q,2)$ such that $\omega(L(q : p_1, ..., p_n)) = L(q : q_1, q_2)$ where $K_q = \{\pm p_1, ..., \pm p_n, \pm q_1, \pm q_2\}$. For $r > 0$, we introduce the family of lens spaces:

$$\mathfrak{L}_r(q,2) = \{L(q : q_1, q_2) \in \mathfrak{L}_0(q,2);\ aq_1 + bq_2 \not\equiv 0 \pmod q,\ 1 \le |a| + |b| \le r+2\},$$

and

$$\mathfrak{L}_r(q,n) = \omega^{-1}(\mathfrak{L}_r(q,2)).$$

THEOREM 13.7. **[Ike89]** *Let q be a prime number ≥ 11 and $q_0 = \frac{q-1}{2} = n+2$. Let L_1 and L_2 be two spaces of $\mathfrak{L}_p(q,n)$, where p is a positive integer number. We have :*

(1) L_1 et L_2 are k-isospectral for $k = 0, 1, ..., p$.
(2) Moreover, if $L_1 \in \mathfrak{L}_{p+1}(q,n)$ but not L_2, then L_1 and L_2 are not $(p+1)$-isospectral.

We introduce the projection $\pi : \mathrm{SO}(2n) \to S^{2n-1} = \mathrm{SO}(2n)/\mathrm{SO}(2n-1)$. Any geodesic of S^{2n-1} is of the form $h.\gamma$ where $h \in \mathrm{O}(2n)$ and:

$$\gamma(t) = \begin{pmatrix} \cos t \\ \sin t \\ 0 \\ \vdots \\ 0 \end{pmatrix}. \tag{13.2}$$

The geodesics of lens space $\Gamma\backslash S^{2n-1}$ are exactly the images of the geodesics of S^{2n-1} by the canonical projection Π of S^{2n-1} to $\Gamma\backslash S^{2n-1}$.

THEOREM 13.8. *Let q be a prime number ≥ 11 and $q_0 = n+2$, where $q_0 = (q-1)/2$. Let $L = L(q : p_1, ..., p_n)$ be a space of $\mathfrak{L}_0(q, n)$. The length spectrum of the lens space L is like this :*

(i) $2l\pi$, $(l \geq 1)$, *of multiplicity 1.*
(ii) $2l\pi/q$, $(l \geq 1)$, *of multiplicity $2n$.*

In particular, if L_1 and L_2 belong to $\mathfrak{L}_0(q, n)$, then they have the same length spectrum.

PROOF. Let $h.\gamma$ be a geodesic of the sphere and σ the matrix defined by (13.1). If $h.\gamma$ is not invariant by σ, it is not so by any non-trivial element of Γ since q is prime. His image by Π in the lens space $\Gamma\backslash S^{2n-1}$ is a geodesic of length 2π with a multiplicity 1.

If the geodesic $h.\gamma$ is invariant by σ, then it is so by all the σ^k, $0 \leq k \leq q-1$. Equivalently, for all $0 \leq k \leq q-1$, there exists a ω_k and a $1 \leq i \leq n$, such that $\cos\omega_k = \cos k\theta_i$. For $0 \leq k \leq q-1$, the smallest positive number ω_k such that $\cos\omega_k = \cos k\theta_i$ is $2\pi/q$.

The multiplicity of $2\pi/q$ is calculated as follows: since q is prime, then for all $p \in \{\pm p_1, ..., \pm p_n\}$, there exists a unique $1 \leq k \leq q-1$ such that $kp \equiv 1 \pmod{q}$. These values of k are distincts. This implies that the number of free homotopy classes which contain a geodesic of length $2\pi/q$, is $2n$. Taking the iterates of every geodesic, we have the proof. □

The k-dimensional totally geodesic submanifolds of the lens space $L(q : p_1, ..., p_n)$ are exactly the projections of the spheres $S^k \subset S^{2n-1}$ on $L(q : p_1, ..., p_n)$. The next proposition characterizes the invariance by Γ of the totally geodesic submanifolds of S^{2n-1}. It is not difficult to be established.

PROPOSITION 13.9. *Let Γ be the group generated by the matrix σ gived by (13.1). Let $X = h.S^r$ be a totally geodesic submanifold of S^{2n-1}, where $h \in O(2n)$. We write h in the following form:*

$$h = \begin{pmatrix} H_{11} & H_{12} \\ \vdots & \vdots \\ H_{n1} & H_{n2} \end{pmatrix},$$

H_{i1} is a $(2, r+1)$ matrix and H_{i2} a $(2, 2n-r-1)$ one $(1 \leq i \leq n)$.

- *When r is even, X is not invariant by any element of Γ.*
- *When r is odd, X is invariant by Γ if and only if*

$$\forall\, 1 \leq i \leq n, \quad H_{i1} = 0 \text{ or } H_{i2} = 0. \tag{13.3}$$

This condition is equivalent to the fact that there exists a unique i with $H_{i1} \neq 0$.

PROPOSITION 13.10. *The volumes of totally geodesic submanifolds $\Pi(h.S^r)$ of $L(q : p_1, ..., p_n)$, where $h \in O(2n)$ and $r \in \{1, \ldots, 2n-1\}$, are given by:*

(i) The volume of $\Pi(h.S^{2m})$ is equal to the volume of S^{2m}.
(ii) The volume of $\Pi(h.S^{2m-1})$ is equal to $\frac{1}{q}$ times the volume of S^{2m-1}) if h verifies (13.3) and to the volume of S^{2m-1} if not.

The multiplicity of this volumes is infinity.

In particular, if L_1 and L_2 are in $\mathfrak{L}_0(q,n)$, then the volumes of totally geodesic submanifolds of L_1 and L_2 are identical.

REMARK 13.11. Similarly, we can calculate the volume of some minimal submanifolds of lens spaces, but we can not determine all the minimal submanifolds of this spaces. Consequently, we can not determine all the volume spectrum.

13.3. Spherical space forms with type 1 fundamental groups

In this section, we introduce the isospectral and non-isometric spherical space forms which are the quotients of the sphere by finite non cyclic and fixed point free groups of type 1. We determine theirs length spectra and we find they are not identical in general.

DEFINITION 13.12. A finite fixed point free group Γ is said to be of type 1 if all the Sylow subgroups of Γ are cyclic.

If a and b are two integers, we denote by (a,b) the greater common divisor of a and b. Let m, n and d be positive integers and r an integer prime with m verifying:

$$\begin{cases} ((r-1)n, m) = 1 \\ \text{the class of } r \text{ in } K_m \text{ is of order } d \\ n = d^2. \end{cases}$$

With these integers, we defined the finite group $\Gamma_d(m,n;r)$ of order $N = mn$ generated by two elements A and B such that:

$$A^m = B^n = 1 \quad \text{and} \quad BAB^{-1} = A^r.$$

We denote by Γ the group $\Gamma_d(m,n;r)$; it is of type 1. we remind some properties of Γ ([**Ike83**] or [**Wol84**]). The fixed point free irreducible representations of Γ are described as follows :

Let k and l be two integers such that $(k,m) = (l,n) = 1$. Let $\pi_{k,l}$ be the real representation of degree $2d$ of Γ, given by:

$$\pi_{k,l}(A) = \operatorname{diag}\left(R(2\pi k/m), R(2\pi kr/m), \ldots, R(2\pi kr^{d-1}/m)\right)$$

and

$$\pi_{k,l}(B) = \left(\begin{array}{c|c} (0) & I_{2d-2} \\ \hline R(2\pi l/d) & (0) \end{array}\right).$$

The space $\pi_{k,l}(\Gamma)\backslash S^{2n-1}$ is isometric to $\pi_{k',l'}(\Gamma)\backslash S^{2n-1}$ if and only if there are integers $\varepsilon = \pm 1$, $c \in \{0, ..., d-1\}$, s and t such that:

$$\begin{cases} (s,m) = (t,n) = 1 \\ t \equiv 1 \pmod{d} \\ k' \equiv \varepsilon s k r^c \pmod{m} \\ l' \equiv \varepsilon t l \pmod{d}. \end{cases}$$

In particular, the spaces $\pi_{1,l}(\Gamma)\backslash S^{2n-1}$ and $\pi_{1,1}(\Gamma)\backslash S^{2n-1}$ are not isometric.

Next, we describe the elements of $\pi_{1,l}(\Gamma)$. For $0 \le a, a' \le m-1$ and $0 \le b, b' \le n-1$, we have:

$$A^a B^b . A^{a'} B^{b'} = A^{a+a' r^b} B^{b+b'}.$$

Then, every element of $\pi_{1,l}(\Gamma)$ can be written $\sigma = \pi_{1,l}(A^a B^b)$. Writing $b = v + wd$, with $0 \leq v, w \leq d-1$, we have:

- If $v = 0$:

$$\sigma = \sigma_{a,v,w} = \operatorname{diag}\left(R(\theta_1), R(\theta_2), \ldots, R(\theta_d)\right), \tag{13.4}$$

  where $\theta_i = 2\pi(a\frac{r^{i-1}}{m} + w\frac{l}{d})$, for all $1 \leq i \leq d$.

- If $1 \leq v \leq d-1$:

(13.5)

$$\sigma = \sigma_{a,v,w} = \left(\begin{array}{c|c} (0) & \operatorname{diag}\left(R(\theta_1), R(\theta_2) \ldots, R(\theta_{d-v})\right) \\ \hline \operatorname{diag}\left(R(\alpha_{d-v+1}), \ldots, R(\alpha_d)\right) & (0) \end{array}\right),$$

where

$$\begin{cases} \theta_i = 2\pi(a\frac{r^{i-1}}{m} + w\frac{l}{d}) & \text{for } \ 1 \leq i \leq d-v, \\ \alpha_i = 2\pi(a\frac{r^{i-1}}{m} + (w+1)\frac{l}{d}) & \text{for } \ d-v+1 \leq i \leq d. \end{cases}$$

PROPOSITION 13.13. [**Ike83**] *For $(l,d) = 1$ and $(l-1,d) = 1$, the groups $\pi_{1,l}(\Gamma)$ and $\pi_{1,1}(\Gamma)$ are almost conjugate (but not conjugate) by:*

$$\begin{array}{cccc} \varphi & : & \pi_{1,l}(\Gamma) & \to & \pi_{1,1}(\Gamma) \\ & & \pi_{1,l}(A^a B^b) & \mapsto & \pi_{1,1}(A^a B^{lb}) \end{array}$$

where $0 \leq a \leq m-1$ and $0 \leq b \leq n-1$.

Consequently, using the Sunada's theorem we have that the non-isometric spaces $\pi_{1,l}(\Gamma)\backslash S^{2d-1}$ and $\pi_{1,1}(\Gamma)\backslash S^{2d-1}$ with $(l,d) = 1$ and $(l-1,d) = 1$ are isospectral for the differential forms of all orders and have the same lengths of closed geodesics.

To calculate the length spectrum of the spaces $\pi_{1,l}(\Gamma)\backslash S^{2d-1}$, we begin by calculating the eigenvalues of elements of the group $\pi_{1,l}(\Gamma)$ (lemma 13.3).

NOTATIONS 13.14. We consider $1 \leq v \leq d-1$. We denote by $v_1 = (v,d)$, $\delta = d/v_1$,

$$\begin{cases} \gamma_i = \theta_i & \text{if} \quad i \leq d-v \\ \gamma_i = \alpha_i & \text{if not} \end{cases}$$

and

$$S_{a,v,w,\mu} = \gamma_\mu + \gamma_{v_1+\mu} + \ldots + \gamma_{(\delta-1)v_1+\mu}.$$

We have:

$$\begin{aligned} S_{a,v,w,\mu} &\equiv 2\pi\left(\frac{ar^{\mu-1}}{m}(1 + r^{v_1} + \ldots + r^{(\delta-1)v_1}) + (\delta w + \delta - \nu)\frac{l}{d}\right) \pmod{2\pi} \\ &\equiv 2\pi\left(\frac{ar^{\mu-1}(1-r^d)}{m(1-r^{v_1})} + (\delta w + \delta - \nu)\frac{l}{d}\right) \pmod{2\pi}, \end{aligned}$$

where $\nu = \dfrac{d-v}{v_1}$.

We show without difficulty the lemma:

LEMMA 13.15. *For $0 \leq a \leq m-1$, $1 \leq v \leq d-1$ and $0 \leq w \leq d-1$, the eigenvalues of $\sigma_{a,v,w}$ are $e^{\pm i\omega}$, when ω is a solution of one of these equations:*

$$e^{i\delta\omega} = e^{iS_{a,v,w,\mu}}, \quad 1 \leq \mu \leq v_1. \tag{13.6}$$

The eigenvalues of $\sigma_{a,v,w} + {}^t\sigma_{a,v,w}$ *are double of the form* $2\cos\omega$*, when* ω *is given by (13.6). The corresponding eigenvectors are the real and imaginary parts of the eigenvectors of* $\sigma_{a,v,w}$ *associated to the eigenvalues of* $e^{i\omega}$.

Using the correspondence between the free homotopy classes and the conjugacy ones of the elements of the group $\pi_{1,l}(\Gamma)$, the next lemma describes the free homotopy classes of closed curves of the space $\pi_{1,l}(\Gamma)\backslash S^{2d-1}$. It could be useful for the calculation of the multiplicities in the length spectrum (theorem 13.18).

LEMMA 13.16. *Let* m *be a prime and let* $0 \le a \le m-1$ *and* $0 \le b \le n-1$*. We write* $b = v + wd$ *with* $0 \le v, w \le d-1$*. The free homotopy class* $[\sigma_{a,v,w}]$ *of* $\sigma_{a,v,w}$ *is given by:*

(1) $[\sigma_{0,0,w}] = \{\sigma_{0,0,w}\}$, *for all* $0 \le w \le d-1$.
(2) $[\sigma_{a,0,w}] = \{\sigma_{ar^t,0,w};\ 0 \le t \le d-1\}$, *for all* $1 \le a \le m-1$ *and* $0 \le w \le d-1$.
(3) $[\sigma_{a,v,w}] = \{\sigma_{c,v,w};\ 0 \le c \le m-1\}$, *for all* $0 \le a \le m-1$, $1 \le v \le d-1$ *and* $0 \le w \le d-1$.

REMARK 13.17. For m prime, we have:

- The free homotopy classes $[\sigma_{a,0,w}]$ and $[\sigma_{m-a,0,w}]$ are the same if and only if d is even or $a = 0$.
- $S_{a,v,w,\mu}$ depends only on v and w. We denote by:

$$S_{v,w} = S_{a,v,w,\mu} \equiv 2\pi\frac{l}{d}(\delta w + \delta - \nu) \pmod{2\pi}.$$

The next theorem determines the multiplicities of the obtained lengths.

THEOREM 13.18. *Let* m *be a prime number and let* ω *be the length of a geodesic of the free homotopy class* $[\sigma_{a,v,w}]$ *in the space* $\pi_{1,l}(\Gamma)\backslash S^{2d-1}$*, where* $0 \le a \le m-1$*,* $0 \le v \le d-1$ *and* $0 \le w \le d-1$*. We set* $v_1 = (d, v)$ *and* $\delta = d/v_1$.

I- First case: d *is an odd number:*

(1) *If* $a = v = w = 0$*, then* $\omega = 2\pi$ *has multiplicity 1.*
(2) *If* $a \ne 0$ *and* $v = 0$*, then* ω *has multiplicity 2.*
(3) *If* $v = 0$ *and* $w \ne 0$*, then* ω *has multiplicity 2.*
(4) *If* $v \ne 0$*, then* ω *has multiplicity* 2δ.

II- Second case: d *is an even number:*

(1) *If* $a = v = w = 0$*, then* $\omega = 2\pi$ *has multiplicity 1.*
(2) *If* $v = 0$ *and* $w \notin \{0, d/2\}$*, then* ω *has multiplicity 2.*
(3) *If* $a \ne 0$ *and* $v = w = 0$*, then* ω *has multiplicity 1.*
(4) *If* $v = 0$ *and* $w = d/2$*, then* ω *has multiplicity 1.*
(5) *If* $v \notin \{0, d/2\}$*, then* ω *has multiplicity* 2δ.
(6) *If* $v = d/2$ *and if* $2w+1$ *is not multiple of* $d/2$*, then* ω *has multiplicity* $2\delta = 4$ *(in this case* $v_1 = d/2$*).*
(7) *If* $v = d/2$ *and if* $2w+1$ *is multiple of* $d/2$*, then* ω *has multiplicity* $\delta = 2$.

PROOF. Let ω be the length of a closed geodesic in the free homotopy class $[\sigma_{a,v,w}]$, where $0 \le a \le m-1$, $0 \le v \le d-1$ and $0 \le w \le d-1$.

- $v = 0$:
In this case, there exists an $\alpha \in \{1, ..., d\}$ such that $\cos\omega = \cos\theta_\alpha$. If $[\sigma_{a',v',w'}]$ $(0 \leq a' \leq m-1,\ 0 \leq v' \leq d-1,\ 0 \leq w' \leq d-1)$ is another free homotopy class which contains a closed geodesic of the same length ω, then $v' = 0$. Indeed, if $v' \neq 0$, we have $\cos\delta\omega = \cos S_{v',w'}$ (lemma 13.15). Then, $\delta\theta_\alpha \equiv \pm S_{v',w'} \pmod{2\pi}$. This implies that d divides $\delta w \pm (\delta w' + (\delta - \nu))$ where $\nu = \frac{d-v'}{(d,v')}$. Consequently, δ divides ν, which is impossible since $1 \leq \nu \leq \delta - 1$.
Hence, we search $0 \leq a' \leq m-1$ and $0 \leq w' \leq d-1$ such that $[\sigma_{a',0,w'}]$ contains a closed geodesic of length ω. In this case, there exists an $\alpha' \in \{1, ..., d\}$ such that $\cos\theta_\alpha = \cos\omega = \cos\theta'_{\alpha'}$. This means that md divides
$$(ar^{\alpha-1} \pm a'r^{\alpha'-1})d + (w \pm w')lm,$$
i.e., m divides $ar^{\alpha-1} \pm a'r^{\alpha'-1}$ and d divides $w \pm w'$.
  - d divides $w - w'$ if and only if $w = w'$, and m divides $ar^{\alpha-1} - a'r^{\alpha'-1}$ if and only if there exists $0 \leq t \leq d-1$ such that $a - a'r^t$ or $ar^t - a'$ is multiple of m.
  Then we find $[\sigma_{a,0,w}] = [\sigma_{a',0,w}]$ (lemma 13.16).
  - d divides $w + w'$ if and only if $w' = d - w$ (modulo d), and m divides $ar^{\alpha-1} + a'r^{\alpha'-1}$ if and only if m divides $ar^t + a'$ or $a + a'r^t$ for a $0 \leq t \leq d-1$.
  This means $w' = d - w$ and m divides $ar^t - (m - a')$ or $a'r^t - (m - a)$. Thus $[\sigma_{a',0,w'}] = [\sigma_{m-a,0,d-w}]$ (lemma 13.16).
- $v \neq 0$:
We obtain $\cos\delta\omega = \cos S_{v,w}$ where $\delta = d/(v,d)$. If $[\sigma_{a',v',w'}]$ is another free homotopy class which contains a closed geodesic of length ω, then $v' \neq 0$. Hence $\cos\delta' S_{v,w} = \cos\delta S_{v',w'}$ where $\delta' = d/(v',d)$. If $\nu = \frac{d-v}{(d,v)}$ and $\nu' = \frac{d-v'}{(d,v')}$, then the last equation implies that d divides $\delta'(\delta w + (\delta - \nu)) \pm \delta(\delta' w' + (\delta' - \nu'))$, which means that δ divides $\delta'\nu$ and δ' divides $\delta\nu'$. We have $\delta = \delta'$ since $(\delta, \nu) = (\delta', \nu') = 1$. Then $\cos S_{v,w} = \cos S_{v',w'}$. This condition is equivalent to the fact:
$$d \quad \text{divides} \quad (\delta w + (\delta - \nu)) \pm (\delta w' + (\delta - \nu')).$$
  - In the case when d divides $(\delta w + (\delta - \nu)) - (\delta w' + (\delta - \nu'))$, we have δ divides $\nu - \nu'$, then $\nu' = \nu$, which implies that d divides $\delta(w - w')$ and consequently (v, d) divides $w - w'$. Let $w - w' = \alpha(v, d)$, then $\frac{w+1}{(v,d)} - \delta \leq \alpha \leq \frac{w}{(v,d)}$. We have δ integer values of α, then δ free homotopy classes containing a geodesic of length ω.
  - In the case when d divides $(\delta w + (\delta - \nu)) + (\delta w' + (\delta - \nu'))$, we have δ divides $\nu + \nu'$, then $\nu' = \delta - \nu$ (modulo δ). This implies that (v, d) divides $w + w' + 1$. Let $w + w' + 1 = \beta(v, d)$ then $\frac{w+1}{(v,d)} \leq \beta \leq \frac{w}{(v,d)} + \delta$. Hence, we obtain also in this case δ free homotopy classes containing a geodesic of length ω.

So w' can take the same value in the two cases if and only if (v, d) divides $2w + 1$.

The proof is a simple consequence of the precedent calculation. □

THEOREM 13.19. *We consider the group* $\Gamma = \Gamma_d(m, n; r)$ *with* m *prime,* $(l, d) = 1$ *and* $(l - 1, d) = 1$. *Then the spaces* $\pi_{1,l}(\Gamma)\backslash S^{2d-1}$ *and* $\pi_{1,1}(\Gamma)\backslash S^{2d-1}$ *have the same length spectrum when* d *is odd and different length spectrum when* d *is even.*

PROOF. Let ω be the length of a closed geodesic in $[\pi_{1,l}(A^a B^{v+wd})]$ of the space $\pi_{1,l}(\Gamma)\backslash S^{2d-1}$. There exists a closed geodesic in $[\pi_{1,1}(A^a B^{v'+w'd})]$ of the space $\pi_{1,1}(\Gamma)\backslash S^{2d-1}$ with $v' \equiv lv \pmod d$ and $w' \equiv lw \pmod d$. In fact, $\pi_{1,1}(A^a B^{v'+w'd})$ and $\pi_{1,l}(A^a B^{v+wd})$ are conjugate in $O(2d)$ (proposition 13.13), so they have the same eigenvalues.

We have $v = 0$ (resp. $w = 0$) if and only if $v' = 0$ (resp. $w' = 0$) because $(l, d) = 1$. In the case d even, $v = d/2$ (resp. $w = d/2$) if and only if $v' = d/2$ (resp. $w' = d/2$); moreover, when $v = v' = d/2$, the integers $2w+1$ and $2w'+1$ can't be simultaneously multiple of $d/2$ because $(l-1, d) = 1$.

Using theorem 13.18, we find that ω has the same multiplicity in the two spaces $\pi_{1,l}(\Gamma)\backslash S^{2d-1}$ and $\pi_{1,1}(\Gamma)\backslash S^{2d-1}$ when d is odd or even with $v = v' \neq d/2$. If $v = v' = d/2$, ω has different multiplicities in $\pi_{1,l}(\Gamma)\backslash S^{2d-1}$ and $\pi_{1,1}(\Gamma)\backslash S^{2d-1}$. □

Conclusion : *Knowing the p-spectrum of the laplacian for all p does not determine the length spectrum even for spaces satisfying the conditions of the Sunada's theorem.*

FIDA EL CHAMI, UNIVERSITÉ LIBANAISE, FACULTÉ DES SCIENCES II, BP 90656 , JDEIDET-EL-MATEN, LIBAN *and* UNIVERSITÉ SAINT-JOSEPH, FACULTÉ DES SCIENCES, BP 17 - 5208, MAR MIKAEL, BEYROUTH 1104 2020, LIBAN.
E-mail address: fida.sayah@fs.usj.edu.lb

CHAPTER 14

Towards a Basic Index Theory

AZIZ EL KACIMI[1]

ABSTRACT. In this note, we give and explain the statement of the Hodge decomposition theorem for a transversely elliptic operator on a Riemannian foliation proved in [**ElK90**]. Such an operator is Fredholm and therefore admits an index. This brings us to formulate the following problem : can this index be computed in terms of (topological) transverse invariants like in the Atiyah-Singer Index theorem? Some examples of transversely elliptic operators whose indices are topologically invariant are given, in particular the basic Euler-Poincaré number and the signature of a Riemannian foliation.

Contents

All the objects considered are assumed to be of class C^∞. Let M be a manifold of dimension $d+n$; for simplicity M will be connected and orientable. Differential forms (and in particular functions) will take their values in the field of complex numbers $\mathbb{C}$. If α is a form, $\overline{\alpha}$ will denote its complex conjugate; we say that α is *real* if $\overline{\alpha} = \alpha$.

14.1. Preliminaries

DEFINITION 14.1. A codimension n *foliation* $\mathcal{F}$ on M is given by an open cover $\{U_i\}_{i\in I}$ and submersions $f_i : U_i \longrightarrow T$ over an n-dimensional manifold T and, for each $U_i \cap U_j \neq \emptyset$, a diffeomorphism $\gamma_{ij} : f_i(U_i \cap U_j) \longrightarrow f_j(U_i \cap U_j)$ such that

$$f_j = \gamma_{ij} \circ f_i. \tag{14.1}$$

[1]I would like to thank Oussama HIJAZI for his kind invitation to participate to this workshop. Thanks also to Ken RICHARDSON for reading the initial version and suggesting me some corrections.

Every fibre of f_i is called a *plaque* of the foliation. Condition (14.1) says that, on the intersection $U_i \cap U_j$, the plaques defined respectively by f_i and f_j are the "same". The manifold M is decomposed into a family of disjoint immersed connected submanifolds of dimension d; each of these submanifolds is called a *leaf* of $\mathcal{F}$. We say that $\mathcal{F}$ is *transversely orientable* if T can be given an orientation which is preserved by all the γ_{ij}.

We denote by $T\mathcal{F}$ the tangent bundle to $\mathcal{F}$ and $\Gamma(\mathcal{F})$ the space of its global sections *i.e.* vector fields tangent to $\mathcal{F}$. We say that a differential form α is *basic* if it satisfies $i_X\alpha = L_X\alpha = 0$ for every $X \in \Gamma(\mathcal{F})$. (This notion extends in obvious way to differential forms with values in a vector space.) Basic forms define a differential complex $(\Omega^*(M/\mathcal{F}), d)$; its homology $H^*(M/\mathcal{F})$ is called the *basic cohomology* of $\mathcal{F}$. A *basic function* is a function constant on the leaves; such functions form an algebra which we denote by A. The quotient $\nu\mathcal{F} = TM/T\mathcal{F}$ is the normal bundle of $\mathcal{F}$. A vector field $Y \in \chi(M)$ is said to be *foliated* if, for every $X \in \Gamma(\mathcal{F})$, the bracket $[X, Y] \in \Gamma(\mathcal{F})$. The space $\chi(M, \mathcal{F})$ of foliated vector fields is an algebra and a A-module for which $\Gamma(\mathcal{F})$ is an ideal. The quotient $\chi(M/\mathcal{F}) = \chi(M, \mathcal{F})/\Gamma(\mathcal{F})$ is called the algebra of *basic vector fields* on M; it is also a A-module.

In all this paper a system of local coordinates means coordinates $(x_1, \cdots, x_d, y_1, \cdots, y_n)$ on an open set U on which the foliation is trivial and defined by the equations $dy_1 = \cdots = dy_n = 0$. If $\mathcal{F}$ is transversely holomorphic (see definition 14.2 below) $y_1, \cdots, y_n$ will be complex numbers.

DEFINITION 14.2. A *transverse structure* to $\mathcal{F}$ is a geometric structure on T invariant by all the local diffeomorphisms γ_{ij}.

A transverse structure can be considered as a geometric structure on the leaf space $M/\mathcal{F}$ (which is not a manifold in general). Let us give some examples.

EXAMPLES 14.3. (1) Suppose that T is a Riemannian manifold and all the γ_{ij} are isometries; then $\mathcal{F}$ is said to be *Riemannian*. This means that the normal bundle $\nu\mathcal{F}$ is equipped with a Riemannian metric which is "invariant along the leaves".

(2) Suppose that T is a complex manifold and all the γ_{ij} are biholomorphic maps; then we say that $\mathcal{F}$ is *transversely holomorphic*. In that case, any transversal to $\mathcal{F}$ inherts a complex structure.

(3) Suppose that T is a Hermitian manifold and all the γ_{ij} preserve the Hermitian structure; then we say that $\mathcal{F}$ is *Hermitian*. (The γ_{ij} are in particular biholomorphic maps and isometries.) The complexified normal bundle $\nu = \nu\mathcal{F} \otimes_{\mathbb{R}} \mathbb{C}$ is equipped with a Hermitian metric "invariant along the leaves".

(4) If T is a Kählerian manifold and all the γ_{ij} preserve the Kähler structure we say that $\mathcal{F}$ is *transversely Kählerian*. In particular such a foliation is Hermitian. This is equivalent to the existence of a Hermitian metric h on the normal bundle $\nu\mathcal{F}$ which can be writen in a transverse local system of coordinates $(y_1, \cdots, y_n)$ in the form $h = \sum_{k,\ell=1}^n h_{k\ell}(y) dy_k \otimes d\overline{y}_\ell$ such that its skew-symmetric part $\omega = \frac{i}{2} \sum_{k,\ell=1}^n h_{k\ell}(y) dy_k \wedge d\overline{y}_\ell$ is closed (ω is a basic 2-form called the *Kähler basic form* of $\mathcal{F}$.) It is easy to see that every Riemannian transversely orientable foliation of codimension 2 can be given a structure of a complex codimension 1 transversely Kählerian foliation.

(5) If T is a Spin manifold and all the γ_{ij} preserve the Spin structure, we say that $\mathcal{F}$ is a *Spin foliation*. Of course, a foliation by points on a Spin manifold is a Spin foliation.

(6) If T is parallelizable and the parallelism is preserved by all the γ_{ij}, we say that $\mathcal{F}$ is *transversely parallelizable*. This is equivalent to say that the A-module $\chi(M/\mathcal{F})$ is free of rank n.

14.2. Foliated bundles

Let $\mathcal{P} : G \hookrightarrow P \stackrel{\iota}{\longrightarrow} M$ be a principal bundle with structural group $G \subset \mathrm{GL}(n, \mathbb{C})$. The group G acts on P on the right and on its Lie algebra $\mathcal{G}$ by the adjoint representation Ad *i.e.*, for $g \in G$ and $X \in \mathcal{G}$, $Ad_g(X) = gXg^{-1}$. Denote by $\mathcal{V}$ the vector bundle whose fibre V_z at a point $z \in P$ is the tangent space at z of the fibre of $\mathcal{P}$.

Let $\mathcal{E} : E \longrightarrow M$ be a complex vector bundle defined by a cocycle $\{U_i, g_{ij}, G\}$ where $\{U_i\}$ is an open cover of M and $g_{ij} : U_i \cap U_j \longrightarrow G \subset \mathrm{GL}(n, \mathbb{C})$ are the transition functions. To such a vector bundle we can always associate a principal bundle $G \longrightarrow P \longrightarrow M$ whose fibre is the group G and the transition functions are exactly the g_{ij} (viewed as translations on G).

There are different ways to define a connection on a vector bundle $\mathcal{E}$: on $\mathcal{E}$ directly or by using the associated principal bundle. We shall make use of all these possibilities.

Definition 14.4. A *connection* on $\mathcal{P}$ is a subbundle $\mathcal{H}$ of TP such that

(a) - for every $z \in \mathcal{P} : T_zP = V_z \oplus H_z$ (H_z is the fibre of $\mathcal{H}$ at z),

(b) - for every $g \in G$ and every $z \in P : H_{zg} = (R_g)_* H_z$ where R_g is the right action of g on P.

Definition 14.5. A *connection* on $\mathcal{P}$ is a subbundle $\mathcal{H}$ given by the kernel of a G-invariant 1-form ξ on P with values in $\mathcal{G}$. The G-invariance of ξ means: $(R_g)^*(\xi) = Ad_{g^{-1}}(\xi)$ *i.e.* for $z \in P$, $X \in T_zP$ and $g \in G$, $\xi_{zg}((R_g)_*(X)) = g^{-1}\xi_z(X)g$ (where $(R_g)_*$ is the derivative of R_g).

Definition 14.6. A *linear connection* on the vector bundle $\mathcal{E}$ is a map $\nabla : (X, \alpha) \in \chi(M) \times C^\infty(\mathcal{E}) \longmapsto \nabla_X\alpha \in C^\infty(\mathcal{E})$ satisfying the following properties

(c) - ∇ is $C^\infty(M)$-linear on the first factor, that is, for $\alpha \in C^\infty(\mathcal{E})$, $X, Y \in \chi(M)$ and functions $f, g \in C^\infty(M)$, we have $\nabla_{fX+gY}\alpha = f\nabla_X\alpha + g\nabla_Y\alpha$,

(d) - for $\alpha \in C^\infty(\mathcal{E})$, $X \in \chi(M)$ and $f \in C^\infty(M)$ we have $\nabla_X(f\alpha) = f\nabla_X\alpha + (Xf)\alpha$ where Xf is the derivative of the function f in the direction of the vector field X.

In fact, the map ∇ is the *covariant derivative* of the connection. The *curvature* of this connection is the 2-form $\mathcal{R}$ with values in $\mathrm{End}(\mathcal{E})$ (the space of endomorphisms of $\mathcal{E}$) defined by

$$\mathcal{R}(X, Y) = \nabla_X\nabla_Y - \nabla_Y\nabla_X - \nabla_{[X,Y]}.$$

Now, suppose that we are given a connection $\mathcal{H}$ (like in the first definition or the second one) on the principal bundle $\mathcal{P}$. It is easy to see that the restriction of ι_* (the derivative of ι) to H_z is an isomorphism onto $T_{\iota(z)}M$. Let $\tau = \iota_*^{-1}(T\mathcal{F})$. We say that $\mathcal{P}$ is *foliated* if τ is integrable. In this case, τ defines a foliation $\widetilde{\mathcal{F}}$ on P such that

(e) - $\dim(\widetilde{\mathcal{F}}) = \dim(\mathcal{F})$,

(f) - $\widetilde{\mathcal{F}}$ is invariant under the action of G.

DEFINITION 14.7. We say that the connection $\mathcal{H}$ is *basic* if ξ is basic; a foliated bundle $\mathcal{E}$ is said to be a *$\mathcal{F}$-bundle* if it admits a basic connection. We say that $\mathcal{E}$ is a *$\mathcal{F}$-bundle* if the associated principal bundle is a $\mathcal{F}$-bundle.

A vector bundle $\mathcal{E}$ with a linear connection is foliated if, and only if, its curvature form $\mathcal{R}$ satisfies $\mathcal{R}(X,Y) = 0$ for $X, Y \in \Gamma(\mathcal{F})$; $\mathcal{E}$ is a $\mathcal{F}$-bundle if, and only if, $i_X\mathcal{R} = 0$ for $X \in \Gamma(\mathcal{F})$ (*cf.* [**KT75**]).

The foliation $\mathcal{F}_{\widehat{E}}$ on $\widehat{E} = P \times \mathbf{C}^n$ whose leaves are of the form (leaf of $\widetilde{\mathcal{F}}$) $\times$ (point of $\mathbb{C}^n$) is invariant by the diagonal action of G ; so it induces a foliation $\mathcal{F}_E$ on $E = P \times_G \mathbb{C}^n$.

A *$\mathcal{F}$-morphism* $\varphi : (\mathcal{E}, \xi) \longrightarrow (\mathcal{E}', \xi')$ between two $\mathcal{F}$-bundles is a morphism of vector bundles such that $\xi = \varphi^*(\xi')$.

The collection of $\mathcal{F}$-bundles and $\mathcal{F}$-morphisms on M is a category. So we can define the group $K(M, \mathcal{F})$ of *foliated K-theory* as in the classical case.

EXAMPLES 14.8.

(1) Suppose that we are given a Riemann metric on M. Let $T\mathcal{F}^\perp$ be the subbundle of TM orthogonal to $\mathcal{F}$ and $\Gamma(T\mathcal{F}^\perp)$ the space of its sections. Every $X \in \chi(M)$ can be uniquely writen $X = X_\mathcal{F} + X_\nu$ where $X_\mathcal{F} \in \Gamma(\mathcal{F})$ and $X_\nu \in \Gamma(T\mathcal{F}^\perp)$. Let $\pi : TM \longrightarrow \nu\mathcal{F}$ be the canonical projection. For every section Y of the bundle $\nu\mathcal{F}$ we denote by $\widetilde{Y}$ a vector field on M which projects on Y. For every $X_\mathcal{F} \in \Gamma(\mathcal{F})$ and every $Y \in C^\infty(\nu\mathcal{F})$, $\pi([X_\mathcal{F}, \widetilde{Y}])$ is independant of the choice of the $\widetilde{Y}$. Let $\widehat{\nabla}$ be any linear connection on $\nu\mathcal{F}$. So we can define a linear connection on the vector bundle $\nu\mathcal{F}$
$$\nabla : \chi(M) \times C^\infty(\nu\mathcal{F}) \longrightarrow C^\infty(\nu\mathcal{F}),$$
by
$$\nabla_X Y = \pi([X_\mathcal{F}, \widetilde{Y}]) + \widehat{\nabla}_{X_\nu} Y.$$
It is called a *Bott connection* of $\mathcal{F}$. A simple calculation, using the integrability of the subbundle $T\mathcal{F}$ and the Jacobi identity, shows that the curvature form $\mathcal{R}$ satifies the equation $\mathcal{R}(X,Y) = 0$ for $X, Y \in \Gamma(\mathcal{F})$; this implies that the vector bundle $\nu\mathcal{F}$ is foliated.

(2) Every flat vector bundle $\mathcal{E} : E \longrightarrow M$ (*i.e.* the transition functions of $\mathcal{E}$ are constant) is a $\mathcal{F}$-bundle.

(3) Let $\mathcal{E} : E \longrightarrow M$ be a $\mathcal{F}$-bundle. Then the dual bundle $\mathcal{E}^*$ and all of its exterior and symmetric powers $\Lambda^*\mathcal{E}^*$ and $\mathcal{S}^*\mathcal{E}^*$ are $\mathcal{F}$-bundles; also $\mathcal{H}^2\mathcal{E} = \{\text{Hermitian forms on } \mathcal{E}\}$ is a $\mathcal{F}$-bundle.

14.3. Transversely Elliptic Operators

Let $\mathcal{E}$ be a $\mathcal{F}$-bundle and denote by $C^\infty(\mathcal{E})$ the space of its global sections; let ∇ denotes the covariant derivative $\chi(M) \times C^\infty(\mathcal{E}) \longrightarrow C^\infty(\mathcal{E})$ associated to the connection.

DEFINITION 14.9. We say that a section $\alpha \in C^\infty(\mathcal{E})$ is *basic* if it satisfies the condition $\nabla_X \alpha = 0$ for every $X \in \Gamma(\mathcal{F})$.

The space $C^\infty(\mathcal{E}/\mathcal{F})$ of basic sections of $\mathcal{E}$ is an A-module. Let $\mathcal{E}$ and $\mathcal{E}'$ be two $\mathcal{F}$-bundles with ranks respectively N and N'.

DEFINITION 14.10. A *basic differential operator* of order $m \in \mathbf{N}$ from $\mathcal{E}$ to $\mathcal{E}'$ is a linear map $D : C^\infty(\mathcal{E}/\mathcal{F}) \longrightarrow C^\infty(\mathcal{E}'/\mathcal{F})$ such that, on local coordinates $(x_1, \ldots, x_d, y_1, \ldots y_n)$, D has the expression

$$D = \sum_{|s|\le m} a_s(y) \frac{\partial^{|s|}}{\partial y_1^{s_1} \ldots \partial y_n^{s_n}} \tag{14.2}$$

where $s = (s_1, \ldots, s_n) \in \mathbf{N}^n$, $|s| = s_1 + \cdots + s_n$ and a_s are $N \times N'$-matrices whose coefficients are basic functions.

The *principal symbol* of D at the point z and the covector $\zeta \in T_z^*M$ is the linear map $\sigma(D)(z,\zeta) : E_z \longrightarrow E'_z$ defined by

$$\sigma(D)(z,\zeta)(\eta) = \sum_{|s|=m} \zeta_1^{s_1} \cdot \cdots \cdot \zeta_n^{s_n} a_s(y)(\eta).$$

We say that D is *transversely elliptic* if $\sigma(D)(z,\zeta)$ is an isomorphism for every $z \in M$ and every transverse covector ζ different from 0. If $\mathcal{F}$ is Riemannian, its conormal bundle $\nu^*\mathcal{F}$ is a $\mathcal{F}$-bundle and is equipped with a foliation $\mathcal{F}^*$. If in addition M is compact, $\sigma(D)(z,\zeta)$ defines an element $[D]$ in the group $K(\nu^*\mathcal{F}, \mathcal{F}^*)$.

A *Hermitian metric* on $\mathcal{E}$ is a positive definite section h of $\mathcal{H}^2\mathcal{E}$. If h is basic we say that $\mathcal{F}$ is a *Hermitian $\mathcal{F}$-bundle.* If $(\mathcal{E}, h)$ is a Hermitian $\mathcal{F}$-bundle and $D : C^\infty(\mathcal{E}/\mathcal{F}) \longrightarrow C^\infty(\mathcal{E}/\mathcal{F})$ is a basic operator of order $m = 2\ell$ we can define a quadratic form A on $\mathcal{E}$ by

$$A_z(\eta) = (-1)^\ell h\big(\sigma(D)(z,\zeta)(\eta), \eta\big)$$

where $\eta \in E_z$. We say that D is *strongly transversely elliptic* if A is positive definite for every $z \in M$ and every transverse covector ζ different from zero. Of course, a strongly transversely elliptic operator is transversely elliptic.

Let $\{\mathcal{E}^r\}$ $(r \in \{0, \cdots, n\})$ be a sequence of Hermitian $\mathcal{F}$-bundles and basic operators $D_r : C^\infty(\mathcal{E}^r/\mathcal{F}) \longrightarrow C^\infty(\mathcal{E}^{r+1}/\mathcal{F})$ such that the sequence
(14.3)

$$0 \longrightarrow C^\infty(\mathcal{E}^0/\mathcal{F}) \xrightarrow{D_0} \cdots \xrightarrow{D_{r-1}} C^\infty(\mathcal{E}^r/\mathcal{F}) \xrightarrow{D_r} C^\infty(\mathcal{E}^{r+1}/\mathcal{F}) \xrightarrow{D_{r+1}} \cdots \xrightarrow{D_{n-1}} C^\infty(\mathcal{E}^n/\mathcal{F}) \longrightarrow 0$$

is a differential complex. Let $\sigma_r = \sigma(D_r)(z,\zeta) : E_z^r \longrightarrow E_z^{r+1}$ denotes the principal symbol of D_r at the point $z \in M$ and the transverse covector ζ. We say that the complex (14.3) is *transversely elliptic* if the sequence

$$0 \longrightarrow E_z^0 \xrightarrow{\sigma_0} \cdots \xrightarrow{\sigma_{r-1}} E_z^r \xrightarrow{\sigma_r} E_z^{r+1} \xrightarrow{\sigma_{r+1}} \cdots \xrightarrow{\sigma_{n-1}} E_z^n \longrightarrow 0$$

is exact for every $z \in M$ and every non zero transverse covector ζ. On each $C^\infty(\mathcal{E}^r/\mathcal{F})$ we can define an inner product given by the formula (14.4) below. Let D^* be the formal of D which is a basic operator from $C^\infty(\mathcal{E}^{r+1}/\mathcal{F})$ to $C^\infty(\mathcal{E}^r/\mathcal{F})$. Then $L_r = DD^* + D^*D$ is a selfadjoint operator on $C^\infty(\mathcal{E}^r/\mathcal{F})$. We can easily show that the differential complex (14.3) is transversely elliptic if, and only if, for every $r \in \{0, \cdots, n\}$, L_r is strongly transversely elliptic.

From now on we suppose that M is compact and connected. Assume that the foliation $\mathcal{F}$ is Riemannian transversely oriented. Let

$$G = \mathrm{SO}(n) \longrightarrow M^\# \xrightarrow{p} M$$

be the principal bundle of the orthonormal direct frames transverse to $\mathcal{F}$. Then, the foliation $\mathcal{F}$ lifts to a transversely parallelizable foliation $\mathcal{F}^\#$ on $M^\#$ of the same dimension and invariant under the action of the group G. Moreover, the leaf closures of $\mathcal{F}^\#$ are the fibres of a locally trivial fibration $F \longrightarrow M^\# \longrightarrow W$ where W is a compact manifold called the *basic manifold* of $\mathcal{F}$ (*cf.* [**Mol88**]). Let $\mathcal{E}^\#$ be the pullback by p of the bundle $\mathcal{E}$; then $\mathcal{E}^\#$ is a G-bundle and a Hermitian $\mathcal{F}^\#$-bundle with respect to a Hermitian metric $h^\#$. The basic sections of $\mathcal{E}$ are canonically identified to basic sections of $\mathcal{E}^\#$ which are invariant under the action of G. In particular, if $f : M \longrightarrow \mathbb{C}$ is a basic function, $f \circ p$ is a basic function on $M^\#$ (with respect to $\mathcal{F}^\#$); moreover $f \circ p$ is invariant by the action of G. Because $f \circ p$ is continuous, it is constant on the leaf closures of $\mathcal{F}^\#$, so it induces a G-invariant C^∞ function on the basic manifold W. We can prove, by the converse process, that any G-invariant C^∞ function on the basic manifold W defines a C^∞ basic function on M; so, the algebra A of basic functions on M is canonically isomorphic to the algebra $A_G(W)$ of functions on W invariant by G. The transverse metric on $M^\#$ (which makes $\mathcal{F}^\#$ Riemannian) induces a Riemannian metric on W for which G acts by isometries. Let μ be the measure on W associated to this metric (μ is a volume form if W is orientable, otherwise it is just a density).

On $C^\infty(\mathcal{E}/\mathcal{F})$ we define an inner product as follows. Let α and β be two elements of $C^\infty(\mathcal{E}/\mathcal{F})$. The function $z \in M \longmapsto h_z(\alpha(z), \beta(z)) \in \mathbb{C}$ is basic; so it defines a G-invariant function $\Theta(\alpha, \beta)$ on W. We set

$$\langle \alpha, \beta \rangle = \int_W \Theta(\alpha, \beta)(w) d\mu(w). \tag{14.4}$$

For any transversely elliptic operator D from a Hermitian $\mathcal{F}$-bundle $\mathcal{E}$ to a Hermitian $\mathcal{F}$-bundle $\mathcal{E}'$, denote by $N(D)$ its kernel D and $R(D)$ its range. Let D^* be the formal adjoint of D; D^* is a basic operator from $\mathcal{E}'$ to $\mathcal{E}$ and it is transversely elliptic.

THEOREM 14.11. *The kernel $N(D)$ of D is finite dimensional, the range $R(D^*)$ of D^* is closed and finite codimensional and we have an orthogonal decomposition*

$$C^\infty(\mathcal{E}/\mathcal{F}) = N(D) \oplus R(D^*). \tag{14.5}$$

The proof of this theorem is very long; it can be found in [**ElK90**]. This theorem says in particular that the operator D is Fredholm and then it has an index defined as usual by the formula

$$\mathrm{ind}_{\mathcal{F}}(D) = \dim \mathrm{Ker} D - \dim \mathrm{Ker} D^* \in \mathbf{Z}. \tag{14.6}$$

PROBLEM 14.12. *Compute this integer in terms of invariants of the bundles $\mathcal{E}$ and $\mathcal{E}'$ and transverse topological invariants of $\mathcal{F}$. More precisely, is there an Atiyah-Singer Index Theorem for a transversely elliptic operator on a Riemannian foliation on a compact manifold?*

14.4. Examples

14.4.1. The basic de Rham complex. We suppose as in Theorem 14.11 that $\mathcal{F}$ is Riemannian of codimension n, transversely oriented and that M is compact. For every $r \in \{0, \cdots, n\}$, let $\mathcal{E}^r$ denotes the vector bundle $\Lambda^r(\nu^*\mathcal{F})$. Then $\mathcal{E}^r$ is a Hermitian $\mathcal{F}$-bundle. Its basic sections are exactly the basic forms of degree r which form a vector space denoted $\Omega^r(M/\mathcal{F})$. The exterior differential $d : \Omega^r(M/\mathcal{F}) \longrightarrow$

$\Omega^{r+1}(M/\mathcal{F})$ is a basic differential operator of order 1. The differential complex (14.7)

$$0 \longrightarrow \Omega^0(M/\mathcal{F}) \xrightarrow{d} \cdots \xrightarrow{d} \Omega^r(M/\mathcal{F}) \xrightarrow{d} \Omega^{r+1}(M/\mathcal{F}) \xrightarrow{d} \cdots \xrightarrow{d} \Omega^n(M/\mathcal{F}) \longrightarrow 0$$

is called the *basic de Rham complex* of $\mathcal{F}$; its homology is the basic cohomology $H^*(M/\mathcal{F})$ we have already defined in section 1.

To make things more simple, we suppose that $\mathcal{F}$ is *homologically orientable*, that is, the vector space $H^n(M/\mathcal{F})$ is non trivial, so it is necessarily one dimensional (*cf.* [**EH86**]). This condition is equivalent to the existence of a (real) volume form on the leaves χ which is $\mathcal{F}$-relatively closed, that is, $d\chi(X_1, \cdots, X_d, Y) = 0$ for $X_1, \cdots, X_d \in \Gamma(\mathcal{F})$ (*cf.* [**Mas92**]). (In that case, we can complete the transverse metric by a Riemannian metric along the leaves to obtain a Riemann metric on the whole manifold for which the leaves are minimal and χ is associated to this metric.) This hypothesis will enable one to define an inner product on $\Omega^r(M/\mathcal{F})$ without using the basic manifold W. As in the classical case we define the Hodge star operator

$$* : \Omega^*(M/\mathcal{F}) \longrightarrow \Omega^*(M/\mathcal{F}) \tag{14.8}$$

in the following way. Let U be an open set on which the foliation is trivial. Let $\theta_1, \cdots, \theta_n$ be real 1-forms such that $(\theta_1, \cdots, \theta_n)$ is an orthonormal basis of the free module $\Omega^1(U/\mathcal{F})$ (over the algebra of basic functions on U). Then define $*$ by

$$*(\theta_{i_1} \wedge \cdots \wedge \theta_{i_r}) = \varepsilon \theta_{j_1} \wedge \cdots \wedge \theta_{j_{n-r}}$$

where $\{j_1, \cdots, j_{n-r}\}$ is the increasing complementary sequence of $\{i_1, \cdots, i_r\}$ in $\{1, \cdots, n\}$ and ε is the signature of the permutation $\{i_1, \cdots, i_r, j_1, \cdots, j_{n-r}\}$. A straightforward calculation shows that $*$ satisfies the identity $** = (-1)^{r(n-r)}\mathrm{id}$. On $\Omega^r(M/\mathcal{F})$ we define a Hermitian product by

$$\langle \alpha, \beta \rangle = \int_M \alpha \wedge \overline{*\beta} \wedge \chi. \tag{14.9}$$

Then it is easy to see that the operator $\delta : \Omega^r(M/\mathcal{F}) \longrightarrow \Omega^{r-1}(M/\mathcal{F})$ defined by the formula $\delta = (-1)^{n(r-1)-1} * d *$ is the formal adjoint of $d : \Omega^{r-1}(M/\mathcal{F}) \longrightarrow \Omega^r(M/\mathcal{F})$ *i.e.* for every $\alpha \in \Omega^{r-1}(M/\mathcal{F})$ and every $\beta \in \Omega^r(M/\mathcal{F})$ we have $\langle d\alpha, \beta \rangle = \langle \alpha, \delta\beta \rangle$. Indeed

$$\begin{aligned} d(\alpha \wedge *\beta \wedge \chi) &= d\alpha \wedge *\beta \wedge \chi + (-1)^{r-1}\alpha \wedge d(*\beta) \wedge \chi + (-1)^{n-1}\alpha \wedge *\beta \wedge d\chi \\ &= d\alpha \wedge *\beta \wedge \chi + (-1)^{(2-r)(r-1)-1}\alpha \wedge *(\delta\beta) \wedge \chi + (-1)^{n-1}\alpha \wedge *\beta \wedge d\chi \\ &= d\alpha \wedge *\beta \wedge \chi - \alpha \wedge *(\delta\beta) \wedge \chi + (-1)^{n-1}\alpha \wedge *\beta \wedge d\chi. \end{aligned}$$

Integrating the two members and using the fact that χ is $\mathcal{F}$-relatively closed, we obtain the desired equality. In the more general case in which the leaves are not minimal, the formula for the adjoint has a correction term involving the mean curvature of the foliation (*cf.* [**ÁL92**], [**Ton97**] or [**PR96**]). Let $\Delta_b : \Omega^r(M/\mathcal{F}) \longrightarrow \Omega^r(M/\mathcal{F})$ be the operator $\Delta_b = \delta d + d\delta$; Δ_b is selfadjoint; it is called the *basic Laplacian* (on the basic r-forms); a simple calculation, using local coordinates, proves that Δ_b is strongly transversely elliptic and therfore the complex (14.7) is transversely elliptic. Let

$$\mathcal{H}^r(M/\mathcal{F}) = \mathrm{Ker}\Delta_b = \{\alpha \in \Omega^r(M/\mathcal{F}) : d\alpha = 0 \text{ and } \delta\alpha = 0\}.$$

An element of $\mathcal{H}^r(M/\mathcal{F})$ is called a *basic harmonic form* (of degree r). Then, applying Theorem 14.11, we obtain

(i) $\dim \mathcal{H}^r(M/\mathcal{F}) < +\infty$;

(ii) *we have orthogonal decompositions*

$$\Omega^r(M/\mathcal{F}) = \mathcal{H}^r(M/\mathcal{F}) \oplus R(\Delta_b) = \mathcal{H}^r(M/\mathcal{F}) \oplus R(d) \oplus R(\delta). \tag{14.10}$$

As a consequence, the basic cohomology $H^r(M/\mathcal{F})$ is finite dimensional and is represented by $\mathcal{H}^r(M/\mathcal{F})$. Moreover the Hermitian map $(\alpha, \beta) \in \Omega^r(M/\mathcal{F}) \times \Omega^{n-r}(M/\mathcal{F}) \longmapsto \int_M \alpha \wedge \overline{\beta} \wedge \chi \in \mathbb{C}$ induces a non degenerate pairing

$$Q : H^r(M/\mathcal{F}) \times H^{n-r}(M/\mathcal{F}) \longrightarrow \mathbb{C} \tag{14.11}$$

i.e. the basic cohomology $H^*(M/\mathcal{F})$ satisfies the Poincaré duality.

These results were originally obtained by B. Reinhart in [**Rei59**] without the assumption that $H^n(M/\mathcal{F})$ is nonzero. But in 1981, Y. Carrière [**Car84**] constructed an example of a foliation whose basic cohomology does not satisfy Poincaré duality; this makes false a part of Reinhart's theorem. One year later F. Kamber and P. Tondeur [**KT82**] proved the same result as B. Reinhart for Riemannian foliations with minimal leaves (this is equivalent by [**Mas92**] to $H^n(M/\mathcal{F}) \neq \{0\}$). We can easily observe that, with this hypothesis, Reinhart's proof is still valid. The general case (without any assumption) was completely established in [**EH86**]. But as we have already pointed, these results are direct consequences of Theorem 3.3.

If $n = 2k = 4\ell$, Q defines a non degenerate quadratic form on $H^k(M/\mathcal{F})$; its signature is called the *signature* of $\mathcal{F}$ and is denoted $\mathrm{Sign}(\mathcal{F})$.

Now let $\mathcal{E}$ and $\mathcal{E}'$ be the vector bundles

$$\mathcal{E} = \bigoplus_{i \geq 0} \Lambda^{2i}(\nu^*\mathcal{F}) \quad \text{and} \quad \mathcal{E}' = \bigoplus_{i \geq 0} \Lambda^{2i+1}(\nu^*\mathcal{F}).$$

They are Hermitian $\mathcal{F}$-bundles and we have

$$C^\infty(\mathcal{E}/\mathcal{F}) = \bigoplus_{i \geq 0} \Omega^{2i}(M/\mathcal{F}) \quad \text{and} \quad C^\infty(\mathcal{E}'/\mathcal{F}) = \bigoplus_{i \geq 0} \Omega^{2i+1}(M/\mathcal{F}).$$

The operator $d + \delta : C^\infty(\mathcal{E}/\mathcal{F}) \longrightarrow C^\infty(\mathcal{E}'/\mathcal{F})$ is basic and transversely elliptic, then it is a Fredholm operator. Its index

$$\mathrm{ind}_{\mathcal{F}}(d + \delta) = \sum_{i=0}^{n} (-1)^i \dim H^i(M/\mathcal{F}). \tag{14.12}$$

is the *basic Euler-Poincaré number* $\chi(M/\mathcal{F})$ of the foliation $\mathcal{F}$. As in the classical case, it is an obstruction to the existence of a nonsingular foliated vector field transverse to $\mathcal{F}$ (*cf.* [**BPR**]).

14.4.2. The basic Dolbeault complex. We suppose that $\mathcal{F}$ is Hermitian and, for simplicity, homologically orientable. Let ν be the complexified normal bundle $\nu\mathcal{F} \otimes_{\mathbb{R}} \mathbb{C}$ of $\nu\mathcal{F}$. Let J be the automorphism of ν associated to the complex structure; J satistfies $J^2 = -\mathrm{id}$ and then has two eigenvalues i and $-i$ with associated eigensubbundles respectively denoted ν^{10} and ν^{01}. We have a splitting $\nu = \nu^{10} \oplus \nu^{01}$ which gives rise to a decomposition

$$\Lambda^r \nu^* = \bigoplus_{p+q=r} \Lambda^{p,q}$$

where $\Lambda^{p,q} = \Lambda^p \nu^{10*} \otimes \Lambda^q \nu^{01*}$. Basic sections of $\Lambda^{p,q}$ are called *basic forms of type* (p,q). They form a vector space denoted $\Omega^{p,q}(M/\mathcal{F})$. We have

$$\Omega^r(M/\mathcal{F}) = \bigoplus_{p+q=r} \Omega^{p,q}(M/\mathcal{F}).$$

The exterior differential decomposes into a sum of two operators

$$\partial : \Omega^{p,q}(M/\mathcal{F}) \longrightarrow \Omega^{p+1,q}(M/\mathcal{F}) \quad \text{and} \quad \overline{\partial} : \Omega^{p,q}(M/\mathcal{F}) \longrightarrow \Omega^{p,q+1}(M/\mathcal{F})$$

as in the classical case of a complex manifold. We have $\overline{\partial}^2 = 0$; so we obtain a differential complex
(14.13)

$$0 \to \Omega^{p,0}(M/\mathcal{F}) \xrightarrow{\overline{\partial}} \cdots \xrightarrow{\overline{\partial}} \Omega^{p,q}(M/\mathcal{F}) \xrightarrow{\overline{\partial}} \Omega^{p,q+1}(M/\mathcal{F}) \xrightarrow{\overline{\partial}} \cdots \xrightarrow{\overline{\partial}} \Omega^{p,n}(M/\mathcal{F}) \to 0$$

called the *basic Dolbeault complex* of $\mathcal{F}$; its homology $H^{p,q}(M/\mathcal{F})$ is the *basic Dolbeault cohomology* of the foliation $\mathcal{F}$: even though the leaf space is bad, it can be considered as a "complex manifold" whose Dolbeault cohomology is $H^{p,*}(M/\mathcal{F})$!

The star operator defined in (14.8) induces an isomorphism $* : \Omega^{p,q}(M/\mathcal{F}) \longrightarrow \Omega^{n-q,n-p}(M/\mathcal{F})$. Moreover the restriction of the operator δ to the space $\Omega^{p,q}(M/\mathcal{F})$ decomposes into a sum of two operators $\delta' = - * \overline{\partial} *$ and $\delta'' = - * \partial *$ respectively of types $(-1,0)$ and $(0,-1)$. We can easily verify that δ'' is the formal adjoint of $\overline{\partial}$ for the inner product (14.9). Then the operator $\Delta''_b = \delta''\overline{\partial} + \overline{\partial}\delta''$ is selfadjoint; a simple computation in local coordinates, like for the basic Laplacian, shows that Δ''_b is strongly transversely elliptic and that the complex (14.13) is transversely elliptic. Let

$$\mathcal{H}^{p,q}(M/\mathcal{F}) = \mathrm{Ker}\Delta''_b = \{\alpha \in \Omega^{p,q}(M/\mathcal{F}) : \overline{\partial}\alpha = 0 \text{ and } \delta''\alpha = 0\}.$$

Applying Theorem 3.3, we obtain

(i) dim $\mathcal{H}^{p,q}(M/\mathcal{F}) < +\infty$;
(ii) *we have orthogonal decompositions*

$$\Omega^{p,q}(M/\mathcal{F}) = \mathcal{H}^{p,q}(M/\mathcal{F}) \oplus R(\Delta''_b) = \mathcal{H}^{p,q}(M/\mathcal{F}) \oplus R(\overline{\partial}) \oplus R(\delta''). \tag{14.14}$$

Consequently, the basic Dolbeault cohomology $H^{p,q}(M/\mathcal{F})$ is finite dimensional and is represented by $\mathcal{H}^{p,q}(M/\mathcal{F})$. Moreover the star operator induces an unitary isomorphism (of real vector spaces) $\overline{*} : \alpha \in \mathcal{H}^{p,q}(M/\mathcal{F}) \longmapsto \overline{*\alpha} \in \mathcal{H}^{n-p,n-q}(M/\mathcal{F})$ and then an isomorphism

$$\overline{*} : H^{p,q}(M/\mathcal{F}) \longrightarrow H^{n-p,n-q}(M/\mathcal{F}) \tag{14.15}$$

i.e. the basic Dolbeault cohomology $H^{*,*}(M/\mathcal{F})$ satisfies the Serre duality.

Suppose now that $\mathcal{F}$ is transversely Kählerian with Kähler form ω (it is a basic differential form of degree 2; it is closed and non degenerate). In this case, we can prove that $\Delta_b = 2\Delta''_b$. Because of the decomposition

$$\Omega^r(M/\mathcal{F}) = \bigoplus_{p+q=r} \Omega^{p,q}(M/\mathcal{F}),$$

every basic differential r-form can be uniquely writen $\alpha = \sum_{p+q=r} \alpha_{pq}$ where $\alpha_{pq} \in \Omega^{p,q}(M/\mathcal{F})$. Then we have the following assertions.

(iii) *α is Δ_b-harmonic if, and only if, each component α_{pq} is Δ''_b-harmonic. So we have a direct decomposition*

$$H^r(M/\mathcal{F}) = \bigoplus_{p+q=r} H^{p,q}(M/\mathcal{F}). \tag{14.16}$$

(iv) *The complex conjugacy induces an isomorphism (of real vector spaces)*

$$H^{p,q}(M/\mathcal{F}) \simeq H^{q,p}(M/\mathcal{F}).$$

(v) *For every odd $r \in \{0, \cdots, 2n\}$, the dimension of the space $H^r(M/\mathcal{F})$ is even. In particular if $n = 1$ we have $b_1(M/\mathcal{F}) = 2\dim H^{01}(M/\mathcal{F})$.*

The integer dim $H^{01}(M/\mathcal{F})$ will be denoted $g(\mathcal{F})$ and called the *genus* of the foliation $\mathcal{F}$. It is similar to the genus of a compact Riemann surface; it counts the number of linearly independant basic holomorphic 1-forms.

(vi) *Por every $p \in \{0, \cdots, n\}$ the differential form $\omega^p = \omega \wedge \cdots \wedge \omega$ (wedge product p times) is harmonic. So, the space $H^{p,p}(M/\mathcal{F})$ is non zero.*

14.5. Signature and Index

The manifold M is supposed to be compact oriented and equipped with a Riemannian foliation $\mathcal{F}$ of even codimension $n = 2k$ and with minimal leaves. Let ν denotes the complexified of the normal bundle of $\mathcal{F}$. Let $\tau : \Omega^*(M/\mathcal{F}) \longrightarrow \Omega^*(M/\mathcal{F})$ be the isomorphism defined on $\alpha \in \Omega^r(M/\mathcal{F})$, by $\tau(\alpha) = i^{r(r-1)+k} * \alpha$. It satisfies $\tau^2 =$id *i.e.* τ is an involution; so it admits 1 and -1 as eingenvalues. Let $\Omega^*_+(M/\mathcal{F})$ and $\Omega^*_-(M/\mathcal{F})$ be the eigensubspaces associated respectively to these eigenvalues. Then we have a decomposition

$$\Omega^*(M/\mathcal{F}) = \Omega^*_+(M/\mathcal{F}) \oplus \Omega^*_-(M/\mathcal{F}).$$

Let $D = d + \delta$. It is clear that D is a transversely elliptic selfadjoint operator and its kernel $N(D)$ is identified, by Theorem 3.3, to $H^*(M/\mathcal{F})$. It is obvious to see that $D\tau = -\tau D$. Let

$$D_+ : \Omega^*_+(M/\mathcal{F}) \longrightarrow \Omega^*_-(M/\mathcal{F}) \text{ and } D_- : \Omega^*_-(M/\mathcal{F}) \longrightarrow \Omega^*_+(M/\mathcal{F})$$

be the restrictions of D respectively to the spaces $\Omega^*_+(M/\mathcal{F})$ and $\Omega^*_-(M/\mathcal{F})$. These operators are transversely elliptic and are adjoint to each other. By Theorem 3.3, the operators D_+ and D_- are Fredholm. Then each of them admits an index.

DEFINITION 14.13. D_+ is called the *signature operator* of the foliation $\mathcal{F}$; its index is the integer

$$\mathrm{ind}_{\mathcal{F}}(D_+) = \dim\, (\mathrm{Ker}\ D_+) - \dim\, (\mathrm{Ker}\ D_-). \tag{14.17}$$

Let $\Delta_b = D^2$ be the basic Laplacian; it is immediate to see that Δ_b and D have the same kernel which, by the basic Hodge Theorem (*cf.* subsection 4.1), is $\mathcal{H}^*(M/\mathcal{F})$ and which is identified to the space $H^*(M/\mathcal{F})$. Because $D\tau = -\tau D$, τ restricts to an involution $\tau : \mathrm{Ker}D \longrightarrow \mathrm{Ker}D$ which defines decompositions

$$H^*(M/\mathcal{F}) = H^*_+(M/\mathcal{F}) \oplus H^*_-(M/\mathcal{F}).$$
$$\mathcal{H}^*(M/\mathcal{F}) = \mathcal{H}^*_+(M/\mathcal{F}) \oplus \mathcal{H}^*_-(M/\mathcal{F}).$$

So

$$\mathrm{ind}_{\mathcal{F}}(D_+) = \dim\, \mathcal{H}^*_+(M/\mathcal{F}) - \dim\, \mathcal{H}^*_-(M/\mathcal{F}) = \dim\, H^*_+(M/\mathcal{F}) - \dim\, H^*_-(M/\mathcal{F}).$$

Let V_r be the space $H^r(M/\mathcal{F}) \oplus H^{2k-r}(M/\mathcal{F})$ where $r \in \{0, \dots, k-1\}$. Then the subspaces $H^r(M/\mathcal{F})$ and $H^{2k-r}H(M/\mathcal{F})$ are permuted by the action of τ on $H^*(M/\mathcal{F})$; so V_r is globally invariant by τ. This gives

$$\mathrm{ind}_{\mathcal{F}}(D_+) = \dim \mathcal{H}^k_+(M/\mathcal{F}) - \dim \mathcal{H}^k_-(M/\mathcal{F}) = \dim H^k_+(M/\mathcal{F}) - \dim H^k_-(M/\mathcal{F}).$$

We will regard indifferently τ as an involution on the basic cohomology $H^*(M/\mathcal{F})$ or on the space $\mathcal{H}^*(M/\mathcal{F})$ of basic harmonic forms.

REMARKS 14.14. i)If k is odd we have $\tau = i*$, then $\tau(\alpha) = -\overline{\tau(\alpha)}$ where $\alpha \in \Omega^k(M/\mathcal{F})$; this implies $** = -\mathrm{id}$. The morphism $\tau : \mathcal{H}^k_+(M/\mathcal{F}) \longrightarrow \mathcal{H}^k_-(M/\mathcal{F})$ induced by τ is an isomorphism (of real vector spaces); consequently $\mathrm{ind}_{\mathcal{F}}(D_+) = 0$.

ii)Suppose $n = 2k = 4\ell$. Recall that we have defined in the subsection 4.1 a non degenerate quadratic form Q on $H^k(M/\mathcal{F}) \simeq \mathcal{H}^k(M/\mathcal{F})$ whose signature is $\mathrm{Sign}(\mathcal{F})$. We verify easily that $\tau = *$ on $\mathcal{H}^k(M/\mathcal{F})$. Then, for non zero basic harmonic k-forms $\alpha \in \mathcal{H}^k_+(M/\mathcal{F})$ and $\beta \in \mathcal{H}^k_-(M/\mathcal{F})$, we have

$$Q(\alpha, \alpha) = \int_M \alpha \wedge \overline{\alpha} \wedge \chi = \int_M \alpha \wedge \tau\overline{\alpha} \wedge \chi = \int_M \alpha \wedge *\overline{\alpha} \wedge \chi = \langle \alpha, \alpha \rangle > 0$$

and

$$Q(\beta, \beta) = \int_M \beta \wedge \overline{\beta} \wedge \chi = -\int_M \beta \wedge \tau\overline{\beta} \wedge \chi = -\int_M \beta \wedge *\overline{\beta} \wedge \chi = -\langle \beta, \beta \rangle < 0$$

i.e. Q is positive definite on $\mathcal{H}^k_+(M/\mathcal{F})$ and negative definite on $\mathcal{H}^k_-(M/\mathcal{F})$. So, if we set, by convention, $\mathrm{Sign}(\mathcal{F}) = 0$ if the integer k is odd, we finally obtain the following theorem.

THEOREM 14.15. *Let $\mathcal{F}$ be a homologically orientable Riemannian foliation of even codimension $n = 2k$ on a compact connected oriented manifold M. Then*

$$\mathit{Sign}(\mathcal{F}) = \mathit{ind}_{\mathcal{F}}(D_+).$$

14.6. As a conclusion to this paper

In [**EN93**] it was shown that the basic cohomology $H^*(M/\mathcal{F})$ is invariant by homeomorphism in the category of Riemannian complete foliations. So the signature of $\mathcal{F}$ and the integers defined respectively in (14.12) and 14.4.2(v) are topological invariants. This reinforces the idea that it is certainly interesting to attack 14.12.

AZIZ EL KACIMI ALAOUI, LAMATH, LE MONT HOUY, UNIVERSITÉ DE VALENCIENNES, 59313 VALENCIENNES CEDEX 9, FRANCE.
E-mail address: aziz.elkacimi@univ-valenciennes.fr

CHAPTER 15

Reilly-type Spinorial Inequalities

NICOLAS GINOUX[1]

ABSTRACT. We obtain a new upper bound for the smallest eigenvalue of the classical Dirac operator of an hypersurface. We show that the geodesic spheres are the only hypersurfaces in the hyperbolic space on which this upper bound is sharp.

Contents

Introduction

Let $(\widetilde{M}, g)$ be an $m+1$-dimensional Riemannian spin manifold and M be an immersed (closed) compact connected and oriented hypersurface in $\widetilde{M}$. We assume $m \geq 2$. We consider the classical Dirac operator acting on spinors on M for the induced metric and spin structure, and we want to obtain upper bounds for the smallest eigenvalue of this operator in terms of extrinsic quantities given by the immersion.

For the Laplace operator acting on functions on M, $\triangle$, with smallest positive eigenvalue $\lambda_1(\triangle)$, the following results are well-known ([**Rei77b**]): if H is the mean curvature of M in $\widetilde{M}$, then

$$\begin{aligned} \lambda_1(\triangle) &\leq \frac{m}{\mathrm{Vol}(M)} \int_M H^2 v_g & \text{if } \widetilde{M} = \mathbb{R}^{m+1}, \\ \text{and } \lambda_1(\triangle) &\leq \frac{m}{\mathrm{Vol}(M)} \int_M (H^2+1) v_g & \text{if } \widetilde{M} = S^{m+1}. \end{aligned}$$

[1]The author thanks the École Doctorale IAEM-Lorraine for its support, as well as the Center for Advanced Mathematical Sciences and the American University of Beirut for their hospitality.

These upper bounds were obtained using the coordinate functions of the immersion as "test-functions" in the Rayleigh-quotient.
If the ambient manifold $\widetilde{M}$ is the hyperbolic space $\mathbb{H}^{m+1}$, no sharp estimate in terms of the L^2-norm of the mean curvature was available until El Soufi and Ilias proved in 1992 ([**SI92**], Théorème 1):

$$\lambda_1(\triangle) \leq \frac{m}{\mathrm{Vol}(M)} \int_M (H^2 - 1) v_g,$$

with equality only achieved for geodesic spheres. This was discovered using the preceding technique and the fact that all the model spaces can be *conformally embedded* in the round sphere of dimension $m+1$.

Analogous results to Reilly's have been proved for the smallest eigenvalue $\lambda_1(D)$ of the Dirac operator D of M by Bär in 1998 ([**Bär98**], Corollaries 4.2 and 4.3):

$$\begin{aligned} (\lambda_1(D))^2 &\leq \frac{m^2}{4\mathrm{Vol}(M)} \int_M H^2 v_g & \text{if } \widetilde{M} = \mathbb{R}^{m+1}, \\ \text{and } (\lambda_1(D))^2 &\leq \frac{m^2}{4\mathrm{Vol}(M)} \int_M (H^2+1) v_g & \text{if } \widetilde{M} = S^{m+1}. \end{aligned}$$

Both results rely on the existence of *real* Killing spinors on the ambient manifold, which are used as "test-spinors" in the Rayleigh-quotient. Refining this argument, Agricola and Friedrich could prove that, for a surface $M^2 \hookrightarrow \mathbb{R}^3$, one has ([**AF99**]):

$$(\lambda_1(D))^2 \leq \lambda_1(\triangle + H^2),$$

where $\lambda_1(\triangle + H^2)$ is the smallest eigenvalue of the Schrödinger operator $\triangle + H^2$. Unfortunately, the hyperbolic space doesn't admit real but *imaginary* Killing spinors, which don't have as nice properties as the real ones for the computation of the Rayleigh-quotient. In this case, Bär obtained ([**Bär98**], Corollary 4.5):

$$|\lambda_1(D)| \leq \frac{m}{2}(1 + ||H||_\infty),$$

which is not sharp, since inequality is strict for geodesic spheres of radius r, even if we have:

$$\frac{2|\lambda_1(D)|}{m(1+||H||_\infty)} \underset{r \to 0}{\longrightarrow} 1.$$

We shall prove that an analogous upper bound to El Soufi and Ilias' one can be derived for the Dirac operator.

15.1. Restriction of spinors to a hypersurface

Let $\iota : M \longrightarrow (\widetilde{M}, g)$ be an immersion. Denote by g the restriction of g on M. Let $\nu : M \longrightarrow NM$ be the normal unitary section such that, for every positively-oriented orthonormal basis $(X_1, \ldots, X_m)$ of TM, $(X_1, \ldots, X_m, \nu)$ is a positively-oriented orthonormal basis of $T\widetilde{M}_{|M}$.

Fact 1: The vector field ν induces a spin structure on M.

Let $\Sigma\widetilde{M}$ be the spinor bundle of $(\widetilde{M}, g)$ and $\Sigma\widetilde{M}|_M$ be its restriction on M. If ΣM is the spinor bundle of (M, g), set

$$\Sigma := \begin{cases} \Sigma M & \text{if m is even} \\ \Sigma M \oplus \Sigma M & \text{if m is odd} \end{cases}$$

FACT 2: There exists an isomorphism

$$\begin{array}{ccc} \Sigma & \longrightarrow & \Sigma\widetilde{M}|_M \\ \psi & \longmapsto & \Psi, \end{array}$$

such that:

(1) $|\Psi| = |\psi|$ where $|\Psi|^2 = <\Psi, \Psi>$ and $< \cdot, \cdot >$ is a natural hermitian inner product on $\Sigma\widetilde{M}$.
(2) For every vector field X on M and for every section ψ of Σ, the spinor field $X \underset{M}{\cdot} \psi$ is mapped by this isomorphism onto $X \cdot \nu \cdot \Psi$, where "$\underset{M}{\cdot}$" (resp. "$\cdot$") denotes Clifford multiplication on Σ (resp. on $\Sigma\widetilde{M}|_M$).
(3) If ∇ (resp. $\widetilde{\nabla}$) is the Levi-Civita connection on TM (resp. on $T\widetilde{M}$), denote by ∇ (resp. $\widetilde{\nabla}$) the associated Levi-Civita connection on Σ (resp. $\Sigma\widetilde{M}|_M$). Then, identifying $\nabla_X\psi$ with its image in $\Gamma(\Sigma\widetilde{M}|_M)$, we have:
Gauß relation:

$$\widetilde{\nabla}_X\Psi = \nabla_X\psi + \frac{1}{2}B(X) \cdot \nu \cdot \Psi,$$

where $B(X) = -\widetilde{\nabla}_X\nu$ is the Weingarten endomorphism.

Let $(X_j)_{1\leq j\leq m}$ be a local orthonormal basis of TM, and define the following operators:

$$* \quad \begin{array}{rcl} D : \Gamma(\Sigma) & \longrightarrow & \Gamma(\Sigma) \\ \psi & \longmapsto & D\psi := \sum_{j=1}^{m} X_j \underset{M}{\cdot} \nabla_{X_j}\psi \end{array}$$

$$D = \begin{cases} D_M & \text{if m is even} \\ D_M \oplus -D_M & \text{if m is odd} \end{cases}$$

where D_M is the classical Dirac operator of M.

$$* \quad \begin{array}{rcl} \widetilde{D} : \Gamma(\Sigma\widetilde{M}|_M) & \longrightarrow & \Gamma(\Sigma\widetilde{M}|_M) \\ \psi & \longmapsto & \widetilde{D}\Psi := \sum_{j=1}^{m} X_j \cdot \widetilde{\nabla}_{X_j}\Psi, \end{array}$$

the Witten operator. Gauß relation yields, for every section ψ of Σ,

$$\widetilde{D}\Psi = \nu \cdot \Big(D\psi - \frac{mH}{2}\Psi\Big),$$

where $H = \frac{1}{m}tr(B)$ is the mean curvature of ι. Then we can obtain the following relation between the squares of the operators D and $\widetilde{D}$, which will be very useful in our computations:

$$\boxed{\widetilde{D}^2\Psi = D^2\psi - \tfrac{m}{2}dH \cdot \nu \cdot \Psi - \tfrac{m^2H^2}{4}\Psi.} \tag{15.1}$$

FACT 3: If $\overline{g} = e^{2u}g$, $u \in C^\infty(M)$, is a conformal change of the metric on M, there exists an isometry

$$\begin{array}{ccc} \Sigma M & \longrightarrow & \Sigma_{\overline{g}} M \\ \psi & \longmapsto & \overline{\psi} \end{array}$$

such that:

- $\overline{X \cdot \psi} = e^{-u} X \,\overline{\cdot}\, \overline{\psi}$ ("$\overline{\cdot}$"is Clifford multiplication on $\Sigma_{\overline{g}} M$)
- If $\overline{\nabla} := \nabla^{\overline{g}}$, then $\overline{\nabla}_X \overline{\psi} = \overline{\nabla_X \psi} - \frac{1}{2}\overline{X \cdot du \cdot \psi} - \frac{1}{2} du(X)\overline{\psi}$.

The same is true for $\widetilde{M}$. Hence we obtain:

$$\boxed{\widetilde{D}_{\overline{g}} \overline{\Psi} = e^{-u}\Big(\overline{\widetilde{D}\Psi} + \tfrac{m-1}{2}\overline{du \cdot \Psi} + \tfrac{m}{2} du(\nu)\overline{\nu \cdot \Psi}\Big)} \tag{15.2}$$

where $du = d(u_{|M})$ and $\widetilde{D}_{\overline{g}}$ is the Witten operator of $(M, \overline{g})$.

15.2. Extrinsic upper bound in case the ambient manifold admits a twistor-spinor

15.2.1. General estimate. The manifold M being assumed to be compact, the spectrum of D is real and forms a discrete sequence of eigenvalues,

$$0 \leq |\lambda_1| \leq \ldots \leq |\lambda_k| \leq |\lambda_{k+1}| \leq \ldots \longrightarrow +\infty.$$

In order to obtain upper bounds for the smallest eigenvalues, we shall apply the following well-known proposition:

PROPOSITION 15.1 (Min-Max principle). *For every integer $k \geq 1$,*

$$(\lambda_k)^2 = \underset{E_k}{Min} \left\{ \underset{\psi \in E_k \setminus \{0\}}{Max} \frac{\int_M < D^2\psi \,, \psi > v_g}{\int_M |\psi|^2 v_g} \right\},$$

where the minimum is taken over all the k-dimensional vector subspaces E_k of $\Gamma(\Sigma)$.

Let's state our main theorem:

THEOREM 15.2. *Assume that, for $\overline{g} = e^{2u}g$, $(\widetilde{M}, \overline{g})$ admits a real Killing spinor. Then*

$$(\lambda_1)^2 \leq \frac{m^2}{4\mathrm{Vol}(M)} \int_M (H^2 + R(\iota))v_g + \frac{1}{4\mathrm{Vol}(M)} ||du||^2_{L^2(M)}, \tag{15.3}$$

where $R(\iota) = \frac{1}{m(m-1)} \sum_{1 \leq i \neq j \leq m} \widetilde{K}(X_i, X_j)$ and $\widetilde{K}$ is the sectional curvature of $(\widetilde{M}, g)$.

PROOF. The assumption says that there exists a section $\overline{\Psi}$ of $\Gamma(\Sigma_{\overline{g}}\widetilde{M})$ and a real constant α such that, for every vector field X on $\widetilde{M}$,

$$\widetilde{\nabla}^{\overline{g}}_X \overline{\Psi} = \alpha X \,\overline{\cdot}\, \overline{\Psi}.$$

Let's compute the Rayleigh-quotient $\frac{\int_M < D^2\psi, \psi > v_g}{\int_M |\psi|^2 v_g}$. Formulas (15.1) and (15.2) yield

$$D^2\psi = \frac{m^2}{4} e^{2u}(H^2_{\overline{g}} + 4\alpha^2)\Psi + \frac{(m-1)^2}{4}|du|^2\Psi - \frac{m-1}{2}(\triangle u)\Psi + X_0 \cdot \Psi + Y_0 \cdot \nu \cdot \Psi,$$

where $H_{\overline{g}}$ is the mean curvature of $\iota : M \longrightarrow (\widetilde{M}, \overline{g})$, and X_0 and Y_0 are two vector fields on M that can be given explicitly in terms of u, α and H (see [**Gin02**]). Note that the term $4\alpha^2$ is just the term $R(\iota, \overline{g})$ for $\iota : M \longrightarrow (\widetilde{M}, \overline{g})$.
Now we know from El Soufi and Ilias that the quantity $H^2 + R(\iota)$ behaves kindly when rescaling the metric ([**SI92**], Proposition 2):

$$e^{2u}(H_{\overline{g}}^2 + 4\alpha^2) = H^2 + R(\iota) - \frac{m-2}{m}|du|^2 + \frac{2}{m}\triangle u.$$

We deduce from this formula that

$$\widetilde{D^2\psi} = \Big\{\frac{m^2}{4}(H^2 + R(\iota)) + \frac{1}{4}|du|^2 + \frac{1}{2}\triangle u\Big\}\Psi + X_0 \cdot \Psi + Y_0 \cdot \nu \cdot \Psi.$$

If we take the inner product of this equality with Ψ, we obtain

$$Re(< D^2\psi, \psi >) = \Big\{\frac{m^2}{4}(H^2 + R(\iota)) + \frac{1}{4}|du|^2 + \frac{1}{2}\triangle u\Big\}|\psi|^2,$$

since $< X_0 \cdot \Psi, \Psi >$ and $< Y_0 \cdot \nu \cdot \Psi, \Psi >$ are purely imaginary numbers. Besides, Ψ being a *real* Killing spinor, its norm is *constant* on $\widetilde{M}$, and therefore on M. Integrating the preceding equality on M and applying the Min-Max principle lead straightforward to inequality (15.3). □

REMARKS 15.3.

(1) If $(\widetilde{M}, g)$ admits a real Killing spinor, then $u = 0$, $R(\iota) = 4\alpha^2$ and inequality (15.3) is just Bär's inequality in [**Bär98**] (theorem 4.1) .
(2) If $(\widetilde{M}, \overline{g})$ admits more than one Killing spinor for the constant α (for example, if $(\widetilde{M}, g)$ is conformally equivalent to an open subset of (S^{m+1}, can)), then more than one eigenvalue can be bounded by the right-hand term in (15.3).

Theorem 15.2 can be applied to manifolds admitting twistor-spinors without zeros.

DEFINITION 15.4. A non-zero section Ψ of $\Sigma\widetilde{M}$ is called a *twistor-spinor* if and only if for every vector field X on $\widetilde{M}$,

$$\widetilde{\nabla}_X \Psi + \frac{1}{m+1} X \cdot D_{\widetilde{M}}\Psi = 0,$$

where $D_{\widetilde{M}}$ is the classical Dirac operator of $\widetilde{M}$.

One can prove that the set of twistor-spinors is a finite-dimensional vector space, and that every non-trivial twistor-spinor can only vanish on a discrete subset of $\widetilde{M}$. Nowhere-vanishing twistor-spinors deserve a particular attention concerning our problem:

LEMMA 15.5 (Lichnerowicz, [**Lic90b**]). *Assume that there exists a twistor-spinor Ψ without zeros on $(\widetilde{M}, g)$. Then for the conformal change of the metric given by $\overline{g} = \frac{1}{|\Psi|^4} g$, the manifold $(\widetilde{M}, \overline{g})$ admits a real Killing spinor.*

Hence we obtain the following corollary:

COROLLARY 15.6. *Assume that $(\widetilde{M}, \overline{g})$ admits a twistor-spinor Ψ without zeros. Then*

$$(\lambda_1)^2 \le \frac{m^2}{4\mathrm{Vol}(M)} \int_M (H^2 + R(\iota)) v_g + \frac{1}{4\mathrm{Vol}(M)} \Big|\Big| \frac{d|\psi|^2}{|\psi|^2} \Big|\Big|_{L^2(M)}^2.$$

An interesting question is to know whether the estimate obtained in theorem 2.2 is sharp or not. It can be proved that, if the manifold $(\widetilde{M},\overline{g})$ admits a *maximal* number of Killing spinors for a *non-zero* real constant α, i.e. if the vector bundle $\Sigma\widetilde{M}$ can be trivialized by such spinors, then equality is achieved in (15.3) if and only if $du=0$ and $H^2+R(\iota)$ is constant and equal to $\frac{4(\lambda_1)^2}{m}$ ([**Gin02**], Theorem 3). In the case where $(\widetilde{M},g)$ admits a real Killing spinor ($u=0$), this just means that equality in (15.3) implies H being constant, which is a well-known fact (see [**Bär98**]). We will see in next section that, for manifolds $\widetilde{M}$ admitting imaginary Killing spinors, such as the hyperbolic space, the condition $du=0$ is not empty and there exist hypersurfaces realizing equality in (15.3).

15.2.2. Application to hypersurfaces of manifolds admitting imaginary Killing spinors. Assume that there exists a Killing spinor Ψ for an *imaginary* constant α on $(\widetilde{M},g)$, i.e. a non-zero section Ψ of $\Sigma\widetilde{M}$ such that, for every vector field X on $\widetilde{M}$,

$$\widetilde{\nabla}_X\Psi=\alpha X\cdot\Psi.$$

Such a spinor is obviously a twistor-spinor, and has no zeros since it's a parallel section for the covariant derivative $\sum_j X_j^*\otimes(\widetilde{\nabla}_{X_j}-\alpha X_j\cdot)$ on $\Sigma\widetilde{M}$. It can be proved that if such a spinor exists on $\widetilde{M}$, then $(\widetilde{M},g)$ is Einstein with scalar curvature equal to $4m(m+1)\alpha^2$; hence $R(\iota)=4\alpha^2=-4|\alpha|^2$. Applying theorem 1, we obtain for a hypersurface M in $\widetilde{M}$:

$$(\lambda_1)^2\le\frac{m^2}{4\mathrm{Vol}(M)}\int_M(H^2-4|\alpha|^2)v_g+\frac{1}{4\mathrm{Vol}(M)}\|\frac{d|\psi|^2}{|\psi|^2}\|^2_{L^2(M)}.\tag{15.4}$$

Note that this inequality depends on Ψ, which is a given spinor on the ambient manifold, and whose norm can be explicitly computed in certain cases (see [**Bau89a**] and example).

If one assumes besides that, for the conformal change of metric as above, there are real non-parallel Killing spinors in maximal number, then equality in (15.4) is in fact equivalent to $d|\psi|^2=0$, i.e. $|\psi|$ is constant on M. One can then prove the following proposition (see [**Gin02**] or [**Bau89b**]):

PROPOSITION 15.7. *Assume that there exists an imaginary Killing spinor Ψ on $\widetilde{M}$ such that $|\psi|$ is contant on M. Then the immersion ι is totally umbilical with constant mean curvature and the manifold (M,g) admits a real Killing spinor. Besides, we have*

$$\frac{mS}{4(m-1)}=(\lambda_1)^2=-m^2|\alpha|^2+\frac{m^2H^2}{4},$$

where S is the scalar curvature of (M,g).

EXAMPLE 15.8. Let's consider hypersurfaces of the hyperbolic space $(\mathbb{H}^{m+1},g)$, which can be seen as the upper half-sphere sitting in $\mathbb{R}^{m+2}$ with metric $g=\frac{g_0}{x^2_{m+2}}$, where g_0 is the standard metric on S^{m+1}. One can compute explicitly $|du|^2$ on a hypersurface M:

$$|du|^2=g_0(e^T_{m+2},e^T_{m+2}),$$

where e_{m+2}^T is the orthogonal projection of the vector $e_{m+2} = (0, \dots, 0, 1)$ on the tangent space of M. Therefore we have in this case:

$$(\lambda_1)^2 \leq \frac{m^2}{4\mathrm{Vol}(M)} \int_M (H^2 - 1)v_g + \frac{1}{4\mathrm{Vol}(M)} \int_M g_0(e_{m+2}^T, e_{m+2}^T)v_g,$$

and equality holds if and only if the function x_{m+2} is constant on M, i.e. M is a geodesic sphere centered at e_{m+2}. Since the group of isometries of $(\mathbb{H}^{m+1}, g)$ acts transitively on $\mathbb{H}^{m+1}$, and since positive isometries between simply-connected manifolds don't change the spin-structure, we can deduce that every geodesic sphere in $\mathbb{H}^{m+1}$ satisfies:

$$(\lambda_1)^2 = \frac{m^2}{4}(H^2 - 1).$$

REMARK 15.9. If the ambient manifold admits an imaginary Killing spinor Ψ for the constant α, one can directly prove the following estimate:

$$(\lambda_1)^2 \leq -m^2|\alpha|^2 + \frac{m^2 C_\Psi}{4\mathrm{Vol}(M)} \int_M H^2 v_g,$$

where $C_\Psi = \frac{\underset{M}{\mathrm{Sup}}(|\psi|^2)}{\underset{M}{\inf}(|\psi|^2)}$. This can be done computing the Rayleigh-quotient with ψ and using relation (15.1). Besides, equality is achieved if and only if $|\psi|$ is constant on M.

NICOLAS GINOUX, UNIVERSITÄT POTSDAM, INSTITUT FÜR MATHEMATIK, PF 60 15 53, D-14415 POTSDAM, GERMANY.
E-mail address: ginoux@math.uni-potsdam.de

CHAPTER 16

Dirac Operators and Hypersurfaces

SEBASTIÁN MONTIEL[1]

ABSTRACT. In this survey paper, upper and lower bounds of the spectrum of the scalar Laplacian and that of the Dirac operator on hypersurfaces are given. Different boundary conditions for the Dirac operator are examined. As a consequence, some geometric results for hypersurfaces are obtained.

Contents

16.1. Introduction

This paper is an expanded version of my talk at the Workshop *Dirac Operators: Yesterday and Today* held at the American University of Beirut in the summer of 2001. It aims to present some ideas and results appearing as interaction field between the topic of the Workshop, Dirac operators, and the Geometry of Submanifolds. Most of the results are part of different joint works with Oussama Hijazi, Xiao Zhang and my student Antonio Roldán at the University of Granada.

Also I would like to thank the organizers of the conference, especially O. Hijazi and A. Chamseddine, for their hospitality and for giving us the opportunity to enjoy such good mathematics in a such beautiful country.

16.2. Spectral Results on the Scalar Laplacian

First of all I want to mention that spectral properties of submanifolds in Riemannian manifolds have been extensively studied since the seventies. These spectral

[1]Research partially supported by a MCyT grant No. BFM2001-2967.

properties are usually referred to the Laplace operator associated with the metric on the submanifold induced from the ambient Riemannian structure. They mainly deal with two classes of problems:

First, obtaining information about the spectrum of the Laplacian on a compact submanifold M immersed in a given Riemannian ambient space $\overline{M}$, which is in general, a Riemannian manifold with a well-behaved curvature: Euclidean space, sphere, projective space and other symmetric spaces. One looks for information given in terms of the extrinsic geometry of the submanifold, that is, in terms of its second fundamental form, its mean curvature, etc. From a physical point of view, one tries to hear the shape of the submanifold (see [**Kac66, GWW92**]), that is, one would like to know the relation between the pure sounds given off by the submanifold and the way in which it is immersed in the ambient space.

Second, characterizing some distinguished submanifolds (with a nice geometry or second fundamental form) by a limiting behaviour of eigenvalues, multiplicities or eigenfunctions of natural differential operators.

To be precise two examples of such results will be stated. The first result is due to Bleecker and Weiner in 1977 and, under a different form, to Reilly [**Rei77b**] and to Chen [**Che79**].

THEOREM 16.1 ([**BW76**]). *The first non-zero eigenvalue λ_1 of the Laplacian Δ of an n-dimensional compact submanifold M immersed into the $(n+k)$-dimensional Euclidean space $\mathbb{R}^{n+k}$ satisfies the following inequality*

$$\lambda_1 \leq \frac{n}{\mathrm{vol}(M)} \int_M |H|^2 \, dV,$$

where $\mathrm{vol}(M)$ *is the volume of the submanifold and H its mean curvature vector, that is, the average of the second fundamental form at each point. The equality occurs if and only if the submanifold lies minimally in a (round) hypersphere.*

Note that this result, in the case of least codimension, characterizes the round hyperspheres among all compact hypersurfaces in the Euclidean space. It is worth to point out that, from Theorem 16.1, one can immediately deduce a corresponding result for submanifolds of a sphere. The analogue for the hyperbolic space is not so easy. In fact it was not obtained until 1992 by El Soufi and Ilias in [**SI92**].

Another interesting example is the following : It is an easy and old observation by Takahashi [**Tak66**] that the position vector function $\varphi : M \to \mathbb{S}^n \subset \mathbb{R}^{n+1}$ of a minimal hypersurface immersed into the unit sphere is an eigenfunction of the Laplacian of its induced metric corresponding to the eigenvalue $n = \dim M$. Yau and Ogiue (see [**Ogi73, Yau82**]) asked in 1982 the following question: what can one say about compact minimal hypersurfaces immersed into a unit sphere for which their dimension n is just the first non-zero eigenvalue? This type of hypersurfaces are usually called *minimal hypersurfaces immersed by their first eigenfunctions.* Yau even conjectured that this must be always the case when the hypersurface is embedded. Up to now, it is not known whether this conjecture is true. The best approach is by Choi and Wang [**CW83**] who showed in 1983 that the first non-zero eigenvalue of the Laplacian on an embedded minimal hypersurface in the $(n+1)$-dimensional unit sphere is bigger than $n/2$.

A. Ros and myself studied in 1986 the case of the three-sphere. It was a well-known result by Almgren [**Alm66**] that a minimal surface in $\mathbb{S}^3$ with genus zero must be an equator. We proved the following result.

THEOREM 16.2 ([**MR86b**]). *Let M be a genus one compact surface immersed in $\mathbb{S}^3$ by the first eigenfunctions. Then it is congruent to the so called Clifford torus*

$$\{(x_1, x_2, x_3, x_4) \in \mathbb{S}^3 \mid x_1^2 + x_2^2 = x_3^2 + x_4^2 = 1/2\}.$$

In this way we saw that, for tori, Yau's conjecture was equivalent to another well-known conjecture of Lawson [**Law70**] which claims that this Clifford torus is the only minimal torus in the three-sphere without self-intersections. Nothing is known for higher genera, but El Soufi and Ilias obtained a version of our theorem for tori immersed by the first eigenfunctions in spheres of arbitrary dimension. In this way they characterized the Clifford torus and the so-called *equilateral* torus (see [**SI00**]).

I think that, at least when the the Laplacian is involved, these problems and results can help to understand the interest in spectral theory of submanifolds.

16.3. Dirac Operator on Hypersurfaces

In the last years, some geometers of submanifolds (like myself [**HMZ01a, HMZ01b, HMZ02**] or the group of the Technische Universität of Berlin: Pinkall, Ferus, Taimanov and others [**BFL⁺02, FLPP01, Tai**]) have been led to study diverse Dirac operators which arise from the extrinsic geometry of surfaces, and some Riemannian geometers (like Bär, Friedrich and others) have started to consider the particular features of this operator on submanifolds [**Amm00b, Amm, Bär98, AF99**].

Our interest here concerns the relation between the spectral properties of the classical Dirac operator and the geometry of submanifolds, and particularly hypersurfaces, in the sense of the previous results on the Laplacian. So, we will assume from now on that the manifold M is a hypersurface immersed in an $(n+1)$-dimensional Riemannian manifold $\overline{M}$. The second fundamental form of M is the geometrical object which controls the extrinsic geometry of the immersion. Its eigenvalues are the *principal curvatures* of the hypersurface and their average is the *mean curvature* function, *denote* by H. The mean curvature is perhaps the most important invariant of extrinsic geometry of hypersurfaces. It is a measure of the variation of the area with respect to local deformations of the hypersurface, in such a way that, for example, the hypersurfaces of constant mean curvature are exactly the first order solutions of the isoperimetric problem. In other words, they are critical points of the area functional with respect to variations preserving the enclosed volume.

If the ambient manifold $\overline{M}$ is spin (and this is the case for the most common ambient spaces in the Theory of Submanifolds) and the hypersurface M is orientable, then M is also a spin manifold and so we will have on M the spinor bundle $\mathbb{S}M$, the Levi-Civita spinorial connection, the Clifford multiplication and the (intrinsic) Dirac operator D (see [**BFGK91, BHMM, LM89**]). The point is to relate this spinorial structure on the hypersurface with the restriction to the hypersurface itself of the spinorial structure of the ambient manifold. Without entering into details (see [**Bär98, Tra95**]) we can check the following. Let $\mathbb{S}\overline{M}$ be the spinor bundle of $\overline{M}$, γ its Clifford multiplication, $\overline{\nabla}$ its Levi-Civita connection and $\overline{D}$ its Dirac operator and denote by

$$\mathbf{S}M = \mathbb{S}\overline{M}_{|M}$$

the restriction to the hypersurface M of the spinor bundle of the ambient space. If the dimension of M is even, then this Hermitian bundle can be identified with the intrinsic spinor bundle $\mathbb{S}M$ of M. Instead, if the dimension of M is odd, it is identified with two copies of $\mathbb{S}M$. As soon as we made these identifications, one can prove that the first order operator $\mathbf{D}$ defined on sections of $\mathbf{S}M$ by

$$\mathbf{D}\psi = \frac{n}{2}H\psi - \gamma(N)\sum_{i=1}^{n}\gamma(e_i)\overline{\nabla}_{e_i}\psi, \tag{16.1}$$

where $\{e_1, \ldots, e_n\}$ is an orthonormal frame tangent to the hypersurface, coincides with the intrinsic Dirac operator D of the hypersurface, if its dimension is even, and with the pair $(D, -D)$, if the dimension is odd. Then the spectrum of $\mathbf{D}$ is nothing but the symmetrization with respect to zero of the spectrum of D.

16.4. Upper Bounds on Hypersurfaces

Looking at Equality (16.1) and accepting that the first order operator $\mathbf{D}$ can be thought of as the intrinsic Dirac operator on the hypersurface, we can see that the mean curvature function H must play an important rôle to control the eigenvalues of the Dirac operator of the hypersurface M. In fact, a first consequence is the following. Suppose that, the ambient space $\overline{M}$, has a non-trivial parallel spinor ψ_0. This is the case when $\overline{M}$ is Euclidean, but it is not the only case. Spin manifolds having non-trivial parallel spinor fields were classified by Wang [**Wan89**] in 1989. They include Calabi-Yau manifolds, hyper-Kähler manifolds and other 8- and 7-dimensional Riemannian manifolds. Then its restriction to the hypersurface (denoted by the same symbol) satisfies

$$\mathbf{D}\psi_0 = \frac{n}{2}H\psi_0. \tag{16.2}$$

This means that, if the mean curvature H of the hypersurface is constant, then ψ_0 is an eigenspinor. In the general case, one can use such a spinor ψ_0 on M as a *test spinor* in the variational characterization (Rayleigh quotient) of the lowest eigenvalues to obtain

$$\lambda_1^2 \leq \frac{\displaystyle\int_M |\mathbf{D}\psi_0|^2\,dV}{\displaystyle\int_M |\psi_0|^2\,dV},$$

where λ_1 is the eigenvalue of the intrinsic Dirac operator with lowest absolute value. From this inequality, since parallel spinors have constant length, we obtain the following result which has been recently established by Bär and can be seen as the analogue of the inequality for the Laplacian obtained by Bleecker and Weiner in 1977, quoted in Theorem 16.1.

THEOREM 16.3 ([**Bär98**]). *Let M be a compact orientable hypersurface immersed into the Euclidean space $\mathbb{R}^{n+1}$ and let λ_1 be the eigenvalue of the intrinsic Dirac operator of M with the smallest absolute value. Then*

$$\lambda_1^2 \leq \frac{n^2}{4\,\mathrm{vol}(M)}\int_M H^2\,dV.$$

In the case of the Laplacian, the limiting-case is achieved only by the round spheres. However, it is an open problem to know when the equality holds in the Bär inequality. We know that in this case, the test spinor field ψ_0 is an eigenspinor for $\mathbf{D}^2$ and so, from (16.2), we have

$$\lambda_1^2\psi_0 = \mathbf{D}^2\psi_0 = \frac{n^2}{4}H^2\psi_0 + \frac{n}{2}\gamma(\nabla H)\psi_0$$

and hence the hypersurface has constant mean curvature. But this is not enough to get further information, even if the ambient space is Euclidean. In fact, the central result for this class of hypersurfaces in the Euclidean space is a celebrated theorem by Alexandrov. He proved [**Ale62**] in 1962 that a compact hypersurface in the Euclidean space with constant mean curvature must be a hypersphere, provided that it is *embedded*, that is, when we can ensure that it has no self-intersections. After this theorem and for some years specialists thought that perhaps the spheres were the only hypersurfaces immersed into the Euclidean space with constant mean curvature (see [**Hop83**]). But since 1983, we have many examples of this class of hypersurfaces which are not round spheres constructed by Hsiang, Ten and Yu [**HTY83**] when the dimension of the ambient space is even and by Wente [**Wen86**] and Kapouleas [**Kap91, Kap95**] in the three-dimensional space.

Hence, if equality is attained in the Bär inequality, the mean curvature H of the hypersurface M is constant and so, as we pointed out before, the restriction of the parallel spinor field ψ_0 is an eigenspinor corresponding to the eigenvalue $(n/2)H$. So an interesting question is:

PROBLEM 16.4. *Is the equality in the Bär inequality achieved only by the spheres? If the answer is not, what can we say about constant mean curvature compact hypersurfaces of the Euclidean space* immersed by the first eigenspinors, *that is, those for which $nH/2$ is the first eigenvalue of the Dirac operator?*

This question should be viewed as the spinorial version of the problem posed by Yau and Ogiue for the Laplacian of minimal hypersurfaces of the sphere. As in the case of the Laplacian operator, inequalities analogous to that in Theorem 16.3 can be obtained, with the same method, for hypersurfaces in spheres (see [**Bär98**]). For hypersurfaces immersed in hyperbolic spaces, the best approach has been found by N. Ginoux (see [**Gin02**]).

16.5. Estimates Under Boundary Conditions

Now we establish some lower bounds for the eigenvalues of the intrinsic Dirac operator on compact hypersurfaces of Riemannian spin manifolds together with certain geometrical applications. The estimates work only for hypersurfaces M which bound a compact domain Ω in the ambient spin manifold $\overline{M}$. For example, according to the Jordan-Brower separation theorem, this happens when M is embedded and the ambient manifold $\overline{M}$ is the Euclidean space or, in general, a manifold with vanishing first Betti number. So, from now on we assume that the oriented hypersurface M is the boundary of a compact manifold with boundary Ω.

In this situation, the basic tool is the Schrödinger–Lichnerowicz integral formula

$$\int_\Omega \left(\frac{1}{4}\overline{R}|\psi|^2 + |\overline{\nabla}\psi|^2 - |\overline{D}\psi|^2\right) = \int_M \left(\langle \mathbf{D}\psi, \psi\rangle - \frac{n}{2}H|\psi|^2\right),$$

where $\overline{R}$ is the scalar curvature of the ambient space and ψ is any spinor field defined on the domain Ω. This integral formula is different from the well-known case of compact manifolds without boundary (see [**Lic63**]): there is a boundary term where, after a suitable manipulation, the symmetrized intrinsic Dirac operator $\mathbf{D}$ and the mean curvature function H of the hypersurface appear. On the left hand side we will use the following elementary pointwise Schwarz inequality

$$|\overline{D}\psi|^2 \leq (n+1)|\overline{\nabla}\psi|^2$$

where the equality is achieved only when the spinor ψ is a twistor spinor. Then one gets (for details see, for example, [**HMZ01a**]) the following integral inequality

$$\int_\Omega \left(\frac{1}{4}\overline{R}|\psi|^2 - \frac{n}{n+1}|\overline{D}\psi|^2\right) \leq \int_M \left(\langle \mathbf{D}\psi, \psi\rangle - \frac{n}{2}H|\psi|^2\right) \tag{16.3}$$

where only the geometries of the domain and of the boundary and the two Dirac operators remain.

One can look at this integral inequality in two different ways. First, one emphasizes on the left hand side. From this point of view, it is useful to get information on the Dirac operator $\overline{D}$ of the ambient manifold, under suitable boundary conditions. It seems that the study of the natural boundary conditions which can be imposed in order to solve first order equations of Dirac type was started by Lopatinsky and Shapiro [**Hör85, Lop53**] in the fifties, continued with the theory of pseudo-differential operators by Calderón in the sixties and systematized later by Seeley (see [**See66, BBW93**]). They gave precise tools to detect when a given boundary condition for the Dirac operator on the domain Ω is *elliptic* and, as a consequence, the corresponding boundary problem is of Fredholm type, to conclude the existence and the regularity for its solutions. Using this machinery, we have studied in [**HMZ01b, HMR02**] the spectrum of $\overline{D}$ under four different boundary conditions: the global Atiyah-Patodi-Singer (APS) condition (see [**APS75**]) associated with the spectral resolution of the intrinsic Dirac operator on the boundary hypersurface (see also [**FS98**]); the local condition associated with a chirality (CHI) operator (see [**FS98**]) on the manifold (for example, when its dimension is even or when it is a space-like hypersurface in a Lorentz manifold); the Riemannian version of the so-called (local) MIT bag condition (see [**CJJT74, CJJ**$^+$**74, Joh75**]); and finally a new global boundary condition obtained by modifying the APS condition (mAPS), introduced in [**HMR02**]. It is shown that these four conditions satisfy ellipticity criteria and the corresponding boundary problems are *well-posed* in the sense of Seeley. Under the four boundary conditions, provided that the mean curvature of the boundary hypersurface is non-negative, we establish the same lower estimate in terms of the scalar curvature as for the Friedrich [**Fri80**] inequality in the closed case. That is,

THEOREM 16.5 ([**HMZ01b, HMR02**]). *Let Ω be an $(n+1)$-dimensional compact Riemannian spin manifold whose non-empty boundary $M = \partial\Omega$ has non-negative mean curvature (w.r.t. the inner normal). Then*

- *Under any of the three boundary conditions APS, CHI or mAPS, the spectrum of the (classical) Dirac operator $\overline{D}$ of Ω is a sequence of unbounded real numbers $\{\lambda_k \mid k \in \mathbb{Z}\}$ which satisfy*

$$\lambda_k^2 \geq \frac{n+1}{4n}\inf_\Omega \overline{\mathrm{R}}.$$

The equality holds only for the CHI and the mAPS conditions and in these cases Ω is respectively isometric to a half-sphere or it carries a non-trivial real Killing spinor and has minimal boundary.

- *Under the MIT condition, the spectrum of D is an unbounded discrete set of complex numbers λ^{MIT} with positive imaginary part satisfying*

$$|\lambda^{MIT}|^2 > \frac{n+1}{4n}\inf_\Omega \overline{\mathrm{R}}.$$

Interestingly enough, in the limitting-case the four conditions behave differently and our main motivation to study these conditions was just the observation that equality is never achieved for the APS and MIT conditions. However, equality occurs for the CHI boundary condition if and only if the manifold is a half-sphere and it is achieved for the mAPS condition if and only if the manifold supports a non-trivial real Killing spinor field and the boundary is a minimal hypersurface (this was a principal reason to look for and introduce this new boundary condition). For example, all the domains in a sphere enclosed by embedded minimal hypersurfaces have the same first eigenvalue for the Dirac operator obeying to the mAPS condition.

16.6. Lower Bounds for Hypersurfaces

A second way to make use of the integral inequality (16.3) is to emphasize on its r.h.s in order to get information on the geometry of the boundary hypersurface M. To do this, we start with a given spinor φ on the boundary M, for example, an eigenspinor for the intrinsic Dirac operator $\mathbf{D}$, and apply inequality (16.3) to a spinor field ψ on the bounded domain Ω which is solution to the boundary problem

$$\begin{cases} \overline{D}\psi &= 0 \qquad \text{on } \Omega \\ B\psi_{|M} &= B\varphi \qquad \text{along } \partial\Omega = M, \end{cases}$$

where B is one of the elliptic boundary conditions above. We know that, in fact, there is a smooth solution to this problem if we prove that 0 does not belong to the spectrum of $\overline{D}$, with the chosen boundary condition. This technique consisting in putting the solution of a boundary problem on the domain, determined by a hypersurface, in the boundary version of the Lichnerowicz integral formula, to get information on the hypersurface itself, was the trick used by Witten in [**Wit81, PT82**] to give a spinorial simple proof of the positive mass theorem due to Schoen and Yau. We point out that in [**Rei77a**], Reilly already used this approach in 1977 to study eigenvalues of the Laplacian.

Later, this technique has been used by many authors, for example, by Min-Oo [**MO89**] to obtain a rigidity theorem for asymptotically hyperbolic manifolds, by Ros [**Ros88**] to characterize the spheres as the only compact hypersurfaces embedded in the Euclidean space with constant scalar curvature and by Herzlich [**Her98**] in his approach to the Penrose conjecture.

Now, as a second tool, we can make use of a property of the Dirac operator which has no analogue for the Laplacian: its conformal covariance discovered by Hitchin [**Hit74**]. This can be done by writting the Reilly integral inequality (16.3) and solving the previous boundary problem, not for the original metric, but for a carefully chosen metric in the conformal class. In fact, we have used in [**HMZ02**] the MIT boundary condition which is invariant under conformal changes of the metric. Precisely, take a metric

$$\langle\ ,\ \rangle^\star = f^{\frac{4}{n-1}}\langle\ ,\ \rangle,$$

such that the conformal Laplacian applied to the change of scale f is non-negative

$$\overline{L}f = -\frac{4n}{n-1}\overline{\Delta}f + \overline{R}f \geq 0 \qquad \text{on } \Omega,$$

that is, we choose a metric on the domain Ω in the conformal class of the original metric with non-negative scalar curvature.

In this way one can prove the following result which generalizes several theorems in [**HMZ01a, HMZ02**].

THEOREM 16.6. *Let φ be a spinor field on the boundary hypersurface $M = \partial\Omega$ of a compact $(n+1)$-dimensional spin Riemannian manifold Ω, such that $\mathbf{D}\varphi = \alpha\varphi$ for a smooth function α which satisfies*

$$0 \leq \alpha \leq \frac{n}{2} f^{-1} B f,$$

where B is the following linear boundary operator

$$B = -\frac{2}{n-1}\frac{\partial}{\partial N} + H$$

(the conformal normal derivative, which controls the mean curvature of the boundary with respect to the conformally modified metric). Then we have the equality

$$\alpha = \frac{n}{2} f^{-1} B f$$

and φ is the restriction to the boundary M of a spinor field ψ on Ω such that $f^{-1}\psi$ is parallel with respect to the metric $\langle\ ,\ \rangle^\star$.

16.7. Geometrical Consequences

1) A first consequence of the above result can be obtained by taking the spinor φ to be an eigenspinor of the intrinsic Dirac operator of the boundary and hence α is constant. If λ is the corresponding eigenvalue, we have

$$|\lambda| \geq \frac{n}{2}\mathrm{Sup}_{f>0, \overline{L}f=0}\mathrm{inf}_{\mathrm{M}}\mathrm{f}^{-1}\mathrm{Bf},$$

(the absolute value arises because, in case λ is negative, one may consider the spinor field $\gamma(N)\varphi$ instead of φ), which coincides with the first eigenvalue $\nu_1(\Omega)$ of the following eigenvalue linear problem

$$\overline{L}f = -\frac{4n}{n-1}\overline{\Delta}f + \overline{R}f = 0 \qquad Bf = -\frac{2}{n-1}\frac{\partial f}{\partial N} + Hf = \nu f,$$

that is, with a sort of conformal Steklov first eigenvalue (see [**Esc00**]). This eigenvalue boundary problem for the conformal Laplacian was introduced by Escobar [**Esc92**] in 1992 because it naturally appears in the Yamabe problem for manifolds with boundary. This first eigenvalue $\nu_1(\Omega)$ is not necessarily finite, although reasonable geometrical assumptions on the domain Ω, such that the non-negativity of the scalar curvature, imply its finiteness. In this case, Escobar proved that, under conformal changes of the metric, the sign of $\nu_1(\Omega)$ is invariant on the domain and that the associated eigenfunction f is positive. Then the new metric $\langle\ ,\ \rangle^\star$ is a scalar-flat metric on Ω and the boundary M has positive, zero or negative mean curvature function according to the sign of $\nu_1(\Omega)$ is positive, zero or negative. A trivial use of the Hölder inequality relates $\nu_1(\Omega)$ with $\mathcal{Q}(\Omega, M)$, the relative Yamabe number of the domain enclosed by the hypersurface, which was studied also by Escobar in 1994. Hence we get the following result.

THEOREM 16.7 ([**HMZ02**]). *Let Ω be an $(n+1)$-dimensional compact spin Riemannian manifold with boundary $\partial\Omega = M$ and let λ be any eigenvalue of the intrinsic Dirac operator of the boundary hypersurface M. Then*

$$|\lambda|\mathrm{vol}(M)^{\frac{1}{n}} \geq \frac{n}{2}\mathcal{Q}(\Omega, M),$$

where $\mathcal{Q}(\Omega, M)$ (the Yamabe relative invariant) is the infimum over the scalar-flat metrics in the conformal class of the domain of the normalized total mean curvature of its boundary. If equality holds, then Ω carries non-trivial parallel spinor fields, the mean curvature of the boundary hypersurface is constant and the eigenspace corresponding to λ consists of restrictions of parallel spinor fields of the domain.

One can see that the above results are extrinsic versions of corresponding Hijazi's intrinsic theorems (see [**Hij86a, Hij86b**]).

2) Suppose now that the scalar curvature $\overline{R}$ of the ambient manifold is non negative. Then, we can choose $f = 1$ in Theorem 16.6 and so $f^{-1}Bf = H$. Hence we obtain:

THEOREM 16.8 ([**HMZ01a**]). *Let Ω be a compact spin Riemannian manifold of dimension $n+1$ with non negative scalar curvature. Then the lowest non negative eigenvalue λ_1 of the intrinsic Dirac operator of M satisfies*

$$\lambda_1 \geq \frac{n}{2}\mathrm{inf_M H},$$

and equality implies that H is constant and the space of eigenspinors for $\lambda_1 = (n/2)H$ is obtained by restricting to M the space of parallel spinors on the manifold Ω.

This result can be thought of as an extrinsic version of the Friedrich theorem in [**Fri80**] and the equality is attained for any bounding hypersurface with constant mean curvature when the ambient Riemannian manifold carries a non-trivial parallel spinor. For example, besides spheres in Euclidean spaces, for any regular solution to the isoperimetric problem in a Calabi-Yau or hyper-Kähler manifold.

Two remarks could be made on this metric lower estimate. The first one refers to the case where the ambient space is the Euclidean space $\mathbb{R}^{n+1}$. The Gauss formula for the curvature implies that the scalar curvature R of M is

$$R = n^2H^2 - |\sigma|^2,$$

and by using the Schwarz inequality for the length of σ, then

$$R \leq n^2H^2 - |\sigma|^2 \leq n(n-1)H^2.$$

Hence,

$$\left(\frac{n}{2}\right)^2 \mathrm{inf_M H^2} \geq \frac{\mathrm{n}}{4(\mathrm{n}-1)}\mathrm{inf_M R},$$

that is, the extrinsic estimate improves that of Friedrich when $H \geq 0$. Even more, there are embedded hypersurfaces in the Euclidean space with positive (inner) mean curvature and such that $\mathrm{inf_M R}$ is negative. For instance, the revolution tori with generating circle, small enough.

The second observation is that there are no similar estimates in terms of H, $\nu_1(\Omega)$ or $Q(\Omega, M)$ for the eigenvalues of the Laplacian. To see this, consider the following smooth embedded surface: a cylinder of radius one with ends closed by half-spheres of the same radius. This surface has (inner) mean curvature bounded by the mean curvature of the cylinder, that is, by $1/2$ and this fact is independent of the length of the intermediate cylinder. However, when we increase the length of that cylinder, the first non-zero eigenvalue of the Laplacian becomes smaller and smaller.

3) A third consequence of the above theorem is a simple spinorial proof of the Alexandrov theorem for compact constant mean curvature hypersurfaces without self-intersections in the Euclidean space. In fact, for such hypersurfaces the restrictions of the constant Euclidean spinors provide eigenspinors with eigenvalue $(n/2)H$. Then, this result implies that $\lambda_1 = (n/2)H$ and the corresponding eigenspace consists only of these type of restrictions. But, given such a parallel spinor ψ_0, one can construct another spinor on the hypersurface, say

$$\varphi = \gamma(Hp + N)\psi_0$$

and a direct computation from (16.1) implies that φ is also an eigenspinor on the hypersurface corresponding to the same eigenvalue $(n/2)H$. Then it has to be the restriction of an Euclidean constant spinor and so, if u is vector tangent to the hypersurface,

$$0 = \overline{\nabla}_u\varphi = \gamma(Hu + (dN)_p u)\psi_0.$$

Then, the differential of the Gauss map N is proportional to the identity map, that is, the hypersurface is totally umbilical and so a hypersphere.

4) Finally, the last consequences of our lower estimate for the intrinsic Dirac operator on bounding hypersurface are some rigidity theorems for compact Riemannian spin manifolds with non-empty boundary. Assume that the scalar curvature of the ambient manifold and the mean curvature function of the boundary are both non-negative and let φ be a non-trivial Killing spinor field on the boundary hypersurface M (see [**BFGK91**]). The existence of such a spinor implies that equality in the Friedrich estimate is achieved. This fact together with the previous result imply that

$$\lambda_1^2 = \frac{n}{4(n-1)}\mathrm{inf_M R} \geq \frac{\mathrm{n}^2}{4}\mathrm{inf_M H}^2,$$

but, on the other hand, the Gauss equation above says that

$$R = \overline{R} - 2\overline{\mathrm{Ric}}(N, N) + n^2H^2 - |\sigma|^2.$$

Then, if we assume also that the Einstein tensor of the ambient space is non negative along the boundary M, then

$$R \leq n^2H^2 - |\sigma|^2 \leq n(n-1)H^2$$

and so we get equalities everywhere. As a consequence either the spinor φ or its Clifford product $\gamma(N)\varphi$ by the unit normal is the restriction of a parallel spinor on the domain Ω enclosed by the hypersurface M. That is, under these conditions, all the supersymmetries of the boundary come from supertranslations of the bulk manifold. In particular, if the boundary is supposed to be isometric to a sphere, there is a maximal number of parallel spinor fields on the manifold, that is, the manifold is flat. In fact, this reasoning may be strengthened for complete manifolds.

THEOREM 16.9 ([**HM03**]). *Let Ω be a complete Riemannian spin manifold with non negative Ricci curvature, mean-convex boundary $\partial\Omega$ and non negative Einstein tensor along the normal direction of $\partial\Omega$. Then Ω is isometric to a flat disc.*

Two consequences of this result are immediately deduced. The first one says that *the only Ricci-flat metric on the unit $(n+1)$-dimensional disc inducing, along the boundary, the usual metric on the sphere $\mathbb{S}^n$ and with non-negative inner mean curvature, is the standard flat metric.* A local version of this uniqueness theorem has been recently obtained by Jean-Marc Schlenker in [**Sch01**]. The second one asserts that *a complete Ricci-flat manifold admitting a round sphere as an embedded hypersurface must be flat.*

Finally, I would like to point out that we have recently obtained in [**HMR03**] some analogous results for spin manifolds with negative scalar and Ricci curvatures.

Sebastián Montiel, Departamento de Geometría y Topología Facultad de Ciencias, Universidad de Granada, 18071 Granada, Spain.
E-mail address: smontiel@ugr.es

CHAPTER 17

From Kirchberg's Inequality to the Goldberg Conjecture

ANDREI MOROIANU[1]

ABSTRACT. The main result of this note is that a compact Kähler manifold whose Ricci tensor has two distinct constant non-negative eigenvalues is locally the product of two Kähler-Einstein manifolds. The problem of existence of Kähler metrics whose Ricci tensor has two distinct constant eigenvalues is related to the limiting case of Kirchberg's inequality [**Kir86**] for the first eigenvalue of the Dirac operator on compact Kähler manifolds, as well as to the celebrated (still open) conjecture of Goldberg [**Gol69**]. 2000 *Mathematics Subject Classification.* Primary 53B20, 53C25

Contents

17.1. Introduction

In this note we consider compact Kähler manifolds (M, g, J) such that the symmetric endomorphism corresponding to the Ricci tensor Ric via the metric has two distinct constant eigenvalues. Since for Kähler manifolds the Ricci tensor is invariant under the action of J, each eigenvalue is of even multiplicity.

Such manifolds first appeared, to our knowledge, in a conjecture of Lichnerowicz concerning compact Kähler spin manifolds with least possible (compared to the scalar curvature) eigenvalue of the Dirac operator. They carry special spinor fields and it was showed in [**Mor97a**] that their Ricci tensor has two constant non-negative eigenvalues. In Section 2 we sketch the proof of the fact that their Ricci tensor has to be parallel. Then in Section 3 we analyze manifolds with constant eigenvalues of the Ricci tensor using tools coming from almost Kähler geometry, and obtain the following splitting theorem.

[1]The material of the present talk is based on a joint work with Vestislav Apostolov and Tedi Draghici.

THEOREM 17.1. [**ADM01**] *Let (M, g, J) be a compact Kähler manifold whose Ricci tensor has two distinct constant non-negative eigenvalues λ and μ. Then the universal cover of (M, g, J) is the product of two simply connected Kähler–Einstein manifolds of scalar curvatures λ and μ, respectively.*

Note that *local* irreducible examples of Kähler manifolds with eigenvalues of the Ricci tensor equal to 0 and 1 are known to exist in complex dimension two, cf. [**Bry**] and [**AAD**, Remark 1(c)]. Note also that the above result fails if one allows the Ricci tensor to have more than two different eigenvalues, as shown by the compact homogeneous Kähler manifolds (the generalized complex flag manifolds). Another reason for this study came from an a priori unexpected link with the Goldberg conjecture [**Gol69**], which states that any compact *Einstein* almost Kähler manifold is, in fact, Kähler–Einstein. For any Kähler manifold (M, g, J) with Ricci tensor having two distinct constant eigenvalues, one can define another g-orthogonal almost complex structure $\overline{J}$, by changing the sign of J on one of the eigenspaces of Ric. The new almost complex structure $\overline{J}$, which is not integrable in general, commutes with J and has a closed fundamental 2-form, i.e. $(g, \overline{J})$ gives rise to an almost Kähler structure on the manifold. The integrability of $\overline{J}$ holds precisely when the Ricci tensor of g is parallel, or equivalently, when g is locally a product of two Kähler–Einstein metrics; see Lemma 17.8. Moreover, any Kähler structure (g, J) with Ricci tensor having two distinct constant eigenvalues either both positive, or both negative, determines (and is determined by) a certain Einstein almost Kähler structure $(\widetilde{g}, \overline{J})$, see Corollary 17.10. In other words, any example of compact Kähler manifold whose Ricci tensor is non-parallel and has two constant eigenvalues of the same sign, would provide a counterexample to the Goldberg conjecture.

17.2. Limiting manifolds for Kirchberg's inequality

17.2.1. The Ricci tensor of limiting manifolds. We follow here the presentation and notations from [**Mor99**]. Let (M^{2m}, g, J) be a compact Kähler spin manifold with positive scalar curvature s and suppose that its complex dimension m is even : $m = 2l \geq 4$. K.-D. Kirchberg [**Kir86**] showed that every eigenvalue λ of the Dirac operator on M satisfies

$$\lambda^2 \geq \frac{m}{4(m-1)} \inf_M s. \tag{17.1}$$

The manifolds which satisfy the limiting case of this inequality (called *limiting manifolds* for the remaining of this paper), are characterized by the following

THEOREM 17.2. (cf. [**Kir86**]) *M is a limiting manifold if and only if its scalar curvature s is a positive constant and there exists a spinor $\Psi \in \Gamma(\Sigma^{l+1}M)$ such that*

$$\nabla_X \Psi = -\frac{1}{n}(X - iJX) \cdot D\Psi, \ \forall X, \tag{17.2}$$

$$\nabla_X D\Psi = -\frac{1}{4}(\mathrm{Ric}(X) + iJ\mathrm{Ric}(X)) \cdot \Psi, \ \forall X, \tag{17.3}$$

$$\kappa(X - iJX) \cdot \Psi = (\mathrm{Ric}(X) - iJ\mathrm{Ric}(X)) \cdot \Psi, \ \forall X, \tag{17.4}$$

$$\kappa(X - iJX) \cdot D\Psi = (\mathrm{Ric}(X) - iJ\mathrm{Ric}(X)) \cdot D\Psi, \ \forall X, \tag{17.5}$$

where $\kappa = \frac{s}{n-2}$. In particular, (17.2) implies (after a Clifford contraction) that $D\Psi \in \Gamma(\Sigma^l M)$.

These relations correspond to formulas (58), (59), (60) and (74) from [**Kir86**], with the remark that Ψ above and ψ^{l-1} of [**Kir86**] are related by $\Psi = j\psi^{l-1}$. (We recall, for the convenience of the reader, that Σ^l denotes the eigenspace for the Clifford action of the Kähler form on the spin bundle ΣM corresponding to the eigenvalue $i(m-2l)$). In [**Mor97a**] we obtained the following

THEOREM 17.3. (cf. [**Mor97a**], Thm.3.1) *The Ricci tensor of a limiting manifold of even complex dimension has two eigenvalues, κ and 0, the first one with multiplicity $n-2$ and the second one with multiplicity 2.*

COROLLARY 17.4. *The tangent bundle of M splits into a J–invariant orthogonal direct sum $TM = \mathcal{E} \oplus \mathcal{F}$ (where $\mathcal{E}$ and $\mathcal{F}$ are the eigenbundles of TM corresponding to the eigenvalues 0 and κ of* Ric *respectively). Moreover, the distributions $\mathcal{E}$ and $\mathcal{F}$ are integrable.*

PROOF. All but the last statement are clear from Theorem 17.3, so we only prove the integrability of $\mathcal{E}$ and $\mathcal{F}$. Let ρ denote the Ricci form of M, defined by $\rho(X,Y) = \mathrm{Ric}(JX,Y)$, which, of course satisfies $d\rho = 0$. Remark that $X \lrcorner \rho = 0$ for $X \in \mathcal{E}$ and $X \lrcorner \rho = -\kappa JX$ for $X \in \mathcal{F}$. We consider arbitrary vector fields $X, Y \in \mathcal{E}$ and $Z \in \mathcal{F}$ and obtain (σ stands for the cyclic sum)

$$\begin{aligned} 0 &= d\rho(X,Y,Z) = \sigma(X(\rho(Y,Z) - \rho([X,Y],Z)) \\ &= -\rho([X,Y],Z), \end{aligned}$$

so $\mathcal{E}$ is integrable. Similarly, for $X, Y \in \mathcal{F}$ and $Z \in \mathcal{E}$ we have

$$\begin{aligned} 0 &= d\rho(X,Y,Z) = \sigma(X(\rho(Y,Z) - \rho([X,Y],Z)) \\ &= Z(\rho(X,Y)) - \rho([Y,Z],X) - \rho([Z,X],Y) \\ &= \kappa(Z(g(JX,Y)) - g(J[Y,Z],X) - g(J[Z,X],Y)) \\ &= \kappa(g(\nabla_Y Z, JX) - g(\nabla_X Z, JY)) \\ &= -\kappa g([X,Y],JZ), \end{aligned}$$

which proves the integrability of $\mathcal{F}$. □

From (17.2)– (17.5) follows that for every section X of $\mathcal{E}$ we have

$$(X - iJX)\cdot\Psi = (X - iJX)\cdot D\Psi = 0 \qquad (17.6)$$

$$\nabla_X \Psi = \nabla_X D\Psi = 0. \qquad (17.7)$$

17.2.2. Kählerian Killing spinors on Hodge manifolds. The theorem below is crucial for proving that the Ricci tensor of limiting manifolds is parallel. Its proof is based on the theory of projectable spinors and the classification of Spin^c manifolds with parallel spinors [**Mor97b**].

DEFINITION 17.5. Let (N^{4l-2}, g, J) be a Kähler manifold equipped with a Spin^c structure. A *Kählerian Killing spinor* on N is a spinor Ψ satisfying

$$\nabla_X^A \Psi + \alpha X \cdot \Psi + i\alpha(-1)^l JX \cdot \overline{\Psi} = 0, \ \forall X, \qquad (17.8)$$

for some real constant $\alpha \neq 0$.

THEOREM 17.6. *Let (N^n, g, J), $n = 4l-2$ be a simply connected compact Hodge manifold endowed with a Spin^c structure carrying a Kählerian Killing spinor $\Psi \in \Gamma(\Sigma^{l-1}N \oplus \Sigma^l N)$. Then this Spin^c structure on N is actually a spin structure.*

PROOF. The proof is in two steps. We first show that the Kählerian Killing spinor on N induces a Killing spinor on some S^1 bundle over N, and then use the classification of Spin^c manifolds carrying Killing spinors to conclude. The Hodge condition just means that $\frac{r}{2\pi}[\Omega] \in H^2(N,\mathbb{Z})$ for some $r \in \mathbb{R}^*$, and we will fix some r with this property. The isomorphism $H^2(N,\mathbb{Z}) \simeq H^1(N,S^1)$ guarantees the existence of some principal $U(1)$ bundle $\pi : S \to N$ whose first Chern class satisfies $c_1(S) = \frac{r}{2\pi}[\Omega]$. Furthermore, the Thom–Gysin exact sequence shows that S is simply connected if r is chosen in such a way that $\frac{r}{2\pi}[\Omega]$ is not a multiple of some integral class in $H^2(N,\mathbb{Z})$ (cf. [**BFGK91**], p.85). The condition above on the first Chern class of S shows that there is a connection on S whose curvature form G satisfies $G = -ir\pi^*\Omega$. This connection induces a 1–parameter family of metrics on S which turn the bundle projection $\pi : S \to N$ into a Riemannian submersion with totally geodesic fibers. These metrics are given by

$$g_S^t(X,Y) = g(\pi_*(X),\pi_*(Y)) - t^2\,\omega(X)\,\omega(Y) \quad (t>0),$$

where ω denotes the (imaginary valued) connection form on S. By pull–back from N, we obtain a Spin^c structure on S, whose spinor bundle is just $\pi^*\Sigma N$. After a straightforward computation, one finds that for a suitably chosen parameter t, the pull back $\pi^*\Psi$ of our Kählerian Killing spinor Ψ is a Killing spinor of the pull–back Spin^c structure on S. Now, a standard argument shows that $\pi^*\Psi$ induces a parallel spinor Φ on the cone $\overline{S}$ over S, endowed with the pull–back Spin^c structure (see [**Mor97b**]). Since S is compact, a theorem of Gallot ([**Gal79**], Prop.3.1) shows that $\overline{S}$ is an irreducible Riemannian manifold. From ([**Mor97b**], Thm.3.1) we then deduce that either the Spin^c structure of $\overline{S}$ is actually a spin structure, or there exists a Kähler structure I on $\overline{S}$ such that

$$X\cdot\Phi = iI(X)\cdot\Phi,\ \forall X \in T\overline{S}, \tag{17.9}$$

and the Spin^c structure of $\overline{S}$ is the canonical Spin^c structure induced by I (these two cases do not exclude each other). In the first case we are done since the pull-back operation for Spin^c structures on Riemannian submersions is one-to-one. The second case turns out to be impossible, cf. [**Mor99**] for the details. □

We are now ready to complete the proof of:

THEOREM 17.7. *The Ricci tensor of a limiting manifold of even complex dimension is parallel.*

PROOF. We give here the main ideas, the reader is referred to [**Mor99**] for details. Let N be a maximal leaf of the integrable distribution $\mathcal{F}$. An easy calculation shows that the Ricci tensor of N is defined positive. As N is complete, Myers' Theorem implies that N is compact and the theorem of Kobayashi ([**Kob61**], Thm. A) shows that N is simply connected. We shall now consider the restriction Φ^N of $\Phi := \Psi + \frac{2}{\sqrt{n\kappa}}D\Psi$ to N. The relations (17.4) and (17.5) show that Φ^N is a section of the Spin^c structure on N with associated line bundle $\mathcal{E}^{1,0}|_N$. A rather intricate computation yields that Φ^N is a Kählerian Killing spinor satisfying the hypothesis of Theorem 17.6. Moreover, N is a Hodge manifold: if we denote by i the inclusion $N \to M$ and by ρ the Ricci form of M, then $\kappa\,\Omega^N = i^*\rho$, which implies $\kappa[\Omega^N] = i^*(2\pi c_1(M))$, and thus $[\Omega^N]$ is a real multiple of $i^*(c_1(M)) \in H^2(N,\mathbb{Z})$. We then apply Theorem 17.6 and deduce that the Spin^c structure on N has actually to be a *spin* structure (i.e. $\mathcal{E}^{1,0}|_N$ is a flat bundle on N). An explicit computation of the curvature of $\mathcal{E}^{1,0}|_N$ in terms of the curvature tensor of M yields that $TM|_N = \mathcal{E}|_N \oplus \mathcal{F}|_N$ is a parallel decomposition of the vector bundle $TM|_N$ over N, so finally, by choosing N arbitrarily, we deduce that E and F are parallel distributions on M. □

17.3. Almost Kähler manifolds

We now relax the hypothesis in Theorem 17.7 : rather than assuming that M is a limiting manifold, we only suppose that the Ricci tensor of M has two constant non-negative eigenvalues.

17.3.1. The commuting almost Kähler structure. The main idea of the proof of Theorem 17.1 is to construct an almost Kähler structure $\overline{J}$ on (M, g, J), which is compatible with g and commutes with J.

LEMMA 17.8. *Let (M, g, J) be a Kähler manifold whose Ricci tensor has constant eigenvalues $\lambda < \mu$. Denote by E_λ and E_μ the corresponding J-invariant eigenspaces and define a g-orthogonal almost complex structure $\overline{J}$ by setting $\overline{J}_{|E_\lambda} = J_{|E_\lambda}$; $\overline{J}_{|E_\mu} = -J_{|E_\mu}$. Then J and $\overline{J}$ mutually commute and $(g, \overline{J})$ is an almost Kähler structure, i.e. the fundamental form $\overline{\Omega}(\cdot,\cdot) = g(\overline{J}\cdot,\cdot)$ is symplectic. Moreover, $(g, \overline{J})$ is Kähler (i.e. $\overline{J}$ is integrable) if and only if (M, g) is locally product of two Kähler–Einstein manifolds of scalar curvatures λ and μ, respectively.*

Proof. Denote by $\Omega(\cdot,\cdot) = g(J\cdot,\cdot)$ the fundamental form of (g, J) and consider the (1,1)-forms α and β defined by

$$\alpha(X, Y) = \Omega(\mathrm{pr}^\lambda(X), \mathrm{pr}^\lambda(Y)), \ \forall X, Y \in TM;$$

$$\beta = \Omega - \alpha,$$

where pr^λ (resp. pr^μ) denotes the orthogonal projection of the tangent bundle TM onto E_λ (resp. E_μ). The Ricci form $\rho(\cdot,\cdot) = \mathrm{Ric}(J\cdot,\cdot)$ of (M, g, J) is then given by

$$\rho = \lambda\alpha + \mu\beta.$$

As $\Omega = \alpha + \beta$ and ρ are both closed (1,1)-forms, so are the 2-forms α and β. By the very definition of $\overline{J}$, the fundamental form $\overline{\Omega}(\cdot,\cdot) = g(\overline{J}\cdot,\cdot)$ is given by

$$\overline{\Omega} = \alpha - \beta,$$

and hence is closed, i.e. $(g, \overline{J}, \overline{\Omega})$ is an almost Kähler structure; it is Kähler as soon as the Ricci tensor is parallel (equivalently, α and β are parallel), i.e. when (M, g) is locally a product of two Kähler-Einstein manifolds with scalar curvatures λ and μ, respectively. □ Let us consider for a moment the more general context of Kähler

manifolds (M, g, J) which admit a *commuting* almost Kähler structure $\overline{J}$. Any g-orthogonal almost complex structure $\overline{J}$ which commutes with and differs from $\pm J$ gives rise to a g-orthogonal, J-invariant endomorphism $Q = -J \circ \overline{J}$ of TM such that $Q^2 = \mathrm{Id}_{|TM}$; we thus define an orthogonal, J-invariant splitting of the tangent bundle TM

$$TM = E_+ \oplus E_-$$

into the sum of the ± 1-eigenspaces of Q, the (complex) sub-bundles $E_\pm$, respectively. As in the proof of Lemma 17.8, we consider the (1,1)-forms α and β, the restrictions of the fundamental form Ω of (g, J) to the spaces E_+ and E_-, respectively. The fundamental forms Ω and $\overline{\Omega}$ of (g, J) and $(g, \overline{J})$ are then given by

$$\Omega = \alpha + \beta; \quad \overline{\Omega} = \alpha - \beta,$$

proving that α and β are *closed*. Therefore, corresponding to any Kähler metric (g, J) admitting a commuting almost Kähler structure $\overline{J}$, we may consider a natural 1-parameter family g^t of metrics having the same property (see [**Mor99**]):

$$g^t = g_{|_{E_+}} + t g_{|_{E_-}}, \quad t > 0, \tag{17.10}$$

where $g_{|_{E_+}}$ (resp. $g_{|_{E_-}}$) denotes the restriction of g to the eigenspaces E_+ (resp. to E_-).

LEMMA 17.9. *For any $t > 0$, the metric g^t is Kähler with respect to J, almost Kähler with respect to $\overline{J}$ and has the same Ricci tensor as the metric $g = g^1$.*

Proof. The first statements follow from the fact that the fundamental form of (g^t, J) (resp. $(g^t, \overline{J})$) is closed as being equal to $\alpha + t\beta$ (resp. $\alpha - t\beta$), where α and β are constructed as above with respect to $g = g^1$. For the last claim, note that the volume form of the metric g^t is a constant multiple of the volume form of $g = g^1$, so, from the local expression in complex coordinates, the Ricci forms of the Kähler structures (g^t, J) and (g, J) coincide. □ As for Kähler metrics with Ricci tensor

having distinct constant eigenvalues, Lemma 17.9 shows that one can deform any such given metric to one whose Ricci tensor has constant eigenvalues equal to $-1, 0$ or $+1$. In particular, we get

COROLLARY 17.10. *On a complex manifold (M, J) there is a one-to-one correspondence between Kähler metrics with Ricci tensor of constant eigenvalues $\lambda < \mu$ with $\lambda\mu > 0$ and Kähler-Einstein metrics $\widetilde{g}$ of scalar curvature 4λ carrying an orthogonal almost Kähler structure $\overline{J}$ which commutes with and differs from $\pm J$; in this correspondence $\overline{J}$ is compatible also with g and coincides (up to sign) with the almost Kähler structure defined in Lemma 17.8; moreover, $\overline{J}$ is integrable precisely when g (and $\widetilde{g}$) is locally product of two Kähler–Einstein metrics.*

Proof. Let (M, g, J) be a Kähler manifold whose Ricci tensor has constant eigenvalues $\lambda < \mu$ and $\overline{J}$ be the almost Kähler structure commuting with J given by Lemma 17.8. It is easy to see that the ± 1-eigenspaces of the endomorphism $Q = -J \circ \overline{J}$ above are given by $E_+ = E_\lambda$, $E_- = E_\mu$, where, we recall, E_λ and E_μ are the eigenspaces of Ric. By Lemma 2, the metric $\widetilde{g} = g^{\mu/\lambda}$ obtained via (17.10) is Kähler–Einstein with scalar curvature 4λ. Conversely, starting from a Kähler–Einstein structure $(\widetilde{g}, J)$ of scalar curvature 4λ, endowed with an almost Kähler structure $\overline{J}$ commuting with J, the deformation (17.10) provides a Kähler metric (g, J) whose Ricci tensor has constant eigenvalues $\lambda < \mu$, by putting $g^1 = \widetilde{g}$ and $g = g^{\lambda/\mu}$. The almost complex structure $\overline{J}$ is compatible to both g and $\widetilde{g}$. It is clear then that the common Ricci tensor of g and $\widetilde{g}$ is $\overline{J}$-invariant, and therefore, $\overline{J}$ coincides (up to sign) with the almost complex structure defined in Lemma 17.8. By Lemma 17.8 we also conclude that the integrability of $\overline{J}$ is equivalent to g (hence also $\widetilde{g}$) being locally a product of two Kähler–Einstein metrics. □

17.3.2. Curvature obstructions to existence of strictly almost Kähler structures. The proof of Theorem 17.1 will be derived by showing the integrability of the almost Kähler structure obtained in Corollary 1. To do this we first observe that existence of a strictly almost Kähler structure imposes several non-trivial relations between different $U(n)$-components of the curvature. Because the almost Kähler structure will take the center stage in what follows, we drop the bar-notation from the previous sub-section and will even forget for now that in our situation the manifold also admits a Kähler structure. Thus, let (M, g, J) be an almost Kähler manifold of (real) dimension $2n$. We start by reviewing some necessary elements of almost Kähler geometry. The almost complex structure J gives rise to a type decomposition of complex vectors and forms, and accordingly, of any complex tensor field; by convention, J acts on the cotangent bundle T^*M by $Ja(X) = a(-JX)$. We thus have a decomposition of the complexified cotangent bundle

$$T^*M \otimes \mathbb{C} = \Lambda^{1,0}M \oplus \Lambda^{0,1}M,$$

and of the bundle of complex 2-forms

$$\Lambda^2 M \otimes \mathbb{C} = \Lambda^{1,1}M \oplus \Lambda^{2,0}M \oplus \Lambda^{0,2}M.$$

A similar decomposition holds for the complex bundle $S^2M \otimes \mathbb{C}$ of symmetric 2-tensors. When considering *real* sections of $\Lambda^2 M$ (resp. of S^2M), we prefer to introduce the super-scripts $'$ and $''$ for denoting the projections to the real sub-bundles $\Lambda^{1,1}_{\mathbb{R}}M$ (resp. $S^{1,1}_{\mathbb{R}}M$) of J-invariant 2-forms (resp. symmetric 2-tensors) and to $[\![\Lambda^{2,0}M]\!]$ (resp. $[\![S^{2,0}M]\!]$) of J-anti-invariant ones (here and henceforth $[\![\ ]\!]$ stands for the real vector bundle underlying a given complex bundle). Thus, for any section ψ of $\Lambda^2 M$ (resp. of S^2M) we have the splitting $\psi = \psi' + \psi''$, where

$$\psi'(\cdot,\cdot) = \frac{1}{2}(\psi(\cdot,\cdot) + \psi(J\cdot, J\cdot)) \text{ and } \psi''(\cdot,\cdot) = \frac{1}{2}(\psi(\cdot,\cdot) - \psi(J\cdot, J\cdot)).$$

Note that $\Lambda^{1,1}_{\mathbb{R}}M$ can be identified with $S^{1,1}_{\mathbb{R}}M$ via the complex structure J: for any $\alpha \in \Lambda^{1,1}_{\mathbb{R}}M$,

$$A = (J \circ \alpha) := \alpha(J\cdot, \cdot)$$

is the corresponding section of $S^{1,1}_{\mathbb{R}}M$. The real bundle $[\![\Lambda^{2,0}M]\!]$ (resp. $[\![S^{2,0}M]\!]$) inherits a canonical complex structure J, acting by

$$(J\psi)(X, Y) := -\psi(JX, Y),\ \forall \psi \in [\![\Lambda^{2,0}M]\!].$$

(We adopt a similar definition for the action of J on $[\![S^{2,0}M]\!]$). It is well known that the fundamental form $\Omega(\cdot,\cdot) = g(J\cdot,\cdot)$ of an almost Kähler structure is a real harmonic 2-form of type $(1,1)$, i.e. satisfies:

$$\Omega(J\cdot, J\cdot) = \Omega(\cdot,\cdot)\ ,\mathrm{d}\Omega = 0 \text{ and } \delta\Omega = 0,$$

where d and δ are the differential and co-differential operators acting on forms. Moreover, if ∇ is the Levi-Civita connection of g, then $\nabla\Omega$ (which is identified with the Nijenhuis tensor of J) is a section of the real vector bundle $[\![\Lambda^{1,0}M \otimes \Lambda^{2,0}M]\!]$. We first derive several consequences from the classical Weitzenböck formula for a 2-form ψ:

$$\begin{aligned}(17.11)\qquad \Delta\psi - \nabla^*\nabla\psi &= [\mathrm{Ric}(\Psi\cdot,\cdot) - \mathrm{Ric}(\cdot,\Psi\cdot)] - 2R(\psi)\\ &= \frac{2(n-1)}{n(2n-1)}s\psi - 2W(\psi) + \frac{(n-2)}{(n-1)}[\mathrm{Ric}_0(\Psi\cdot,\cdot) - \mathrm{Ric}_0(\cdot,\Psi\cdot)],\end{aligned}$$

where: $\Delta = \mathrm{d}\delta + \delta\mathrm{d}$ denotes the Riemannian Laplace operator acting on 2-forms, ∇^* denotes the adjoint of ∇ with respect to g; $\mathrm{Ric}_0 = \mathrm{Ric} - \frac{s}{2n}g$ is the traceless part of the Ricci tensor, $s = \mathrm{trace(Ric)}$ is the scalar curvature, Ψ is the skew-symmetric endomorphism of TM identified to ψ via the metric, and R and W are respectively the curvature tensor and the Weyl tensor, considered as endomorphisms of $\Lambda^2 M$ or as sections of $\Lambda^2 M \otimes \Lambda^2 M$, depending on the context. Applying relation (17.11) to the (harmonic) fundamental form Ω of the almost Kähler structure (g, J), we obtain

$$(17.12)\qquad \nabla^*\nabla\Omega = 2R(\Omega) - [\mathrm{Ric}(J\cdot,\cdot) - \mathrm{Ric}(\cdot, J\cdot)]\ .$$

Note that the Ricci tensor of a Kähler structure is J-invariant, but this is no longer true for an arbitrary almost Kähler structure. It will be thus useful to introduce the invariant and the anti-invariant parts of the Ricci tensor with respect to the almost complex structure J, Ric' and Ric'', respectively. We also put

$$\rho = J \circ \mathrm{Ric}'$$

to be the (1,1)-form corresponding the the J-invariant part of Ric, which will be called *Ricci form* of (M, g, J). For Kähler manifolds, ρ is clearly equal to the image of Ω under the action of the curvature R, but this is not longer true for almost Kähler manifolds. In fact,

$$\rho^* = R(\Omega)$$

can be considered as a second (twisted) Ricci form of (M, g, J) which is not, in general, J-invariant (see e.g. [**TV81**]). We will consequently denote by $(\rho^*)'$ and $(\rho^*)''$ the corresponding 2-forms which are sections of the bundles $\Lambda^{1,1}_{\mathbb{R}}M$ and $[\![\Lambda^{2,0}M]\!]$, respectively. With these notations, formula (17.12) is a measure of the difference of the two types of Ricci forms on an almost Kähler manifold:

$$\rho^* - \rho = \frac{1}{2}(\nabla^*\nabla\Omega). \tag{17.13}$$

Taking the inner product with Ω of the relation (17.13) we obtain the difference of the two types of scalar curvatures:

$$s^* - s = |\nabla\Omega|^2 = \frac{1}{2}|\nabla J|^2 \, , \tag{17.14}$$

where, we recall $s = \mathrm{trace}(\mathrm{Ric})$ is the usual scalar curvature of g, and $s^* = 2\langle R(\Omega), \Omega\rangle$ is the so-called *star-scalar curvature* of the almost Kähler structure (g, J). Here and throughout the paper, the inner product induced by the metric g on various tensor bundles over the manifold will be denoted by $\langle\ \rangle$, while the corresponding norm is denoted by $|\ |$; note that $\langle\ \rangle$ acting on 2-forms differs by a factor of 1/2 compared to when it acts on corresponding tensors or endomorphisms. In the present paper $\nabla\Omega$ is viewed as a Λ^2M-valued 1-form, while ∇J is considered as a section of $(T^*M)^{\otimes 2} \otimes TM$, etc. Formulae (17.13) and (17.14) can be interpreted as "obstructions" to the (local) existence of a strictly almost Kähler structure J, compatible with a given metric g; see e.g. [**Arm02**]. We derived these relations by using properties of the 2-jet of J (although eventually (17.14) depends on the 1-jet only), so that (17.13) and (17.14) can be viewed as obstructions to the lifting of the 0-jet of J to the 2-jet. In fact, there is even a more general identity than (17.13), due to Gray [**Gra76**], which could also be interpreted as an obstruction to the lifting of the 0-jet of J to the 2-jet: Starting from the splitting

$$\Lambda^2M = \Lambda^{1,1}_{\mathbb{R}}M \oplus [\![\Lambda^{2,0}M]\!],$$

we denote by $\widetilde{R}$ the component of the curvature operator acting trivially on the first factor, i.e.

$$\widetilde{R}_{X,Y,Z,T} = \frac{1}{4}\Big(R_{X,Y,Z,T} - R_{JX,JY,Z,T} - R_{X,Y,JZ,JT} + R_{JX,JY,JZ,JT}\Big).$$

Thus, $\widetilde{R}$ can be viewed as a section of the bundle $\mathrm{End}_{\mathbb{R}}([\![\Lambda^{2,0}M]\!])$, which in turn decomposes further as

$$\mathrm{End}_{\mathbb{R}}([\![\Lambda^{2,0}M]\!]) = \Big(\mathrm{End}_{\mathbb{R}}([\![\Lambda^{2,0}M]\!])\Big)' \oplus \Big(\mathrm{End}_{\mathbb{R}}([\![\Lambda^{2,0}M]\!])\Big)'' \, ,$$

into the sub-bundles of endomorphisms of $[\![\Lambda^{2,0}M]\!]$ which commute, respectively, anti-commute with the action of J on $[\![\Lambda^{2,0}M]\!]$. Denoting by $\widetilde{R}'$ and $\widetilde{R}''$ the corresponding components of $\widetilde{R}$, Gray's identity is [**Gra76**]

$$\widetilde{R}' = -\frac{1}{4}\sum(\nabla_{e_j}\Omega)\otimes(\nabla_{e_j}\Omega)\ . \tag{17.15}$$

As for the component $\widetilde{R}''$, from its definition we have

$$(\widetilde{R}'')_{X,Y,Z,T} = \frac{1}{8}\Big(R_{X,Y,Z,T} - R_{JX,JY,Z,T} - R_{X,Y,JZ,JT} + R_{JX,JY,JZ,JT}$$
$$+R_{X,JY,Z,JT} + R_{JX,Y,Z,JT} + R_{X,JY,JZ,T} + R_{JX,Y,JZ,T}\Big),$$

showing that

$$(\widetilde{R}'')_{Z_1,Z_2,Z_3,Z_4} = R_{Z_1,Z_2,Z_3,Z_4} = W_{Z_1,Z_2,Z_3,Z_4} \quad \forall Z_i \in T^{1,0}M.$$

Thus, $\widetilde{R}''$ is actually determined by the Weyl curvature of M. The next Weitzenböck-type formula provides a further obstruction, this time to the lift of the 3-jet of J to the 4-jet. We skip the proof, which can be found in [**ADM01**].

PROPOSITION 17.11. *For any almost Kähler structure (g, J, Ω) the following relation holds:*

$$\begin{aligned}(17.16)\ \Delta(s^* - s) \quad = \quad & -4\delta\big(J\delta(J\mathrm{Ric}'')\big) + 8\delta\big(\langle\rho^*, \nabla.\,\Omega\rangle\big) + 2|\mathrm{Ric}''|^2 \\ & -8|\widetilde{R}''|^2 - |\nabla^*\nabla\Omega|^2 - |\varphi|^2 + 4\langle\rho, \varphi\rangle - 4\langle\rho, \nabla^*\nabla\Omega\rangle\ ,\end{aligned}$$

where the semi-positive (1,1)-form φ is given by $\varphi(X, Y) = \langle\nabla_{JX}\Omega, \nabla_Y\Omega\rangle$; δ denotes the co-differential with respect to ∇, acting on 1-forms and 2-tensors.

Integrating (17.16) over the manifold, one obtains an integral formula identical to the one in [**Sek87**, Proposition 3.2], up to some integration by parts. In particular we have:

COROLLARY 17.12. ([**Sek87**]) *For any compact almost Kähler manifold with J-invariant Ricci tensor the following inequality holds:*

$$\int_M \left[4\langle\rho, \varphi\rangle - 4\langle\rho, \nabla^*\nabla\Omega\rangle - |\nabla^*\nabla\Omega|^2 - |\varphi|^2\right] dV_g \geq 0, \tag{17.17}$$

where $dV_g = \frac{1}{n!}\Omega^n$ is the volume form of g.

Remark 3. As shown by Sekigawa, the above inequality gives an obstruction to the (global) existence of strictly almost Kähler structures, when the metric g is Einstein with non-negative scalar curvature. Indeed, in this case $\mathrm{Ric}'' = 0$ and $2\langle\rho, \varphi\rangle = \langle\rho, \nabla^*\nabla\Omega\rangle = \frac{s}{2n}|\nabla\Omega|^2$, so that, by (17.17), $\nabla\Omega = 0$, i.e. J is necessarily Kähler. In dimension 4, other integrability results have been derived from (17.16); see e.g. [**Dra99**], [**AA00**], [**OS98**]. **Proof of Theorem 17.1.** We now turn back

to the notation used in Section 3.1. Thus, (g, J, Ω) denotes the Kähler structure, with Ricci tensor having two non-negative distinct constant eigenvalues $0 \leq \lambda < \mu$, while $(g, \overline{J}, \overline{\Omega})$ is the almost Kähler structure constructed by Lemma 17.8; we shall also use the (1,1)-forms α and β introduced in Section 2.1, so that we have

$$\Omega = \alpha + \beta;\ \overline{\Omega} = \alpha - \beta;\ \rho = \lambda\alpha + \mu\beta;\ \overline{\rho} = \lambda\alpha - \mu\rho, \tag{17.18}$$

where ρ and $\overline{\rho}$ are the Ricci forms of (g, J) and $(g, \overline{J})$, respectively. For proving Theorem 17.1 it is enough to show that $\overline{J}$ is integrable (see Lemma 17.8), or equivalently, that $\nabla\overline{\Omega} = 0$. The latter will be derived from the integral inequality stated in

Corollary 17.12 (see Remark 3). Let $\overline{\varphi}(X,Y) = \langle \nabla_{\overline{J}X}\overline{\Omega}, \nabla_Y\overline{\Omega}\rangle$ be the semi-positive definite (1,1)-form with respect to $\overline{J}$, defined in Proposition 17.11. By (17.18) and using the semi-positivity of the (1,1)-forms α and $\overline{\varphi}$, we get

$$\begin{aligned}(17.19)\quad \langle\overline{\rho},\overline{\varphi}\rangle - \langle\overline{\rho},\nabla^*\nabla\overline{\Omega}\rangle &= (\lambda-\mu)\langle\alpha,\overline{\varphi}\rangle + (\mu-\lambda)\langle\alpha,\nabla^*\nabla\overline{\Omega}\rangle \\ &\quad +\mu\langle\overline{\Omega},\overline{\varphi}\rangle - \mu\langle\overline{\Omega},\nabla^*\nabla\overline{\Omega}\rangle \\ &= (\lambda-\mu)\langle\alpha,\overline{\varphi}\rangle + (\mu-\lambda)\langle\alpha,\nabla^*\nabla\overline{\Omega}\rangle - \frac{\mu}{2}|\nabla\overline{\Omega}|^2 \\ &\le (\mu-\lambda)\langle\alpha,\nabla^*\nabla\overline{\Omega}\rangle - \frac{\mu}{2}|\nabla\overline{\Omega}|^2.\end{aligned}$$

Since $\langle\alpha,\nabla\overline{\Omega}\rangle = 0$ (because α and $\nabla\overline{\Omega}$ are of type $(1,1)$ and $(2,0)+(0,2)$, respectively), we have

$$\langle\alpha,\nabla^*\nabla\overline{\Omega}\rangle = \langle\nabla\alpha,\nabla\overline{\Omega}\rangle = \frac{1}{2}|\nabla\overline{\Omega}|^2\ ,$$

where in the last step we used that $\alpha = \frac{1}{2}(\Omega+\overline{\Omega})$ and Ω is parallel. Substituting into the inequality (17.19), we obtain

$$\langle\overline{\rho},\overline{\varphi}\rangle - \langle\overline{\rho},\nabla^*\nabla\overline{\Omega}\rangle \le -\frac{\lambda}{2}|\nabla\overline{\Omega}|^2.$$

Since by assumption $\lambda \ge 0$, the latter inequality shows that $\langle\overline{\rho},\overline{\varphi}\rangle - \langle\overline{\rho},\nabla^*\nabla\overline{\Omega}\rangle$ is an everywhere non-positive function and Corollary 17.12 then implies that $\nabla^*\nabla\overline{\Omega} = 0$; after multiplying by $\overline{\Omega}$ we reach $\nabla\overline{\Omega} = 0$. □

Andrei Moroianu, CMAT, École Polytechnique, UMR 7640 du CNRS, 91128 Palaiseau, France.
E-mail address: am@math.polytechnique.fr

Bibliography

[AA00] V. Apostolov and J. Armstrong, *Symplectic 4-manifolds with Hermitian Weyl tensor*, Trans. Amer. Math. Soc. **352** (2000), 4501–4513.

[AAD] V. Apostolov, J. Armstrong, and T. Draghici, *Local rigidity of certain classes of almost Kähler 4-manifolds*, math.DG/9911197.

[AB98] B. Ammann and C. Bär, *The Dirac Operator on Nilmanifolds and Collapsing Circle Bundles*, Ann. Global Anal. Geom. **16** (1998), no. 3, 221–253.

[AB00] ———, *Dirac eigenvalues and total scalar curvature*, J. Geom. Phys. **33** (2000), 229–234.

[AB02a] ———, *Dirac eigenvalue estimates on surfaces*, Math. Zeitschrift **240** (2002), 423–449.

[AB02b] I. Avramidi and T. Branson, *A discrete leading symbol and spectral asymptotics for natural differential operators*, J. Funct. Anal. **190** (2002), 292–337.

[ABS64] M.F. Atiyah, R. Bott, and A. Shapiro, *Clifford Modules*, Topology **3** (1964), 3–38.

[ACDS98] D.V. Alekseevsky, V. Cortes, C. Devchand, and U. Semmelmann, *Killing spinors are Killing vector fields in Riemannian supergeometry*, J. Geom. Phys. **26** (1998), no. 1-2, 51–78.

[ADM01] V. Apostolov, T. Draghici, and A. Moroianu, *A splitting theorem for Kähler manifolds with constant eigenvalues of the Ricci tensor*, Int. J. Math. **12** (2001), 769–789.

[AF99] I. Agricola and T. Friedrich, *Upper bounds for the first eigenvalue of the Dirac operator on surfaces*, J. Geom. Phys. **30** (1999), 1–22.

[Agr] I. Agricola, *Connections on naturally reductive spaces, their Dirac operator and homogeneous models in string theory*, arXiv:math.DG/0202094 v1.

[AH70] M.F. Atiyah and F. Hirzebruch, *Spin manifolds and group actions*, Essays in Topology and Related Topics: Mémoires Dediés à Georges de Rham (A. Haefliger and B. Narashmhan, eds.), 1970, pp. 18–28.

[ÁL92] J. Álvarez-López, *The basic component of the mean curvature of Riemannian foliations*, Ann. Global Anal. Geom. **10** (1992), 179–194.

[Ale62] A.D. Alexandrov, *A characteristic property of spheres*, Ann. Mat. Pura Appl. **58** (1962), 303–315.

[Ale68] D.V. Alekseevskii, *Riemannian spaces with unusual holonomy groups*, Funct. Anal. Appl. **2** (1968), 97–105.

[Alm66] F.J. Almgren, *Some interior regularity theorems for minimal surfaces and an extension of Bernstein's theorem*, Ann. of Math. **84** (1966), 277–292.

[Amm] B. Ammann, *Ambient Dirac eigenvalue estimates and the Willmore functional*, in this volume.

[Amm98] ———, *Spin-Strukturen und das Spektrum des Dirac-Operators*, Shaker-Verlag Aachen, 1998, Ph.D. thesis, University of Freiburg, Germany.

[Amm00a] ———, *Spectral estimates on 2-tori*, Preprint, CUNY, Graduate Center, 2000.

[Amm00b] ———, *The Willmore Conjecture for immersed tori with small curvature integral*, Manuscripta Math. **101** (2000), 1–22.

[Amm03] ———, *A spin-conformal lower bound of the first positive Dirac eigenvalue*, Diff. Geom. and its Appl. **18** (2003), no. 1, 21–32.

[APS75] M.F. Atiyah, V.K. Patodi, and I.M. Singer, *Spectral asymmetry and Riemannian geometry, I, II and III*, Math. Proc. Cambridge Phil. Soc. **77, 78, 79** (1975), 43–69, 405–432, 71–99.

[Arm02] J. Armstrong, *An Ansatz for Almost-Kähler, Einstein 4-manifolds*, J. reine angew. Math. **542** (2002), 53–84.

[AS53] A. Ambrose and I. Singer, *A theorem on holonomy*, Trans. Amer. Math. Soc. **75** (1953), 428–443.

[AS68a] M.F. Atiyah and G. Segal, *The index of elliptic operators, II*, Annals of Math. **87** (1968), 531–545.

[AS68b] M.F. Atiyah and I. Singer, *The index of elliptic operators, III*, Annals of Math. **87** (1968), 546–604.

[AS68c] ———, *The index of elliptic operators, saierI*, Annals of Math. **87** (1968), 484–530.

[AS69] ———, *Index theory for skew-adjoint Fredholm operators, IV*, I. H. É. S. **37** (1969), 305–326.

[AS71a] ———, *The index of elliptic operators, IV*, Annals of Math. **93** (1971), 119–138.

[AS71b] ———, *The index of elliptic operators, V*, Annals of Math. **93** (1971), 139–149.

[AS77] M. Atiyah and W. Schmid, *A geometric construction of the discrete series for semisimple Lie groups*, Inventiones Math. **42** (1977), 1–62.

[Ati67] M.F. Atiyah, *K-Theory*, Benjamin Press, New York, 1967.

[Ati70] ———, *Vector fields on manifolds*, Arbeitsgemeinschaft fur Forschung des Landes Nordrhein-Westfalen **200** (1970), 7–23.

[Bai97] P.D. Baier, *Über den Diracoperator auf Mannigfaltigkeiten mit Zylinderenden*, Master's thesis, Diplomarbeit, Universität Freiburg, 1997.

[Bär91] C. Bär, *Das Spektrum von Dirac-Operatoren*, vol. 217, Bonner Math. Schr., 1991, Doktorarbeit, Universität Bonn.

[Bär92a] ———, *Lower eigenvalue estimates for Dirac operators*, Math. Ann. **293** (1992), 39–46.

[Bär92b] ———, *The Dirac operator on homogeneous spaces and its spectrum on 3-dimensional lens spaces*, Arch. Math. **59** (1992), 65–79.

[Bär92c] ———, *Upper eigenvalue estimates for Dirac operators*, Ann. Glob. Anal. Geom. **10** (1992), 171–177.

[Bär93] ———, *Real Killing spinors and holonomy*, Commun. Math. Phys. **154** (1993), 509–521.

[Bär96a] ———, *Metrics with harmonic spinors*, Geom. Func. Anal. **6** (1996), 899–942.

[Bär96b] ———, *The Dirac operator on space forms of positive curvature*, J. Math. Soc. Japan **48** (1996), 69–83.

[Bär98] ———, *Extrinsic bounds for eigenvalues of the Dirac operator*, Ann. Glob. Anal. Geom. **16** (1998), 573–596.

[Bär00] ———, *The Dirac operator on hyperbolic manifolds of finite volume*, J. Diff. Geom. **54** (2000), 438–488.

[Bau89a] H. Baum, *Complete Riemannian manifolds with imaginary Killing spinors*, Ann. Glob. Anal. Geom. **7** (1989), 205–226.

[Bau89b] ———, *Odd-dimensional Riemannian manifolds admitting imaginary Killing spinors*, Ann. Glob. Anal. Geom. **7** (1989), 141–153.

[Bau99] H. Baum, *Lorentzian twistor spinors and CR-geometry*, Diff. Geom. and its Appl. **11** (1999), no. 1, 69–96.

[BBW93] B. Booß-Bavnbek and K.P. Wojciechowski, *Elliptic Boundary Problems for the Dirac Operator*, Birkhäuser, Basel, 1993.

[BC97] I.M. Benn and P. Charlton, *Dirac symmetry operators from conformal Killing-Yano tensors*, Class. Quant. Grav. **14** (1997), 1037–1042.

[BCK97] I.M. Benn, P. Charlton, and J. Kress, *Debye potentials for Maxwell and Dirac fields from a generalisation of the Killing-Yano equation*, J. Math. Phys. **38** (1997), 4504–4527.

[BDS77] D. Burns, K. Diederich, and S. Shnider, *Distinguished curves in pseudoconvex boundaries*, Duke Math. J. **44** (1977), 407–431.

[BE] T.N. Bailey and A.W. Knapp (Editors), *Representation Theory and automorphic forms*, Proceedings of Symposia in Pure Mathematics, vol. 61, American Mathematical Society, Instructional Conference, March 1996, Edinburgh, Scotland.

[BEG94] T.N. Bailey, M.G. Eastwood, and A.R. Gover, *The Thomas structure bundle for conformal, projective and related structures*, Rocky Mountain J. Math. **24** (1994), 1–27.

[Ber55] M. Berger, *Sur les groupes d'holonomie des variétés connexion affine et des variétés riemanniennes*, Bull. Soc. Math. France **83** (1955), 279–330.

[Bér93] P. Bérard, *Transplantation et isospectralité II*, J. London Math. Soc. **48** (1993), no. 3, 565–576.

[Bes87] A. Besse, *Einstein Manifolds*, Ergebnisse der Mathematik und ihrer Grenzgebiete, 3. Folge, Band 10, Springer-Verlag, Berlin, 1987.

[BFGK91] H. Baum, T. Friedrich, R. Grunewald, and I. Kath, *Twistor and Killing Spinors on Riemannian Manifolds*, vol. 108, Teubner–Verlag, Stuttgart/Leipzig, 1991.

[BFL+02] F. Burstall, D. Ferus, K. Leschke, F. Pedit, and U. Pinkall, *Conformal geometry of surfaces in S^4 and quaternions*, Lect. Notes in Math. **1772** (2002), 171–189.

[BFOHP01] M. Blau, J.M. Figueroa-O'Farrill, C. Hull, and G. Papadopoulos, *New maximally supersymmetric background of IIB superstring theory*, hep-th/0110242, 2001.

[BG] T. Branson and A.R. Gover, *Spin-tractors*, in preparation.

[BG72] R. Brown and A. Gray, *Riemannian manifolds with holonomy group* Spin(9), Differential Geometry (in honor of Kentaro Yano), Kinokuniya, Tokyo, 1972, pp. 41–59.

[BGV91] N. Berline, E. Getzler, and M. Vergne, *Heat Kernels and Dirac Operators*, Springer-Verlag, Berlin-New York, 1991.

[BH60] A. Borel and F. Hirzebruch, *Characteristic classes and homogeneous spaces I, II, III*, Amer. J. Math. **80, 81, 82** (1958, 1959, 1960), 458–538, 315–382, 491–504.

[BH97] T. Branson and O. Hijazi, *Vanishing theorems and eigenvalue estimates in Riemannian spin geometry*, International J. Math. **8** (1997), 921–934.

[BH00] ———, *Improved forms of some vanishing theorems in Riemannian spin geometry*, International J. Math. **11** (2000), 291–304.

[BH02] ———, *Bochner-Weitzenböck formulas associated with the Rarita-Schwinger operator*, International J. Math. **13** (2002), 137–182.

[BHMM] J.P. Bourguignon, O. Hijazi, J.-L. Milhorat, and A. Moroianu, *A Spinorial Approach to Riemannian and Conformal Geometry*, Monograph (In Preparation).

[BL] H. Baum and F. Leitner, *The twistor equation in Lorentzian spin geometry*, to appear in Math. Zeit, 2004.

[BL52] A. Borel and A. Lichnerowicz, *Groupes d'holonomie des variétés riemanniennes*, C.R. Acad. Sci. Paris **234** (1952), 1835–1837.

[Boh03] Ch. Bohle, *Killing spinors on Lorentzian manifolds*, J. Geom. Phys., 2003.

[BÓO96] T. Branson, G. Ólafsson, and B. Orsted, *Spectrum generating operators, and intertwining operators for representations induced from a maximal parabolic subgroup*, J. Funct. Anal. **135** (1996), 163–205.

[Bot59] R. Bott, *Stable homotopy of the classical groups*, Ann. of Math. **70** (1959), 313–337.

[Bot69] ———, *Lectures on $K(X)$*, Benjamin Press, New York, 1969.

[Bou68] N. Bourbaki, *Groupes et Algèbres de Lie, Chap. IV, V, VI.*

[Bou71] ———, *Groupes et Algèbres de Lie, I*, vol. 1258, Hermann, Paris, 1971.

[Bou72] ———, *Groupes et Algèbres de Lie, Chap. 2,3*, vol. 1349, Hermann, Paris, 1972.

[BP97] P. Bérard and H. Pesce, *Construction de variétés isospectrales autour du théorème de T. Sunada*, Progress in inverse spectral geometry, Basel: Birkhaeuser, 1997, pp. 63–83.

[BPR] V. Belfi, E. Park, and K. Richardson, *A Hopf index theorem for foliations*, Preprint.

[Bra84] T. Branson, *Intertwining differential operators for spinor-form representations of the conformal group*, Advances in Math. **54** (1984), 1–21.

[Bra96] ———, *Nonlinear phenomena in the spectral theory of geometric linear differential operators*, Proc. Symp. Pure Math. **59** (1996), 27–65.

[Bra97a] ———, *Stein-Weiss operators and ellipticity*, J. Func. Analysis **151** (1997), 334–383.

[Bra97b] ———, *Stein-Weiss operators and ellipticity*, J. Funct. Anal. **151** (1997), 334–383.

[Bra99] ———, *Spectra of self-gradients on spheres*, J. Lie Theory **9** (1999), 491–506.

[Bra00] ———, *Kato constants in Riemannian geometry*, Math. Research Letters **7** (2000), 245–261.

[Bry] R. Bryant, *private communication.*

[Bry87] ———, *Metrics with exceptional holonomy*, Ann. Math. **126** (1987), 525–576.

[Bry96] ———, *Classical, exceptional, and exotic holonomies: a status report*, Actes de la Table Ronde de Géométrie Différentielle (Luminy, 1992), Sémin. Congr., 1, Soc. Math. France, Paris, 1996, pp. 93–165.

[Bry00] ———, *Pseudo-Riemannian metrics with parallel spinor fields and vanishing Ricci tensor. In: Global Analysis and Harmonic Analyis. eds J.-P. Bourguignon, T. Branson, O. Hijazi*, Seminaires et Congress of the Soc. Math. France **4** (2000), 53–94.

[Bun91a] U. Bunke, *The spectrum of the Dirac operator on the hyperbolic space*, Math. Nachr. **153** (1991), 179–190.

[Bun91b] ———, *Upper bounds of small eigenvalues of the Dirac operator and isometric immersions.*, Ann. Global Anal. Geom. **9** (1991), 109–116.

[BvCE+] T. Branson, A. ˇ Cap, M.G. Eastwood, , and A.R. Gover, *Prolongations of geometric overdetermined systems*, Preprint, math.DG/0402100.

[BW76] D. Bleecker and J. Weiner, *Extrinsic bound of λ_1 of Δ on a compact manifold*, Comment. Math. Helv. **51** (1976), 601–609.

[BW80] A. Borel and N. Wallach, *Continuous cohomology, Discrete Subgroups, and Representations of Reductive Groups*, Annals of Math Studies (1980), no. 94.

[Car25] É. Cartan, *La Géometrie des Espaces de Riemann*, Mémorial des Sciences Mathematiques, Fasc. IX, (Especially, Chapitre VII, Section II.), 1925.

[Car26] ———, *Les groupes d'holonomie des espaces généralisés*, Acta. Math. **48** (1926), 1–42.

[Car84] Y. Carrière, *Flots riemanniens*, Astérisque, vol. 116, 1984, Journées sur les structures transverses des feuilletages, Toulouse.

[CFG89] M. Cahen, A. Franc, and S. Gutt, *Spectrum of the Dirac operator on complex projective space $P_{2q-1}(\mathbb{C})$*, Lett. Math. Phys. **18** (1989), 165–176.

[CFG94] ———, *Erratum to "The spectrum of the Dirac operator on complex projective space $P_{2q-1}(\mathbb{C})$"*, Lett. Math. Phys. **32** (1994), 365–368.

[CGH00a] D. Calderbank, P. Gauduchon, and M. Herzlich, *On the Kato inequaliy in Riemannian geometry*, Global Analysis and Harmonic Analysis (Marseille-Luminy, 1999) (J.-P. Bourguignon, T. Branson, and O. Hijazi, eds.), Séminaires & Congrès, Collection Société Mathématique de France, no. 4, 2000, pp. 95–113.

[CGH00b] ———, *Refined Kato inequalities and conformal weights in Riemannian geometry*, J. Funct. Anal. **173** (2000), 214–255.

[CGLS86] M. Cahen, S. Gutt, L. Lemaire, and P. Spindel, *Killing spinors*, Bull. Soc. Math. Belg. Sér. A **38** (1986), 75–102.

[CH96] R. Camporesi and A. Higuchi, *On the eigenfunctions of the Dirac operator on spheres and real hyperbolic spaces*, J. Geom. Phys. **20** (1996), 1–18.

[Cha00] F. El Chami, *Spectre du laplacien sur les formes versus spectre des volumes : le cas des grassmanniennes*, Ph.D. thesis, Thèse de doctorat de l'université Paris-Sud, 2000.

[Che79] B.Y. Chen, *On the total curvature of immersed manifolds IV: Spectrum and total mean curvature*, Bull. Math. Acad. Sinica **7** (1979), 301–311.

[Chr70] E.B. Christoffel, *Über die Transformationen der homogenen Differentialausdrücke zweiten Grades*, Borchardt J. **LXX** (1870), 46–70.

[CJJ+74] A. Chodos, R.L. Jaffe, K. Johnson, C.B. Thorn, and V.S. Weisskopf, *New extended model of hadrons*, Phys. Rev. D **9** (1974), 3471–3495.

[CJJT74] A. Chodos, R.L. Jaffe, K. Johnson, and C.B. Thorn, *Baryon structure in the bag theory*, Phys. Rev. D **10** (1974), 2599–2604.

[CP02] R. Camporesi and E. Pedon, *The continuous spectrum of the Dirac operator on noncompact Riemannian symmetric spaces of rank one*, Proc. Amer. Math. Soc. **130** (2002), 507–516.

[CRC01] T. Levi Civita and G. Ricci-Curbastro, *Méthodes de calcul différentiel absolu et leurs applications*, Math. Ann. **54** (1901), 125–201.

[CW83] H.I. Choi and A.N. Wang, *A first eigenvalue estimate for minimal hypersurfaces*, J. Diff. Geom. (1983), 559–562.

[DG89] D.M. DeTurck and C.S. Gordon, *Isospectral deformations. II: Trace formulas, metrics, and potentials*, Commun. Pure Appl. Math. **42** (1989), no. 8, 1067–1095.

[DK00] J.J. Duistermaat and J.A.C. Kolk, *Lie Groups*, Springer Verlag, Berlin, 2000.

[DNP86] M. Duff, B. Nilsson, and C. Pope, *Kaluza-Klein Supergravity*, Phys. Rep. **130** (1986), 1–142.

[Dra99] T. Draghici, *Almost Kähler 4-manifolds with J-invariant Ricci tensor*, Houston J. Math. **25** (1999), 133–145.

[dV73] Y. Colin de Verdière, *Spectre du laplacien et longueurs des géodésiques périodiques I et II*, Compositio Math. **27** (1973), 83–106 et 159–184.

[EH86] A. ElKacimi and G. Hector, *Décomposition de Hodge basique pour un feuilletage riemannien*, Ann. Inst. Fourier, Grenoble **36** (1986), no. 3, 207–227.

[ElK90] A. ElKacimi, *Opérateurs transversalement elliptiques sur un feuilletage riemannien et applications*, Compositio Mathematica **73** (1990), 57–106.

[EN93] A. ElKacimi and M. Nicolau, *On the topological invariance of the basic cohomology*, Math. Ann. **295** (1993), 627–634.

[ER87] M.G. Eastwood and J.W. Rice, *Conformally invariant differential operators on Minkowski space and their curved analogues*, Commun. Math. Phys. **109** (1987), 207–228, Erratum, Commun. Math. Phys. **144** (1992), 223.

[ES52] S. Eilenberg and N. Steenrod, *Foundations of Algebraic Topology*, Princeton University Press, Princeton, 1952.

[Esc92] J.F. Escobar, *Conformal deformation of a Riemannian metric to a scalar flat metric with constant mean curvature on the boundary*, Ann. Math. **136** (1992), 1–50, (Addendum in **139** (1994),749–750.).

[Esc00] ———, *A comparison theorem for the first non-zero Steklov eigenvalue*, J. Funct. Anal. **178** (2000), 143–155.

[Fef76] C. Fefferman, *Monge–Ampère equations, the Bergman kernel, and geometry of pseudoconvex domains*, Ann. Math. **103** (1976), 395–416.

[Feg76] H. Fegan, *Conformally invariant first order differential operators*, Quart. J. Math.(Oxford) **27** (1976), 371–378.

[Feg87] ———, *The spectrum of the Dirac operator on a simply connected compact Lie group*, Simon Stevin **61** (1987), 97–108.

[FG85] T. Friedrich and R. Grunewald, *On the first eigenvalue of the Dirac operator on 6-dimensional manifolds*, Ann. Global Anal. Geom. **3** (1985), 265–273.

[FK88] T. Friedrich and I. Kath, *Variétés riemanniennes compactes de dimension 7 admettant des spineurs de Killing*, C. R. Acad. Sci. Paris **307** (1988), 967–969.

[FK89] ———, *Einstein manifolds of dimension five with small first eigenvalue of the Dirac operator*, J. Diff. Geom. **29** (1989), 263–279.

[FK90] ———, *7-dimensional compact Riemannian manifolds with Killing spinors*, Commun. Math. Phys. **133** (1990), 543–561.

[FK02] T. Friedrich and K.-D. Kirchberg, *Eigenvalue estimates for the Dirac operator depending on the Weyl tensor*, J. Geom. Phys. **41** (2002), 196–207.

[FLPP01] D. Ferus, K. Leschke, F. Pedit, and U. Pinkall, *Quaternionic holomorphic geometry: Plücker formula, Dirac eigenvalue estimates and energy estimates of harmonic 2-tori*, Invent. Math. **146** (2001), 507–593.

[FOP01] J.M. Figueroa-O'Farrill and G. Papadopoulos, *Homogeneous fluxes, branes and a maximally supersymmetric solution of M-theory*, hep-th/0105308, 2001.

[Fra72] D. Frank, *On the index of a tangent 2-field*, Topology **11** (1972), 245–252.

[Fra87] A. Franc, *Spin structures and Killing spinors on lens spaces*, J. Geom. Phys. **4** (1987), 277–287.

[Fre82] M. Freedman, *The topology of four-dimensional manifolds*, J. Diff. Geom. **17** (1982), 357–453.

[Fri80] T. Friedrich, *Der erste Eigenwert des Diracoperators einer kompakten Riemannschen Mannigfaltigkeit nichtnegativer Skalarkrümmung*, Math. Nachr. **97** (1980), 117–146.

[Fri84] ———, *Zur Abhängigkeit des Dirac-Operators von der Spin-Struktur*, Coll. Math. **48** (1984), 57–62.

[Fri89] ———, *On the conformal relation between twistors and Killing spinors*, Suppl. Rend. Circ. Math. Palermo **II** (1989), no. 22, 59–75.

[Fri93] ———, *The classification of 4-dimensional Kähler manifolds with small eigenvalue of the Dirac operator*, Math. Ann. **295** (1993), 565–574.

[FS98] S. Farinelli and G. Schwarz, *On the spectrum of the Dirac operator under boundary conditions*, J. Geom. Phys. **28** (1998), 67–84.

[Gal79] S. Gallot, *Équations différentielles caractéristiques de la sphère*, Ann. Sci. Ec. Norm. Sup. Paris **12** (1979), 235–267.

[GHL87] S. Gallot, D. Hulin, and J. Lafontaine, *Riemannian Geometry*, Universitext, Springer-Verlag, Berlin Heidelberg New York, 1987.

[Gil84] P. Gilkey, *Invariance Theory, the Heat Equation and the Atyiah-Singer Index Theorem*, Math. Lecture Series, vol. 4, Publish or Perish Press, Berkeley, California, 1984.

[Gin02] N. Ginoux, *Reilly-type spinorial inequalities*, Math. Zeit. **241** (2002), 513–525.

[GL80a] M. Gromov and H.B. Jr. Lawson, *Spin and scalar curvature in the presence of a fundamental group*, Ann. of Math. **111** (1980), 209–230.

[GL80b] ———, *The classification of simply-connected manifolds of positive scalar curvature*, Ann. of Math. **111** (1980), 423–434.

[GL83] ———, *Positive scalar curvature and the Dirac operator on complete riemannian manifolds*, Publ. I.H.E.S. **58** (1983), 295–408.

[GN] A.R. Gover and P. Nurowski, *Obstructions to conformally Einstein metrics in n dimensions*, Preprint, math.DG/0405304.

[Gol69] S.I. Goldberg, *Integrability of almost Kähler manifolds*, Proc. Amer. Math. Soc. **21** (1969), 96–100.

[Gor86] C.S. Gordon, *The Laplace spectra versus the length spectra of Riemannian manifolds*, Nonlinear problems in geometry, Proc. AMS Spec. Sess., 820th Meet., vol. 51, AMS, Contemp. Math., 1986, pp. 63–80.

[Gor94] R. Gornet, *The length spectrum and representation theory on two and three-step nilpotent Lie groups*, Geometry of the spectrum (P. Perry R. Brooks, C.S. Gordon, ed.), vol. 173, AMS, Contemp. Math, 1994, pp. 133–155.

[Gor96] ———, *The marked length spectrum vs. the Laplace spectrum on forms on Riemannian nilmanifolds*, Comment. Math. Helv. **71** (1996), no. 2, 297–329.

[Gra76] A. Gray, *Curvature identities for Hermitian and almost Hermitian manifolds*, Tôhoku Math. J. **28** (1976), 601–612.

[Gra87] C.R. Graham, *On Sparling's characterization of Fefferman metrics*, Amer. J. Math. **109** (1987), 853–874.

[Gro86a] M. Gromov, *Large Riemannian manifolds*, Curvature and topology of Riemannian manifolds (Katata), Lecture Notes in Math., no. 1201, Springer-Verlag, Berlin-New York, 1986, pp. 108–121.

[Gro86b] ———, *Partial Differential Relations*, Springer-Verlag, Berlin-New York, 1986.

[Gro96] ———, *Carnot-Carathéodory spaces seen from within, in* Sub-Riemannian geometry, Progr. Math. **144**, Birkhäuser, Basel, 1996.

[Gru90] R. Grunewald, *Six-dimensional Riemannian manifolds with a real Killing spinor*, Ann. Global Anal. Geom. **8** (1990), 43–59.

[GS02] S. Goette and U. Semmelmann, *The point spectrum of the Dirac operator on noncompact symmetric spaces*, Proc. Amer. Math. Soc. **130** (2002), 915–923.

[GSW87] M. Green, J. Schwartz, and E. Witten, *Superstring Theory*, vol. 2, Cambridge University Press, 1987.

[GWW92] C. Gordon, D.L. Webb, and S. Wolpert, *One cannot hear the shape of a drum*, Bull. Amer. Math. Soc. **27** (1992), 134–138.

[Hab90] K. Habermann, *The twistor equation on Riemannian manifolds*, J. Geom. Phys. **7** (1990), 469–488.

[Hab93] ———, *The graded algebra and the conformal Lie derivative of spinor fields related to the twistor equation*, preprint, 1993.

[Hab94] ———, *Twistor spinors and their zeroes*, J. Geom. Phys. **14** (1994), 1–24.

[Hab96] ———, *The graded algebra and the derivative of spinor fields related to the twistor equation*, J. Geom. Phys. **18** (1996), 131–146.

[Hel78a] S. Helgason, *Differential Geometry, Lie Groups, and Symmetric Spaces*, Academic Press, New York, 1978.

[Hel78b] ———, *Differential Geometry, Lie Groups, and Symmetric Spaces*, Academic Press, New York, 1978.

[Her70] J. Hersch, *Quatre propriétés isopérimétriques de membranes sphériques homogènes*, C. R. Acad. Sci. Paris Sér. A-B **270** (1970), A1645–A1648.

[Her98] M. Herzlich, *A Penrose-like inequality for the mass of Riemannian asymptotically flat manifolds*, Comm. Math. Phys. **188** (1998), 121–133.

[Hij86a] O. Hijazi, *A conformal lower bound for the smallest eigenvalue of the Dirac operator and Killing spinors*, Comm. Math. Phys. **104** (1986), 151–162.

[Hij86b] ———, *Caractérisation de la sphère par les premières valeurs propres de l'opérateur de Dirac en dimensions* 3, 4, 7 *et* 8, C. R. Acad. Sci. Paris Sér. I Math. **303** (1986), 417–419.

[Hij91] ———, *Première valeur propre de l'opérateur de Dirac et nombre de Yamabe*, C. R. Acad. Sci. Paris Sér. I Math. **313** (1991), 865–868.

[Hir66] F. Hirzebruch, *Topological Methods in Algebraic Topology*, Springer-Verlag, Berlin-New York, 1966.

[Hir76] M. Hirsch, *Differential Topology*, Springer-Verlag, Berlin-New York, 1976.

[Hit74] N. Hitchin, *Harmonic spinors*, Adv. in Math. **14** (1974), 1–55.

[HL82] F.R. Harvey and H.B. Jr. Lawson, *Calibrated geometries*, Acta Math. **148** (1982), 47–157.

[HM97] O. Hijazi and J.-L. Milhorat, *Twistor operators and eigenvalues of the Dirac operator on compact quaternionic spin manifolds*, Ann. Global Anal. Geom. **2** (1997), 117–131.

[HM03] O. Hijazi and S. Montiel, *Extrinsic Killing spinors*, Math. Zeit., 2003.

[HMR02] O. Hijazi, S. Montiel, and A. Roldán, *Eigenvalue boundary problems for the Dirac operator*, Commun. Math. Phys. **231, Issue 3** (2002), 375–390.

[HMR03] O. Hijazi, S. Montiel, and A. Roldán, *Dirac operator on hypersurfaces in negatively curved manifolds*, Ann. Global Anal. Geom., 2003.

[HMZ01a] O. Hijazi, S. Montiel, and X. Zhang, *Dirac operator on embedded hypersurfaces*, Math. Res. Lett. **8** (2001), 195–208.

[HMZ01b] ———, *Eigenvalues of the Dirac operator on manifolds with boundary*, Comm. Math. Phys. **221** (2001), 255–265.

[HMZ02] ———, *Conformal lower bounds for the Dirac operator of embedded hypersurfaces*, Asian J. Math. **6** (2002), 23–36.

[Hop83] H. Hopf, *Differential geometry in the large*, Lecture Notes in Math., vol. 1000, Springer-Verlag, Berlin, 1983.

[Hör85] L. Hörmander, *The Analysis of Linear Partial Differential Operators III*, Springer, Berlin, 1985.

[How92] R. Howe, *A Century of Lie Theory*, Mathematics into the Twenty-first Century, AMS Centennial Publications, Providence, RI,VII (1992), 101–320.

[HP02] J.-S. Huang and P. Pandžić, *Dirac cohomology, unitary representations and a proof of a conjecture of Vogan*, J. Amer. Math. Soc. **15** (2002), no. 1, 185–202.

[HTY83] W.Y. Hsiang, Z.H. Teng, and W.C. Yu, *New examples of constant mean curvature immersions of $(2k-1)$-spheres into Euclidean $2k$-space*, Ann. Math. **117** (1983), 609–625.

[Hua01] J.-S. Huang, *Invariant differential operators and eigenspace representations on an affine symmetric space*, Ann. of Math. **2, 154** (2001), no. 3, 703–737.

[Hum72] J. Humphreys, *Introduction to Lie Algebras and Representation Theory*, Springer-Verlag, Berlin, 1972.

[Hus94] D. Husemoller, *Fibre bundles*, Graduate Texts in Mathematics, vol. 20, Springer-Verlag, 1994.

[HZ01a] O. Hijazi and X. Zhang, *Lower bounds for the eigenvalues of the Dirac operator I: The hypersurface Dirac operator*, Ann. Global Anal. Geom. **19** (2001), 355–376.

[HZ01b] ———, *Lower bounds for the eigenvalues of the Dirac operator II: The submanifold Dirac operator*, Ann. Global Anal. Geom. **20** (2001), 163–181.

[Ike80] A. Ikeda, *On lens spaces which are isospectral but not isometric*, Ann. scient. Ec. Norm. Sup. 4e série **13** (1980), 303–315.

[Ike83] ———, *On spherical space forms which are isospectral but not isometric*, J. Math. Soc. Japan **35** (1983), 437–444.

[Ike89] ———, *Riemannian manifolds p–isospectral but not (p+1)–isospectral*, Coll. Pap. 35th Symp., Perspect. Math., vol. 8, 1989, pp. 383–417.

[Joh75] K. Johnson, *The M.I.T. bag model*, Acta Phys. Pol. **B6** (1975), 865–892.

[Joy00] D. Joyce, *Compact manifolds with special holonomy*, Oxford University Press, Oxford, 2000.

[Kac66] M. Kac, *Can one hear the shape of a drum?*, Amer. Math. Monthly **73** (1966), 1–23.

[Kap91] N. Kapouleas, *Compact constant mean curvature surfaces in Euclidean three-space*, J. Diff. Geom. **33** (1991), 683–715.

[Kap95] ———, *Constant mean curvature surfaces constructed by fusing Wente tori*, Invent. Math. **119** (1995), 443–518.

[Kas68] T. Kashiwada, *On conformal Killing tensors*, Nat. Sci. Rep. Ochan. Univ. **19** (1968), 67–74.

[Kat99] I. Kath, *Killing Spinors on Pseudo-Riemannian Manifolds*, Habilitationsschrift Humboldt-Universität Berlin, 1999.

[Kir86] K.-D. Kirchberg, *An estimation for the first eigenvalue of the Dirac operator on closed Kähler manifolds of positive scalar curvature*, Ann. Global Anal. Geom. **4** (1986), 291–325.

[Kir88] ———, *Compact six-dimensional Kähler Spin manifolds of positive scalar curvature with the smallest possible first eigenvalue of the Dirac operator*, Math. Ann. **282** (1988), 157–176.

[Kir90] ———, *The first eigenvalue of the Dirac operator on Kähler manifolds*, J. Geom. Phys. **7** (1990), 449–468.

[Kir02] ———, *A relation between the Ricci tensor and the spectrum of the Dirac operator*, math.DG/0201279, 2002.

[KN69a] S. Kobayashi and K. Nomizu, *Foundations of Differential Geometry*, vol. I and II, Interscience, New York, 1963 and 1969.

[KN69b] ———, *Foundations of Differential Geometry, I & II*, Interscience, New York, 1963 and 1969.

[Kna96] A.W. Knapp, *Lie Groups Beyond an Introduction*, Progress in Mathematics, vol. 140, Birkhaüser, Boston, MA, 1996.

[Kob61] S. Kobayashi, *On compact Kähler Manifolds with Positive Definite Ricci Tensor*, Ann. of Math. **74** (1961), 570–574.

[Kos] B. Kostant, *Dirac cohomology for the cubic Dirac operator*, arXiv:math.RT/0208048.

[Kos99] ———, *A cubic Dirac operator and the emergence of Euler number multiplets of representations for equal rank subgroups*, Duke Math. J. **100** (1999), no. 3, 447–501.

[Kos00] ———, *A generalization of the Bott-Borel-Weil theorem and Euler number multiplets of representations*, Letters in Math. Phys. **52** (2000), 61–78.

[KR94] W. Kühnel and H.-B. Rademacher, *Twistor spinors with zeros*, Int. J. Math. **5** (1994), 887–895.

[KR96] ———, *Twistor spinors and gravitational instantons*, Lett. Math. Phys. **38** (1996), 411–419.

[KR97] ———, *Twistor spinors on conformally flat manifolds*, Illinois J. Math. **41** (1997), 495–503.

[KR98] ———, *Asymptotically Euclidean manifolds and twistor spinors*, Comm. Math. Phys. **196** (1998), 67–76.

[KSW98a] W. Kramer, U. Semmelmann, and G. Weingart, *The first eigenvalue of the Dirac operator on quaternionic Kähler manifolds*, Commun. Math. Phys. **199** (1998), 327–349.

[KSW98b] ———, *The first eigenvalue of the Dirac operator on quaternionic Kähler manifolds*, Commun. Math. Phys. **199** (1998), 327–349.

[KSW99] ———, *Eigenvalue estimates for the Dirac operator on quaternionic Kähler manifolds*, Math. Z. **230** (1999), 727–751.

[KT75] F. Kamber and P. Tondeur, *Foliated Bundles and Characteristic Classes*, Lecture Notes in Math., vol. 493, Springer-Verlag, 1975.

[KT82] ———, *Duality for Riemannian foliations*, Proc. Symp. Pure Math. **4** (1982), 609–618.

[Lan65] S. Lang, *Algebra*, Addison-Wesley, Reading, Mass., 1965.

[Lan00] G.D. Landweber, *Harmonic spinors on homogeneous spaces*, Represent. Theory **4** (2000), 466–473.

[Lan01] ———, *Multiplets of representations and Kostant's Dirac operator for equal rank loop groups*, Duke Math. J. **110** (2001), no. 1, 121–160.

[Law70] H.B. Jr. Lawson, *Complete minimal surfaces in S^3*, Ann. Math. **92** (1970), 334–374.

[Lee86] J.M. Lee, *The Fefferman metric and pseudohermitian invariants*, Trans. AMS **296** (1986), no. 1, 411–429.

[Lei01a] T. Leistner, *Lorentzian manifolds with special holonomy and parallel spinors*, Proceedings of the 21st Winter School on "Geometry and Physics" (Srni), Rend. Circ. Mat. Palermo, 2001.

[Lei01b] F. Leitner, *The Twistor Equation in Lorentzian Spin Geometry*, Ph.D. thesis, Humboldt University Berlin, 2001.

[Lew91] J. Lewandowski, *Twistor equation in a curved spacetime*, Class. Quant. Grav. **8** (1991), 11–17.

[Lic63] A. Lichnerowicz, *Spineurs harmoniques*, C.R. Acad. Sci. Paris **257, Série I** (1963), 7–9.

[Lic88a] ———, *Killing spinors, twistor spinors and Hijazi inequality*, J. Geom. Phys. **5** (1988), 2–18.

[Lic88b] ———, *Les spineurs-twisteurs sur une variete spinoreielle compacte*, C. R. Acad. Sci Paris, Ser. I **306** (1988), 381–385.

[Lic89] ———, *On twistor spinors*, Lett. Math. Phys. **18** (1989), 333–345.

[Lic90a] ———, *La première valeur propre de l'opérateur de Dirac pour une variété kählérienne et son cas limite*, C. R. Acad. Sci. Paris Sér. I Math. **311** (1990), 717–722.

[Lic90b] ———, *Sur les zéros des spineurs-twisteurs*, C.R. Acad. Sci. Paris **310** (1990), 19–22.

[LM80] H.B. Jr. Lawson and M.-L. Michelsohn, *Clifford bundles, spinors and Dirac operators*, J. Diff. Geom. **15** (1980), 237–267.

[LM89] H.B. Jr. Lawson and M.L. Michelsohn, *Spin Geometry*, Princeton Math. Series, vol. 38, Princeton University Press, 1989.

[Lop53] J. Lopatinsky, *On a method for reducing boundary problems for systems of differential equations of elliptic type to regular integral equations*, Ukrain. Math. Z. **5** (1953), 125–151.

[Lot86] J. Lott, *Eigenvalue bounds for the Dirac operator*, Pacific J. Math. **125** (1986), 117–126.

[LY74] H.B. Jr. Lawson and S.-T. Yau, *Scalar curvature, non-abelian groups actions and the degree of symmetry of exotic spheres*, Comm. math. Helv. **49** (1974), 232–244.

[LY82] P. Li and S.-T. Yau, *A new conformal invariant and its applications to the Willmore Conjecture and the first eigenvalue of compact surfaces*, Invent. Math. **69** (1982), 269–291.

[Mas92] X. Masa, *Duality and minimality in Riemannian foliations*, Comment. Math. Helv. **67** (1992), 17–27.

[May85] K.H. Mayer, *Elliptische Differentialoperatoren und ganzzahligkeitsatze für characteristische Zahlen*, Topology **4** (1985), 295–313.

[MH73] J. Milnor and D. Husemoller, *Symmetric Bilinear Forms*, Springer-Verlag, Berlin-New York, 1973.

[Mic80] M.-L. Michelsohn, *Clifford and Spinor cohomology of Kähler manifolds*, Amer. J. Math. **102** (1980), 1083–1146.

[Mil92] J.-L. Milhorat, *Spectre de l'opérateur de Dirac sur les espaces projectifs quaternioniens*, C. R. Acad. Sci. Paris Sér. I Math. **314** (1992), 69–72.

[Mil98] ———, *Spectrum of the Dirac operator on $Gr_2(\mathbb{C}^{m+2})$*, J. Math. Phys. **39** (1998), 594–609.

[MO89] M. Min-Oo, *Scalar curvature rigidity of asymtotically hyperbolic spin manifolds*, Math. Ann. **285** (1989), 527–539.

[Mol88] P. Molino, *Riemannian foliations*, Birkhäuser, 1988.

[Mor95] A. Moroianu, *La première valeur propre de l'opérateur de Dirac sur les variétés kählériennes compactes*, Commun. Math. Phys. **169** (1995), 373–384.

[Mor97a] ———, *On Kirchberg inequality for compact Kähler manifolds of even complex dimension*, Ann. Global Anal. Geom. **15** (1997), 235–242.

[Mor97b] ———, *Parallel and Killing Spinors on Spin^c Manifolds*, Commun. Math. Phys **187** (1997), 417–428.

[Mor99] ———, *Kähler manifolds with small eigenvalues of the Dirac operator and a conjecture of Lichnerowicz*, Ann. Inst. Fourier (Grenoble) **49** (1999), 1637–1659.

[MR86a] S. Montiel and A. Ros, *Minimal immersions of surfaces by the first eigenfunctions and conformal area*, Invent. Math. **83** (1986), 153–166.

[MR86b] ———, *Minimal immersions of surfaces by the first eigenfunctions and conformal area*, Invent. Math. **83** (1986), 153–166.

[MS74] J. Milnor and J. Stasheff, *Characteristic Classes*, no. 76, Princeton University Press, 1974.

[MS89] H. Moscovici and R.J. Stanton, *Eta invariants of Dirac operators on locally symmetric manifolds*, Inventiones Math. **95** (1989), 629–666.

[Nir70] L. Nirenberg, *Pseudo-differential operators*, A.M.S. Proc. of Symp. in Pure Math. **16** (1970), 149–167.

[Ogi73] K. Ogiue, *Some open problems in differential geometry*, Proc. Symp. Pure Math. **XXVII** (1973), 407–411.

[OS89] B. Orsted and I.E. Segal, *A pilot model in two dimensions for conformally invariant particle theory*, J. Funct. Anal. **83** (1989), 150–184.

[OS98] T. Oguro and K. Sekigawa, *Four-dimensional almost Kähler Einstein and $*$-Einstein manifolds*, Geom. Dedicata, **69** (1998), 91–112.

[Pal65] R.S. Palais, *Seminar on the Atiyah-Singer Index Theorem*, no. 57, Princeton University Press, Princeton, 1965.

[Par72] R. Parthasarathy, *Dirac operator and the discrete series*, Annals of Math. **96** (1972), no. 1, 1–30.

[Pfä00] F. Pfäffle, *The Dirac spectrum of Bieberbach manifolds*, J. Geom. Phys. **35** (2000), 367–385.

[PR86] R. Penrose and W. Rindler, *Spinors and space-time*, vol. 1 and 2, Cambridge University Press, Cambridge, 1984 and 1986.

[PR96] E. Park and K. Richardson, *The basic Laplacian for Riemannian foliations*, Am. J. of Math. **118** (1996), 1249–1275.

[PRV67] R. Parthasarathy, R. Ranga Rao, and V.S. Varadarajan, *Representations of complex semisimple Lie groups,*, Annals of Math. **85** (1967), 383–429.

[PSV87] S. Paneitz, I.E. Segal, and D. Vogan, *Analysis in space-time bundles IV: Natural bundles deforming into and composed of the same invariant factors as the spin and form bundles*, J. Funct. Anal. **75** (1987), 1–57.

[PT82] T. Parker and C.H. Taubes, *On Witten's proof of the positive energy theorem*, Commun. Math. Phys. **84** (1982), 223–238.

[PW70] R. Penrose and M. Walker, *On quadratic first integrals of the geodesic equation for type {22} spacetimes*, Comm. Math. Phys. **18** (1970), 265–274.

[RDSS92] R. R. Delanghe, F. Sommen, and V. Souček, *Clifford Algebra and Spinor-Valued Functions*, Kluwer Academic Publishers, Dordrecht, 1992.

[Rei59] B.P. Reinhart, *Harmonic integrals on foliated manifolds*, Am. J. of Math. **81** (1959), 529–586.

[Rei77a] R.C. Reilly, *Applications of the Hessian operator in a Riemannian manifold*, Indiana Univ. Math. J. **26** (1977), 459–472.

[Rei77b] ———, *On the first eigenvalues of the Laplacian for compact submanifolds of Euclidean space*, Comment. Math. Helv. **52** (1977), 525–533.

[Ric98] K. Richardson, *Asymptotic of heat kernels on Riemannian foliations*, Geom. Funct. Ana. **8** (1998), 356–401.

[Ros83] J. Rosenberg, *C^*-algebras, positive scalar curvature and the Novikov Conjecture*, Publ. I.H.E.S. **58** (1983), 197–212.

[Ros88] A. Ros, *Compact hypersurfaces with constant scalar curvature and a congruence theorem*, J. Diff. Geom. **27** (1988), 215–220.

[Ros02] W. Rossmann, *Lie Groups: An Introduction through Linear Groups.* , Oxford Graduate Texts in Mathematics, vol. 5, Oxford University Press, 2002.

[RS95] J. Rosenberg and S. Stolz, *A "stable" version of the Gromov-Lawson conjecture*, The Čech centennial (Boston, MA,1993), vol. 181, Amer. Math. Soc., Providence, RI, 1995, pp. 405–418.

[RS01] ———, *Metrics of positive scalar curvature and connections with surgery*, Surveys on surgery theory, vol. 2, Ann. of Math. Stud., no. 149, Princeton Univ. Press, 2001, pp. 353–386.

[Sal89] S. Salamon, *Riemannian geometry and holonomy groups*, Pitman Research Notes in Math., no. 201, Longman Scientific & Technical, Essex, 1989.

[Sch18] J. Schouten, *On the number of degrees of freedom of the geodetically moving systems*, Proc. Kon. Acad. Wet. Amsterdam **21** (1918), 607–613.

[Sch32] E. Schrödinger, *Diracsches Electron im Schwerefeld I*, Sitzingsbericht der Preussischen Akademie der Wissenschaften Phys.-Math. Klasse 1932, Verlag der Akademie der Wissenschaften, Berlin (1932), 436–460.

[Sch74] R. Schimming, *Riemannsche Räume mit ebenfrontiger und ebener Symmetrie*, Mathematische Nachrichten **59** (1974), 129–162.

[Sch79] J. Scherk, *Extended Supersymmetries and Extended Supergravity Theories*, Cargese Summer Institute on Recent Developments in the Theory of Gravitation, Plenum Press, 1979.

[Sch01] J.-M. Schlenker, *Einstein manifolds with convex boundaries*, Comment. Math. Helv. **76** (2001), 1–28.

[Sch02] M.U. Schmidt, *A proof of the Willmore conjecture*, Preprint, Berlin, ArXiv:math.DG/0203224, 2002.

[See66] R. Seeley, *Singular integrals and boundary problems*, Amer. J. Math. **88** (1966), 781–809.

[See97] L. Seeger, *The Dirac operator on oriented Grassmann manifolds and $G_2/\mathrm{SO}(4)$*, Preprint, Mathem. Inst. Bonn, 1997.

[See99] ———, *The spectrum of the Dirac operator on $G_2/\mathrm{SO}(4)$*, Ann. Global Anal. Geom. **17** (1999), 385–396.

[See01] ———, *Metriken mit harmonischen Spinoren auf geradedimensionalen Sphären*, Shaker Verlag, Aachen, 2001.

[Sek87] K. Sekigawa, *On some compact Einstein almost Kähler manifolds*, J. Math. Soc. Japan **39** (1987), 677–684.

[Sem01] U. Semmelmann, *Conformal Killing forms in Riemannian Geometry*, Habilitationsschrift, Universität München, 2001.

[SI92] A. El Soufi and S. Ilias, *Une inégalité de type Reilly pour les sous-variétés de l'espace hyperbolique*, Comment. Math. Helv **67** (1992), no. 2, 167–181.

[SI00] ———, *Riemannian manifolds admitting isometric immersions by their first eigenfunctions*, Pacific J. Math. **195** (2000), 91–99.

[Sim62] J. Simons, *On the transitivity of holonomy systems*, Ann. of Math. (2) **76** (1962), 213–234.

[Sle85] S. Slebarski, *Dirac operators on a compact Lie group*, Bull. London Math. Soc. **17** (1985), no. 6, 579–583.

[Sle87a] S. Slebarksi, *The Dirac operator on homogeneous spaces and representations of reductive Lie groups I.* , Amer. J. Math. **109** (1987), no. 2, 283–301.

[Sle87b] S. Slebarski, *The Dirac operator on homogeneous spaces and representations of reductive Lie groups II.*, Amer. J. Math. **109** (1987), no. 3, 499–520.

[Sle88] ———, *The Dirac operator on homogeneous spaces and representations of reductive Lie groups. III.*, Amer. J. Math. **110** (1988), no. 3, 451–471.

[Soh85] M. Sohnius, *Introducing Supersymmetry*, Phys. Rep. **128** (1985), no. 2-3, 39–204.

[Spa85] G.A.J. Sparling, *Twistor theory and the characterization of Fefferman's conformal structures*, Preprint Univ. Pittsburg, 1985.

[Spi79] M. Spivak, *Differential Geometry*, vol. I, Publish or Perish, Houston, Texas, 1979.

[SS93] S. Seifarth and U. Semmelmann, *The spectrum of the Dirac operator on the odd dimensional complex projective space* $P^{2m-1}(\mathbb{C})$, Preprint, SFB 288, 1993.

[Ste51] N. Steenrod, *The topology of fibre bundles*, Princeton University Press, Princeton, 1951.

[Sto92] S. Stolz, *Simply connected manifolds of positive scalar curvature*, Ann. of Math. **136** (1992), 511–540.

[Sto95] ———, *Positive scalar curvature metrics—existence and classification questions*, Proceedings of the International Congress of Mathematicians, (Zürich, 1994), Birkhäuser, Basel, 1995, pp. 625–636.

[Str80a] H. Strese, *Spektren symmetrischer Räume*, Math. Nachr. **98** (1980), 75–82.

[Str80b] ———, *Über den Dirac-Operator auf Grassmann-Mannigfaltigkeiten*, Math. Nachr. **98** (1980), 53–59.

[Str88] R. Strichartz, *Linear algebra of curvature tensors and their covariant derivatives*, Canad. J. Math. **40** (1988), 1105–1143.

[Sul79] S. Sulanke, *Berechnung des Spektrums des Quadrates des Dirac-Operators auf der Sphäre*, Ph.D. thesis, Doktorarbeit, HU Berlin, 1979.

[Sun85] T. Sunada, *Riemannian coverings and isospectral manifolds*, Ann. of Math. **121** (1985), 169–186.

[SW68] E. Stein and G. Weiss, *Generalization of the Cauchy-Riemann equations and representations of the rotation group*, Amer. J. Math. **90** (1968), 163–196.

[SY79a] R. Schoen and S.-T. Yau, *Existence of incompressible minimal surfaces and the topology of three dimensional manifolds of non-negative scala curvature*, Ann. of Math. **110** (1979), 127–142.

[SY79b] ———, *On the proof of the positive mass conjecture in general relativity*, Comm. Math. Phy. **65** (1979), 45–76.

[SY79c] ———, *On the structure of manifolds with positive scalar curvature*, Manuscripta Math. **28** (1979), 159–183.

[SY89] ———, *Complete three dimensional manifolds with positive Ricci and scalar curvature*, Seminar on Differential Geometry, no. 102, Princeton University Press, 1989.

[Tai] I.A. Taimanov, *Dirac operator and conformal invariants of tori in 3-space*, preprint, arXiv:math.DG/0005223.

[Tak66] T. Takahashi, *Minimal immersions of Riemannian manifolds*, J. Math. Soc. Japan **18** (1966), 380–385.

[Tan75] N. Tanaka, *A differential geometric study on strongly pseudoconvex manifolds*, Kinokuniya Company Ltd., Tokyo, 1975.

[Tha92] B. Thaller, *The Dirac equation*, Springer-Verlag, Berlin Heidelberg New York, 1992.

[Tho23] G. Thomsen, *Über konforme Geometrie, I: Grundlagen der konformen Flächentheorie*, Abh. Math. Sem. Hamburg **3** (1923), 31–56.

[Tho26] T.Y. Thomas, *On conformal geometry*, Proc. Natl. Acad. Sci. USA **12** (1926), 352–359.

[TK69] S. Tashibana and T. Kashiwada, *On the integrability of Killing-Yano's equation*, J. Math. Soc. Japan **21** (1969), 259–265.

[Ton97] P. Tondeur, *Geometry of Foliations*, Monographs in Mathematics, 90, Birkhaüser-Verlag, 1997.

[Top98] P. Topping, *Mean curvature flow and geometric inequalities*, J. Reine Angew. Math. **503** (1998), 47–61.

[Tra93] A. Trautman, *Spin structures on hypersurfaces and the spectrum of the Dirac operator on spheres*, Spinors, Twistors, Clifford Algebras and Quantum Deformations, Proceedings of the 2nd Max Born symposium, Wroclaw, Poland, September 24–27, 1992. (Zbigniew et al. Oziewicz, ed.), Kluwer Academic Publishers, 1993, pp. 25–29.

[Tra95] ———, *The Dirac operator on hypersurfaces*, Acta Phys. Pol. **26** (1995), 1283–1310.

[TV81] F. Tricerri and L. Vanhecke, *Curvature tensors on almost Hermitian manifolds*, Trans. Amer. Math. Soc. **267** (1981), 365–398.

[vN81] P. van Nieuwenhuizen, *Supergravity*, Phys. Rep. **68** (1981), no. 189.

[Wal88] N.R. Wallach, *Real Reductive Groups, I*, vol. 132, Academic Press, Boston, MA, 1988.

[Wal92] ———, *Real Reductive Groups, II*, vol. 132-II, Academic Press, Boston, MA, 1992.

[Wan89] M.Y. Wang, *Parallel spinors and parallel forms*, Ann. Global Anal. Geom. **7** (1989), 59–68.

[War71] F.W. Warner, *Foundations of differentiable manifolds and Lie groups*, Scott, Foresman and Co., Glenview, London, 1971.

[WB92] J. Wess and J. Bagger, *Supersymmetry and Supergravity*, Princeton University Press, 1992.

[Web78] S.M. Webster, *Pseudohermitian structures on a real hypersurface*, J. Diff. Geom. **13** (1978), 25–41.

[Wei78] J.L. Weiner, *On a problem of Chen, Willmore, et al.*, Ind. Math. J. **27** (1978), 19–35.

[Wei98] S. Weinberg, *Quantum Field Theory*, vol. 1, Cambridge University Press, 1998.

[Wen86] H.C. Wente, *Counterexample to a conjecture of H. Hopf*, Pacific J. Math. **121** (1986), 193–243.

[Wes92] P. West, *Introduction to Supersymmetry*, second ed., World Scientific, 1992.

[Wil65] T.J. Willmore, *Note on embedded surfaces*, An. Sti. Univ. "Al. I.Cuza" **11B** (1965), 493–496.

[Wit81] E. Witten, *A new proof of the positive energy theorem*, Commun. Math. Phys. **80** (1981), 381–402.

[Wol73] J.A. Wolf, *Essential self adjointness for the Dirac operator and its square*, Indiana Univ. Mathematics Journal **22** (1973), no. 7, 611–640.

[Wol84] ———, *Spaces of constant curvature*, Publish or Perish, Boston, 1984.

[Yau82] S.T. Yau, *Problem section*, Ann. Math. Studies **102** (1982), 669–706.

[Zel73] D.P. Zelobenko, *Compact Lie groups and their representations*, Translation Math. Monographs, AMS **40** (1973).

List of participants

Abdelkhalek Selman	American University of Beirut, Lebanon
Abou Rizk Zeina	Université Saint Joseph, Lebanon
Abou-Dina Mustapha	Cairo University, Egypt
Adwan Ziad	American University of Beirut, Lebanon
Aghnatios Fadi	Hagazian University, Lebanon
Agib Youssef	Beirut Arab University, Lebanon
Aissaoui Habib	Université Mentouri Costantine, Algeria
Ammann Bernd	Université Henri Poincaré, France
Ashkar Ihab	American University of Beirut, Lebanon
Ayadi Abdelhamid	Université Mentouri Costantine, Algeria
Ayoub Joseph	École Normale Supérieure de Paris, France
Aysh Jihad	Tripoli, Lebanon
Bär Christian	Universität Potsdam, Germany
Baum Helga	Humboldt-Universität zu Berlin, Germany
Bayard Pierre	Université de Nice, France
Beaufort Éric	Université Henri Poincaré, France
Belkhiat Saad	Université Abbas Farhat de Sétif, Algeria
Berrimi Saïd	Université Abbas Farhat de Sétif, Algeria
Boucenna Ahmad	Université Abbas Farhat de Sétif, Algeria
Boudine Azzedine	Université Mentouri Costantine, Algeria
Bourguignon Jean-Pierre	IHÉS, France
Boutchebak Souray	Université Abbas Farhat de Sétif, Algeria
Branson Thomas	University of Iowa, USA
Bryant Robert	Duke University, USA
Buchholz Volker	Humboldt-Universität zu Berlin, Germany
Chamseddine Ali	American University of Beirut, Lebanon
Charuel Xavier	Université Henri Poincaré, France
Chmeisany Wissam	American University of Beirut, Lebanon
Dahl Mattias	Universtät Hamburg, Germany
Delay Erwann	Université de Tours, France
Denche Mohamed	Université Mentouri Costantine, Algeria
Doma Salah	Alexandiria University, Egypt
El Helou Yousef	Université Libanaise, Lebanon
El Kacimi Aziz	Université de Valenciennes, France
Fox Daniel	Duke University, USA
Ghayour Omid	Chamran University of Ahvaz, Iran
Ginoux Nicolas	Universität Potsdam, Germany
Grosjean Jean-Francois	Université Henri Poincaré, France
Guillarmou Colin	Université de Nantes, France
Hachemi Hacene	Université Abbas Farhat de Sétif, Algeria
Haroun Abdelhalim	Université Abbas Farhat de Sétif, Algeria
Hijazi Oussama	Université Henri Poincaré, France
Hong Doojin	University of Iowa, USA
Humbert Emmanuel	Université Henri Poincaré, France
Ionel Marianty	Duke University, USA
Issa Rasha	American University of Beirut, Lebanon

Jaafar Samar	American University of Beirut, Lebanon
Jaber Silvana	American University of Beirut, Lebanon
Kadem Abdelouhab	Université Abbas Farhat de Sétif, Algeria
Kebiche Zakia	Université Abbas Farhat de Sétif, Algeria
Khuri-Makdisi Kamal	American University of Beirut, Lebanon
Klinker Frank	Universität Leipzig, Germany
Konderak Jerzy	Universita di Bari, Italy
Krump Lukas	Charles University Prague, Czech Republic.
Kumar Ajit	University of Mumbai, India
Lakhdar Chiter	Université Abbas Farhat de Sétif, Algeria
Latrous Chahla	Université Mentouri Costantine, Algeria
Lawson H. Blaine	SUNY, Stony Brook, USA
Leitner Felipe	Humboldt-Universität zu Berlin, Germany
Lyzzaik Abdellah	American University of Beirut, Lebanon
Maamache Mustapha	Université Abbas Farhat de Sétif, Algeria
Maubon Julien	Université Henri Poincaré, France
Merouani Boubaker	Université Abbas Farhat de Sétif, Algeria
Michelsohn Marie-Louise	SUNY, Stony Brook, USA
Mohamad Abid	University of Baghdad, Iraq
Montiel Sebastián	University of Granada, Spain
Moroianu Andrei	École Polytechnique, France
Moukadem Nazih	Université Libanaise, Lebanon
Nükrichner Thomas	Humboldt-Universität zu Berlin, Germany
Pirinccioglu Nurettin	University of Dicle, Turkey
Rababah Abdellah	Jordan University of Science and Technology, Jordan
Rabie Eqab	Jordan University of Science and Technology, Jordan
Raji Wissam	American University of Beirut, Lebanon
Rashid Adel	Beirut Arab University, Lebanon
Saab Jihad	Université Saint Joseph, Lebanon
Sabra Wafic	American University of Beirut, Lebanon
Samarah Salti	Jordan University of Science and Technology, Jordan
Sayah Toni	Université Saint Joseph, Lebanon
Sayah Fida	Université Saint Joseph et Université Libanaise, Lebanon
Shirbisheh Vahid	University of Tehran, Iran
Stanton Robert	Ohio State University, USA
Ugalde William	University of Iowa, USA

Christian Bär

Robert Bryant

Thomas Branson

Robert J. Stanton

Ali Chamseddine

Andrei Moroianu

Pictures from the Conference Room

The Excursion to Ehden

Made in United States
Orlando, FL
04 November 2026

90200890R00175